HOW WELL DO FACTS TRAVEL?

Why write about facts? Facts are everywhere. They litter the utterances of public life as much as the private conversations of individuals. They frequent the humanities and the sciences in equal measure. But their very ubiquity tells us not only why it is difficult to form general but sensible answers in response to seemingly simple questions about facts, but also why it is important to do so.

This book discusses how facts travel, and when and why they sometimes travel well enough to acquire an independent life of their own. Whether or not facts travel in this manner depends not only on their character and ability to play useful roles elsewhere, but also on the labels, packaging, vehicles, and company that take them across difficult terrains and over disciplinary boundaries. These diverse stories of travelling facts, ranging from architecture to nanotechnology and from romance fiction to climate science, change the way we see the nature of facts. Facts are far from the bland and rather boring but useful objects that scientists and humanists produce and fit together to make narratives, arguments, and evidence. Rather, their extraordinary abilities to travel well – and to fly flags of many different colours in the process – show when, how, and why facts can be used to build further knowledge beyond and away from their sites of original production and intended use.

Peter Howlett is an expert on the economic history of World Wars I and II, and contributed the text for the official history *Fighting with Figures* (1995). Dr. Howlett's publications also explore international economic growth and convergence since 1870 and the development of internal labour markets, and have appeared in edited volumes and journals such as the *Economic History Review*, *Explorations in Economic History*, and *Business History*. He teaches at the London School of Economics and is secretary of the Economic History Society.

Mary S. Morgan is professor of history and philosophy of economics at the London School of Economics and the University of Amsterdam. She has published widely on topics ranging from statistics to experiments to narrative, and from social Darwinism in late-nineteenth-century America to game theory in the Cold War. Her major works inclu￼ The History of Econometric Ideas (1990), *The Foundations of Econometric Anal*￼ ￼ndry), and *Models as Mediators* (1999, co-￼ ￼ofessor Morgan's account of scientific modell￼ *Model.* She is currently engaged in the resear￼ ￼ across the Social Sciences" as a British Acac￼

How Well Do Facts Travel?

The Dissemination of Reliable Knowledge

EDITED BY

PETER HOWLETT

London School of Economics

MARY S. MORGAN

London School of Economics

CAMBRIDGE UNIVERSITY PRESS

CAMBRIDGE UNIVERSITY PRESS
Cambridge, New York, Melbourne, Madrid, Cape Town, Singapore,
São Paulo, Delhi, Dubai, Tokyo, Mexico City

Cambridge University Press
32 Avenue of the Americas, New York, NY 10013-2473, USA

www.cambridge.org
Information on this title: www.cambridge.org/9780521159586

© Cambridge University Press 2011

First published 2011

Printed in the United States of America

A catalog record for this publication is available from the British Library.

Library of Congress Cataloging in Publication data
How well do facts travel? : the dissemination of reliable knowledge /
[edited by] Peter Howlett, Mary S. Morgan.
p.   cm.
ISBN 978-0-521-19654-3 (hardback)
ISBN 978-0-521-15958-6 (pbk.)
1. Facts (Philosophy)   I. Howlett, Peter, 1962–   II. Morgan, Mary S.
B105.F3H69   2010
001–dc22   2010031326

ISBN 978-0-521-19654-3 Hardback
ISBN 978-0-521-15958-6 Paperback

Contents

Figures

Contributors

Jon Adams studied philosophy and literature at the universities of Keele and Durham, subsequently working on the "How Well Do 'Facts' Travel?" project at the London School of Economics (LSE) from 2005 to 2009. In 2007, he published *Interference Patterns* about the possibility of making literary criticism into a science. With Edmund Ramsden, he is writing a book about crowding. Having theorised about the process of knowledge popularisation, in 2009, he was offered an opportunity by the LSE to put the theory into practice, and began making short documentary films about academic research.

Rachel A. Ankeny is an associate professor in the School of History & Politics at the University of Adelaide, Australia. From 2000 to 2006 she was director and lecturer/senior lecturer in the Unit for History and Philosophy of Science at the University of Sydney. Her research in the history and philosophy of science includes explorations of the roles of models and case-based reasoning in science, model organisms, the philosophy of medicine, and the history of contemporary life sciences. She also has ongoing research in bioethics and food studies. She was a visiting faculty member associated with the "Facts" project for its duration.

Richard W. Burkhardt, Jr., is professor emeritus of history at the University of Illinois, Urbana-Champaign. His primary research interests are the historical development of evolutionary theory, ethology, zoos, and naturalist voyages. His publications include *The Spirit of System: Lamarck and Evolutionary Biology* (1977, 1995) and *Patterns of Behavior: Konrad Lorenz, Niko Tinbergen, and the Founding of Ethology* (2005). He is writing a book on the early history of the menagerie of the Muséum d'Histoire Naturelle in Paris, the first public zoo of the modern era.

David Boyd Haycock is a Wellcome Research Fellow in the History of Medicine at the University of Oxford; he has held research fellowships at the London School of Economics; The Center for Seventeenth and Eighteenth

Century Studies at the University of California, Los Angeles; and at Wolfson College, Oxford. His research fields include the histories of early modern science and medicine, as well as the history of art; he is the author of *Mortal Coil: A Short History of Living Longer* (Yale University Press 2008) and *William Stukeley: Science, Archaeology and Religion in Eighteenth-Century England* (The Boydell Press 2002).

Peter Howlett is a senior lecturer in the Department of Economic History at the London School of Economics. He has previously studied the economic aspects of World Wars I and II; international economic growth and convergence since 1870; and the development of internal labour markets, publishing on these topics in journals such as the *Economic History Review* and *Explorations in Economic History*. His involvement in the "Facts" project grew out of an interest in the Indian Green Revolution; his Facts Working Paper (24/08) considers this from the perspective of facts travelling across social science disciplines.

Catharina Landström is a post-doctoral research associate at the School of Geography and the Environment, Oxford University, and is working on environmental knowledge controversies and the science and politics of flood risk. She studied theory of science at University of Göteborg. After post-doctoral research in Australia on the biological control of exotic pests, she returned to Sweden to conduct research on gender and technology. Her publications include "Justifiable Bunnycide: Narrating the Recent Success of Australian Biological Control of Rabbits" (*Science as Culture* 2001) and "A Gendered Economy of Pleasure. Representations of Cars and Humans in Motoring Magazines" (*Science Studies* 2006).

Sabina Leonelli is a research Fellow of the ESRC Centre for Genomics in Society (Egenis) based at the Department of Sociology and Philosophy, University of Exeter. Her research spans the fields of history and philosophy of biology, science and technology studies, and general philosophy of science, to investigate the epistemic and regulatory role of experimental practices such as data sharing, modelling, standardizing, and classifying. She co-edited *Scientific Understanding: Philosophical Perspectives* (Pittsburgh University Press 2009). Following her work for the "Facts" project, she is currently exploring the use of e-Science tools, such as bio-ontologies and databases, to carry out data-driven and translational research on model organisms.

Erika Mansnerus received her PhD from the University of Helsinki in 2007 and, after working on the "Facts" project, became a British Academy post-doctoral Fellow, dividing her time between the Centre for Research in the Arts, Social Sciences and Humanities (CRASSH) at the University of Cambridge and LSE Health. Her disciplinary background is in sociology and philosophy of science and science and technology studies. Her current research focuses on how computational tools (models, simulations) are utilised in public health decision-making processes, especially in pandemic preparedness planning. Her most recent publication is "The Lives of Facts in Mathematical Models: A Story of Population-Level Disease Transmission of *Haemophilus influenzae* Type B Bacteria" (*BioSocieties* 2009).

Martina Merz is a professor of the Swiss National Science Foundation (SNSF) at the Institute of Sociology, University of Lucerne, Switzerland. She is also a guest professor at the Technology and Society Lab, Swiss Federal Laboratories for Materials Science and Technology (Empa), St. Gallen. Her area of expertise is primarily in the social studies of science and technology. She currently heads the research project "Epistemic Practice, Social Organization, and Scientific Culture: Configurations of Nanoscale Research in Switzerland" (funded by SNSF), and is a collaborator of the National Centre of Competence in Research "Iconic Criticism: The Power and Meaning of Images."

Mary S. Morgan is professor of history and philosophy of economics at the London School of Economics and the University of Amsterdam. She has published widely on topics ranging from statistics to experiments to narrative, and from social Darwinism in late-nineteenth-century America to game theory in the Cold War. Her major works include *The History of Econometric Ideas* (1990) and *Models as Mediators* (1999 with Margaret Morrison), and her account of scientific modelling is forthcoming in *The World in the Model*. She is currently engaged in the research project "Re-thinking Case Studies Across the Social Sciences" as a British Academy–Wolfson research professor.

Naomi Oreskes is professor of history and science studies at the University of California, San Diego. Her research focuses on the historical development of scientific knowledge, methods, and practices in the earth and environmental sciences. She is best known for her work on climate change (which was cited in the film *An Inconvenient Truth*). Her publications include *The Rejection of Continental Drift* (1999), "The Scientific Consensus on Climate Change" (*Science* 2004), and *Merchants of Doubt* [with Erik M. Conway] (2010); and she co-edited *Plate Tectonics: An Insider's History of the Modern Theory of the Earth* (2001).

Edmund Ramsden is a research Fellow at the Centre for Medical History at the University of Exeter working on a Wellcome Trust–funded project on the history of stress. His research interests are in the history and sociology of the social and biological sciences and their relations, with a particular focus on the behavioural and population sciences. As a result of his post-doctoral work on the "Facts" project, he is currently writing a book with Jon Adams on John B. Calhoun's experiments.

Heather Schell is an assistant professor at George Washington University, where she directs the First-Year Writing Program and teaches courses on popular culture. Her work focuses on the lines of communication between the worlds of popular culture and biological science, examining such topics as the understanding of global pandemics, the effect of using epidemiological models to understand human prehistory, and the increasing popularity of predators.

Lambert Schneider is a classical archaeologist and has recently retired as a professor at Hamburg University; in 1996, he was a scholar at the J. P. Getty Research Institute in Santa Monica, California. His main research fields include theory and methodology in archaeology, gender studies in Greek culture and its modern applications, Thracian and Scythian imagery, Greek sculpture, late Roman imagery, and re-use of ancient Greek cultural forms in modern societies. His publications include *Die Akropolis von Athen* [with Ch. Höcker) (Darmstadt 2001) and *Die ungezähmte Frau. Weibliche Antibilder in Mythos und Bildkunst der Griechen* [with M. Seifert] (Stuttgart 2010).

Simona Valeriani has a background in architecture, the history of architecture, and building archaeology, earning her PhD in Berlin (2006); she joined the "Facts" project in 2005. Currently, she is analysing how scientific and technical knowledge were accumulated and transmitted in late medieval and early modern Europe and how the two spheres interacted as part of the LSE project "Useful and Reliable Knowledge in Global Histories of Material Progress in the East and the West" (URKEW, financed by the European Research Council). Her publications include the book *Kirchendächer in Rom. Zimmermannskunst und Kirchenbau von der Spätantike bis zur Barockzeit* (2006).

Aashish Velkar is an LSE Fellow in the Department of Economic History at the London School of Economics. He has previously studied measurement standards in historical markets and the role of economic groups in setting standards and reducing transaction costs, publishing this work in *Enterprise & Society* and *Business History*. His involvement with the "Facts" project led to a publication about how "facts" about quality travel within grain markets

(*Graduate Journal of Social Science* 2009). He is currently working on the role of measurement systems in the foundations of markets, using the case study of international grain markets in the nineteenth century.

Sarah J. Whatmore studied geography at University College, London, and worked in policy research at the Greater London Council. She is currently professor of environment and public policy at the University of Oxford. Her research addresses the interface between cultural geography, political theory, and science and technology studies. She is the author of *Hybrid Geographies* (Sage 2002) and joint editor, with Bruce Braun, of *Political Matter: Technoscience, Democracy and Public Life* (University of Minnesota Press 2010). Her current research is on environmental knowledge controversies and the science and politics of flood risk.

Alison Wylie is professor of philosophy and anthropology at the University of Washington. She is a philosopher of science who works on philosophical issues raised by archaeological practice and by feminist research in the social sciences: ideals of objectivity, the role of contextual values in research practice, and models of evidential reasoning. She has published widely on these topics, including *Thinking from Things: Essays in the Philosophy of Archaeology* (2002), *Value-free Science?* (2007, edited with Kincaid and Dupré) and the Sage *Handbook of Feminist Research* (2007). She is currently working on a monograph, *Standpoint Matters, in Feminist Philosophy of Science*.

Editors' Preface

Why write about facts? Facts are everywhere. They litter the utterances of public life as much as the private conversations of individuals. They frequent the humanities and the sciences in equal measure. Facts, understood in their everyday sense as bits of knowledge, make their appearances across the terrains of knowledge. As such, facts are not only expressed in verbal claims and counter-claims, but in all sorts of things and in all sorts of ways: in the drawings of insects, in the maps of our globe, in the beams of buildings, or in the shards of our forebears. Facts may be tiny, and on their own seem quite trivial (as a piece of genetic information about a plant), or important and earth-saving (as our temperature measures of climate change). And, of course, as we all know, individual facts may be strong and secure bits of knowledge, or sometimes hard to distinguish from fictions, or shaky to the point of falsehood. But their very ubiquity, in conjunction with the many forms they take and the different qualities they hold, tells us not only why it is difficult to form general but sensible answers in response to seemingly simple questions about facts, but why it is important to do so.

We may take this notion of facts – as pieces of knowledge – for granted. But it is equally pertinent for this book that facts are also recognised to be *separable* bits of knowledge that can be abstracted from their production context and shared with others. And because they are such independent pieces of knowledge, facts have the possibility to travel, and indeed some circulate freely, far and wide. So, how do such bits of knowledge – whatever their appearance and size – circulate while maintaining their integrity as facts? For, of course, it matters that travelling facts do hold their knowledge: They are not just an essential category of the way we talk in modern times, but provide one of the forms of knowledge upon which we act. This is one way we explore our question: "How *well* do facts travel?" *Well enough* to act upon them: Facts need to retain their integrity if we are to act upon them safely. Yet our recognising that facts have travelled well depends on us noticing how certain facts get used again and again, by other communities or for other purposes. This provides our second insight into the problem of understanding *travelling well*. Facts travel well if their travels prove fruitful. So these two senses of travelling well, with integrity and fruitfully, frame our answers to the question "How well do facts travel?"

The essays here do not adopt any *one* theoretical or disciplinary approach. We are not committed *as a group* to any particular sociological theory about knowledge transfer, nor to the establishment of a philosophical test of the truth or falsity of facts, nor to the provision of an epistemic history of facts as a category. The essays here are written by those with disciplinary backgrounds in the natural and social sciences and the humanities (and many of our authors have training in more than one academic discipline) and they take as subjects cases from the sciences, the humanities, and the arts. But like a number of other recent volumes that cross this traditional science–arts divide – such as Lorraine Daston's *Things That Talk* (2004) and Angela Creager et al.'s *Science Without Laws* (2007) – we take a relaxed, open view about how to study knowledge, one that coalesces around a particular object of study, facts, unbounded by a disciplinary framework. So, the reader will find here a variety of resources used for thinking about travelling facts offered from several disciplinary approaches and in a number of different fields and contexts. But we do share common grounds: The essays in this book address a common question, share an understanding of what facts are and a framework for answering questions about what it means for facts to travel well. They all address directly the lives of facts, and only indirectly other aspects of facts, such as their production, their context, their value, and the communities they pass through.

The answers to our question "How well do facts travel?" are given individually and differently by each author, and each writes from their own theoretical focus. But although each author has followed their own stories, they have all contributed to the shared analysis that shapes our answers. Each essay answers our question in a different way, ways that succeed in adding twists to our agenda, or in shaking our framework to offer us new perspectives, or in creatively turning the way we understand the problem of exploring how well facts can travel. The presence – within each essay – of these multiple contributions made the imposition of part headings in the volume somewhat arbitrary, and the essays cannot be as neatly docketed as our headings suggest. Rather, these headings are indicative of something special that we have found in each of those essays. But as the index makes clear, the elements of our shared thinking about what it means to travel well are either centrally found in each essay or else have crept into the subplots or side analyses in informative ways. The book itself has been strongly supported by our editor, Scott Parris, and our publishers, Cambridge University Press, and we thank them and their readers and referees for their penetrating comments on the book proposal and manuscript.

As a research programme, "How Well Do Facts Travel?" grew into one of the most enjoyable and one of the most intellectually exciting experiences of our careers. Between late 2004 and the end of 2009, we were able to work – not all at once, but over the period – with a group of young researchers who became close colleagues: post-docs Simona Valeriani, Edmund Ramsden, Erika Mansnerus, Sabina Leonelli, and Jon Adams; and PhD students Aashish Velkar, Ashley Millar, Julia Mensink, and Albane Forestier. These were all wonderful to work with, but we might just mention the two longest-serving post-docs: Simona Valeriani helped in organising the project and getting this book into production, and Jon Adams created our public face of posters, Web pages, and logo. We would also like to thank Aashish Velkar and Eric Golson for their sterling work in preparing the index for this volume, and Rajashri Ravindranathan for her patience in chaperoning our book through the production process. We were delighted when the project was recognised at the Times Higher Education Awards as being amongst the best "research projects of the year" in 2008. An account of the project, giving its history, descriptions of workshops and a British Academy congress, the development of its logo, and its working papers can be found at: http://www2.lse.ac.uk/economicHistory/Research/facts/Home.aspx.

During those years we enticed a small number of senior visitors to spend time with the project group at LSE and to contribute essays to this volume. We also persuaded a much larger number of participants to attend our seven workshops, and would particularly like to thank Marcel Boumans and Harro Maas, who loyally took part in most of them and contributed some wonderful commentaries on these occasions. We gained huge insight from working with all these visitors: academics, museum curators, and professionals from many fields. We had the good fortune to be strongly supported by our Department of Economic History, and by a particular group of colleagues therein: especially Patrick Wallis, who was as closely engaged with the project as any of us; Tracy Keefe, who calmly and efficiently administered the project; and Max Schulze, the late Stephan (Larry) Epstein, Paul Johnson, and Rick Steckel, who helped us get the project off the ground. We thank the British Academy, who gave us and our sister project on "The Nature of Evidence" at University College London a public space to report the project in 2007. Last, but no means least, we were generously supported by Sir Richard Brook and Sir Geoffrey Allen at the Leverhulme Trust, which (in conjunction with the ESRC) funded the project (grant F/07004/Z).

Their commitment to these "blue skies" programmes is truly admirable. We thank them all, and most heartily.

We hope that the project has proved as exciting to our many visitors as it has been to us, and that it will equally attract the readers of this volume to recognise the challenge posed in our research question and to share our own engagement with the nature of facts. We appreciate the ambition of the Leverhulme Trust, who want their programme grants to "make a difference." At the least, we are confident that the project was instrumental in turning some smart young researchers into a cadre of really good ones. But our broader ambition might be described thus: that all those who make contact with our research – via the project or this volume – will, as a result, come to think differently about those ordinary, but most important, bits of knowledge we know as "facts."

Peter Howlett and Mary S. Morgan, 2010

HOW WELL DO FACTS TRAVEL?

PART ONE

INTRODUCTION

ONE

TRAVELLING FACTS

MARY S. MORGAN

1. The Lives of Travelling Facts

Travelling Fact 1

What a clever idea to stick black silhouettes of birds-of-prey on windows to stop small birds flying into the glass! When Niko Tinbergen and Konrad Lorenz (Nobel Prize winners for their work on animal behaviour) originally showed that certain species of birds on the ground instinctively take cover in the presence of overhead moving silhouettes of such predators, they had no reason to imagine those window stickers as an outcome. Yet, their facts travelled well enough to prompt owners of glass walls around the world to take their own evasive action by sticking these birds-of-prey shapes on their walls. Years of experience later, according to other facts sent out into the public domain by reputable authorities (such as the Audubon Society), it turns out that those silhouettes don't work. Stationary "flying" predators do not scare away genuinely flying birds. (Separating the original scientific facts from their experimental context and reversing that situation subverted that instinctual behaviour.) So even while those scientific facts – still suitably qualified – have travelled well in the scientific communities (albeit with debates about how to interpret them), the efficacy of those black silhouettes turns out to be the scientific equivalent of an "urban legend." The facts travelled far, but not entirely well (Burkhardt, this volume).

Travelling Fact 2

St. Paul's Cathedral dominates the City of London skyline and epitomises the arrival in England of a new aesthetic style from Italy, and we might reasonably assume that construction methods just travelled alongside the new style. Both the extraordinary construction of the building and the career of its architect, Christopher Wren, are well studied, yet the details of how the technical facts required for its construction travelled to England and from where they came (if indeed they travelled from abroad) remain opaque. So, the historian wonders: Did the details of the construction design come through architectural treatises, or through travelling craftsmen, or through Wren's own visual inspections of such buildings elsewhere? And how do the clues left by carpenters in roof beams, joists and joints tell stories about the facts of construction itself? Was the roof built and assembled off-site and reassembled on-site like a giant IKEA flat-pack; or was it built in situ? This is the stuff of history, but a history dependent on the study of real stuff to reveal what facts travelled, raising interesting questions about the nature of facts that travel embedded in artefacts and technologies, and just what it means for such facts to travel well. The building stands – but do we yet understand the travelling facts of how it came to do so (Valeriani, this volume)?

Travelling Fact 3

We all know about climate change from the scientists, but these facts did not travel easily to us. We all know now that the world's climate is getting warmer, but for a long time, we were not very sure what facts we knew: how certain it was, how serious it was, how fast the change was happening, how different bits of evidence fitted together to form a consistent account and how far different scientists were in agreement about it. And we still don't know much about how it will affect different parts of the globe. The facts did not travel easily, perhaps because the information did not form itself into the kinds of definite, separable pieces of knowledge we think of as facts; perhaps because the implications of its human causes and its uncomfortable consequences were too severe to be accepted and perhaps because climate change itself became the subject of fictions in novels and films during this same period. But this is only part of the story, for climate scientists found their facts fiercely resisted by the interests of certain political and business circles, and even countered by facts produced by scientists in other fields. While it is tempting to imagine there is a free market in facts and that good facts will somehow travel freely of their own accord, maybe, just as "bad money drives out good money," bad facts (poorly attested, dubious, fictional) can drive good (well-evidenced) facts out of circulation. Facts require a variety of charismatic companions and good authorities to travel well, and those faced by competition, as in recent climate science, may fail to do so (Oreskes, this volume).

Travelling Fact 4

When several young men in New York consulted their doctors with an unusual coalition of symptoms in 1979, it was not clear what disease they had. But quite soon their condition came to be recognised as an early case of a new disease, the HIV-AIDS syndrome. Their facts travelled effectively through a system of medical case reporting that gives first notice of unusual combinations of facts about symptoms in conjunction with patient characteristics. By packaging the facts together, the case provided the means to recognise and define a new disease, and gave early warning of a disease that would create a major world epidemic (Ankeny, this volume). Another system of medical case reporting exists to carry facts about well-known, and highly infectious, diseases such as measles or flu. Like the "bills of mortality" of earlier times, which kept a head count of plague deaths, the individual cases of our current pandemics are gathered together and repackaged into statistical and mathematical facts that nowadays travel around communities of modellers and systems of simulations. From these, epidemiologists map and predict the spread and outcomes of such diseases and public health authorities decide the best control, treatment or vaccination procedures. Our life expectancy depends on the careful packaging of such facts and their chaperoned travels around a variety of medical establishments (Mansnerus, this volume).

1.1 The Lives of Facts

These brief accounts sketch the life stories of certain facts, life histories that are told in greater depth and detail elsewhere in this book. As they suggest, the possibilities for facts to travel well are important to our lives. We depend upon systems, conventions, authorities and all sorts of good companions to get facts to travel well – in various senses – and danger may lurk when these are subverted or fail to work. The fact that birds-of-prey silhouettes do not work to solve the problem of birds flying into glass windows tells us that a fact about the relationship between birds and their predators has not travelled intact, but it is not one (or at least, not so far as we know) that is dangerous in itself. But if our medical reporting system had not picked up and set travelling some early facts about HIV-AIDS, this could have exacerbated the dangers from the epidemic – as we see in countries that have refused to recognise the travelling facts of the disease as legitimate. Constraints on the travels of facts may be seriously detrimental to our well-being. Yet the free market may be equally problematic. The Internet is such a free market, but one in which – as is well known – it is difficult to distinguish trustworthy facts from untrustworthy ones, an age-old problem of open (or free) product markets that has led to their habitual regulation, for example, to prevent the use of poisonous additives to make bread white, or, in the case of travelling facts, to regulate the claims made for the efficacy of medicines.

But this problem of getting facts to travel well should not be seen as only a question for the public domain of science. Humanists as much as scientists should beware the trickle-down theory that they merely need to supply facts of good character and those facts will find their way where they are needed, to new homes with honest and welcoming users, professional, amateur or public.[1] Historians and novelists, as much as sociologists and economists, or medics and climate scientists, should be careful of the ways that they package their facts for successful travel and, as much as possible, take care about the company they keep while these facts are in their charge. Once facts leave home, it is more difficult to keep them safe. Historians and archaeologists often find themselves rewriting the past by retrieving lost facts that have failed to travel in replacement for better-travelled, but false, facts. Thus, the original construction of St. Pancras Station in London did not have the useful side effect of clearing a slum, as earlier generations have maintained, but of demolishing a respectable working-class neighbourhood.[2]

The life histories of facts that turn out to be false, that become corrupted or that die out make good short stories, stories that often stick better in the memory than those accounts of facts that remain steadfast throughout successful careers.[3] This bias towards revealing falsehood in our histories of facts may be because we expect specific facts not to be forever facts, either through a natural scepticism about the category of fact or because, in our experience, some particular facts turn out not to be facts after all. Not all facts travel well, some travel only to be found out and many hardly travel at all.

---

[1] See Oreskes and Conway (2010).
[2] See Swenson (2006).
[3] For example, the BBC Radio 4 programme "More or Less" about numbers in the public domain is full of such stories of misleading or mangled facts that travel well precisely because their falsity has made them more dramatic than they really are, for example, a misquoted fact about the proportion of women whose life is cut short by domestic violence (15 May 2009: http://news.bbc.co.uk/2/hi/programmes/more_or_less/8051629.stm). The programme hardly mentions those straightforward stories of facts that travel with integrity, and also misses the kinds of extraordinary and successful stories that we find in this volume.

Yet, many facts do travel well, retaining their integrity when they do so, for we all regularly transfer and make use of facts – without subverting them – in new contexts, often without even noticing that we are doing so. The research project that created this book set out to look for those travelling facts that we do not normally notice. When we asked, "How well do facts travel?" and looked for answers in the travels of facts (rather than knowledge flows more generally) and focussed our attention on the facts themselves (rather than on the people and communities through which they pass), we found that many facts do indeed travel far and wide to new users and new uses.[4] And their trajectories were so extraordinarily varied and sometimes so completely unexpected that we feel justified in saying that, just like some experiments and models in science, some facts acquire an independent life of their own.[5]

Even so, as we found in our research, it is not always easy to figure out why those facts that travelled well did travel or, indeed, what exactly travelling well means in any particular context, for the extent of such travel raises its own puzzles. In travelling to other spheres and in being used to address other questions, we found that facts may grow in scope, sharpen or become more rounded; they may acquire new labels and fulfil new functions, even while they maintain a strong hold on their integrity. It is through these processes that facts produced in one locality come to speak with authority to other questions, even to other fields, times and places. By following these independent lives of facts, we not only found answers to the question, "How well do facts travel?" but we began to understand how it is that facts come to play foundational roles in situations beyond those of their production and original usage.

2. "A Fact is a Fact is a Fact"

Facts seem such obvious things: We think of them nowadays as settled pieces of knowledge that we can take for granted. And while individual or particular facts may be seen as important or striking within a particular field, considered as a general category of knowledge, facts seem less problematic than the elements of evidence, theories, hypotheses or causal claims that appear in both our humanities and sciences, and less colourful than the

---

[4] Information about the project funding is found in the Acknowledgements note. A full record of the research project can be found at http://www2.lse.ac.uk/economicHistory/Research/facts/AboutTheProject.aspx

[5] On the life of experiments, see Hacking (1983) and Shapin and Schaffer (1985); on the life of models, see Morgan (forthcoming).

characters or cases that appear in our narratives, histories and philosophies. But facts are not quite such straightforward things as they seem.

First of all, our research led us to take a generous view of what facts are and where facts are found, for they come in a bewildering variety of forms in those various communities of scientists and humanists that use them. Facts may be expressed in linguistic statements or as bits of digitised information; they may appear in pictures, diagrams, models, maps, documents, biographies or novels; they may be found as material facts located in artefacts such as mediaeval swords, or expressed in the behavioural characteristics of crowded rats or the healthy growth of fertigated plants; they may be found in the fragments from ancient civilizations, in the fossils of long-dead nature or as numerical constructions about the future of our overheated planet. And we found that facts can be little (observations on the buds of a specific flowering plant, data points on a weather graph), big (the regularity of business cycles in modern economies), singular (the age of a particular person), come in crowds (infection rates) or be generic (the alpha male in romance fiction or the exit pattern of firms in declining industries).

Indeed, facts come in so many guises and sizes that it proves difficult to produce a sharp description about what counts as a fact, particularly one that would cover the many times, places and fields that we studied. Nor did the presence (or absence) of the word itself give us any natural starting point for our investigations. In some fields, scientists are profligate in their use of the term (as in parts of the life sciences), while in others (as in parts of economics and physics) the term "data" is preferred for something we might label small facts, while still other scientists might refer to a well-attested "phenomenon" for something that we might label a big fact or a generic fact.

Nevertheless, facts are a usable category, for, in our experience, all communities have some kinds of things that they take to be facts or factlike: shared pieces of knowledge that hold the qualities of being autonomous, short, specific and reliable. These are the qualities that make us think "a fact is a fact is a fact" – wherever it is, for whatever purpose it is used. These are the qualities that enable settled pieces of knowledge to travel (assuming that they are communicable or transportable in some way or another) beyond their place of origin to be used in those new contexts.

These qualities of the things that are taken as facts have historical roots of course. The notion of fact began in legal circles in the mediaeval period according to the account by Barbara Shapiro: so that by the early modern period, the actions that occurred (the murders, frauds, etc.) were referred to as "matters of fact," drawn in contrast to "matters of law" and thus making sense of the otherwise strange phrases "before/after the fact."[6] This sense of facts as deeds or actions seeped from law into history as the narration of well-evidenced facts – the deeds of history – in the sixteenth and seventeenth centuries, and thence into reports of newsworthy items. And this idea of facts continued to hold sway into the nineteenth century at least, as we see in the contrast drawn between facts as deeds versus words found in: "Gracious in fact if not in word" (Jane Austen: *Emma*, 1815).[7] From law and history, the notion of facts moved into natural philosophy and modern sciences where, as Steven Shapin and Simon Schaffer recount, "matters of fact" – properly witnessed, experimentally produced events – came to be distinguished from their interpretation.[8] Lorraine Daston takes up the story to recount how facts – as those noteworthy and particular "things" – grew in scientific circles into a "form of experience most sharply distinguished from 'hypothesis' or 'conjecture.'"[9]

These historically formed qualities invade the current sense of facts in ways that are important for our project of understanding why facts can travel well. Facts are "independent" of their explanations – a quality that goes back to their legal sense discussed by Shapiro, where "matters of fact" – deeds or actions – are established independently of their motivations. Just as in law and history, where facts were not to be conflated with the causes of those facts or with the evidence advanced for them, facts in science were, as Daston tells us, "in principle, strictly independent of this or that explanatory framework."[10] These historical roots tell us why facts – as pieces of knowledge in their various senses and guises – are understood to be independent of their explanations, causes and motivations, and so are free to travel without reference to them.

---

[6] Shapiro (2000) and her earlier shorter version of the argument (1994) argues for law as the field within which the notion of facts emerged into a mature idea. See also Poovey (1998) who grounds the notion in early accounting and the collection of essays on the history of facts in Cerutti and Pomata (2001).

[7] See p. 231 of 1971 edition published by Oxford University Press.

[8] See their, 1985, Chapter 2 ; see also Haycock (this volume).

[9] Daston, 2001 (English version p. 6).

[10] Daston goes on, "They can therefore be potentially mobilised in support of competing theories, and, again in principle, endure the demise of any particular way of explaining a phenomena" (Daston, 2001, English version, p. 6). There is an intimate relation between "fact" and "evidence," yet the distinctions and relation of facts and evidence seem to be a matter of local usage varying over time, country and disciplinary usage. Although Barbara Shapiro's (1994 and 2000) concentration on the legal framework in early modern times shows how evidence and witnesses were needed to attest to, and so establish, matters of fact, Lorraine Daston's (1991) writing about early modern sciences portrays facts as the jigsaw of pieces that create evidence for a hypothesis or conjecture.

Facts are also "short" – an epithet that Daston (2001) uses to capture the particularity of facts in early science in a way that is still shared not just across the sciences but into the humanities. This recognisable quality of facts from those earlier times is described more fully by Ernest Gellner when he observes that we use the term facts to refer to "concrete" and "specific" events, objects and findings rather than to those things we describe in "abstract" terms.[11] This particularity has implications for the way that facts travel to find new uses and the new uses to which they are put, but does not make them any less transferable than the abstract ideas, metaphors, stories or theories that also travel well between fields.

These historical notions of facts are clearly manifest in our modern views of facts, where current definitions rely on drawing definitional contrasts, but not necessarily opposites (as we shall see in what follows), just as earlier ones did.[12] Now we find:

- facts versus evidence and inference (in legal fields)
- facts versus fictions (in the humanities)
- facts versus hypotheses, theories or interpretations (in the sciences)
- facts versus the untrue and surmised (in both everyday life and in philosophy).

Common to all those contrasts is the non-conjectural quality of facts. Facts are *not* fictions, theories, inferences or the merely surmised. This non-conjectural sense that facts carry is captured more positively by describing them as "useful and reliable knowledge" (a phrase from historians of technology).[13] Of course, not all facts are especially useful, and not all useful and reliable knowledge has fact-like qualities (for such a phrase equally captures rule-of-thumb knowledge and more general or abstract knowledge). The point is rather that the sense of facts as "useful and reliable" not only helps to rule out both superstition and opinion, as well as the conjectural

---

[11] Gellner (1964), p. 255.

[12] Evidence from the Oxford English Dictionary, online version.

[13] The first usage of this phrase seems to date from Nathan Rosenberg (1974, p. 97) to describe scientific knowledge that could form the basis for technological knowledge. The term is used here in its current and more generally used form, dating from the 1990s' "Achievement Project" sponsored by the Renaissance Trust in which economic historians (particularly Patrick O'Brien and Ian Inkster) and historians and sociologists of science (amongst others) were engaged in figuring out why some technologies were developed in some countries and not in others. (See, e.g., Gouk 1995 and Inkster 2006.) In this context, the phrase "reliable" refers not necessarily to a scientific source but to the usefulness of a technology, where the addition of "reliable" seems to imply tested by patent or experience or the market in many conditions and circumstances.

aspects of knowledge, but also points to a certain steadiness, even sturdiness, in the quality of facts that makes them sufficient for people to act upon them or use them in support of their actions.[14]

Facts are a form of shared knowledge: They have a public or community aspect, as is evident also from their historical roots.[15] And we have relied on the communities we study – that send, receive and use facts – to reveal what counts to them as facts (even where the word itself is not used), that is, as pieces of knowledge established according to their standards of evidence of discipline, time and place. Such matters of facts should be understood then, *not* as an expression of that community's *belief*, or *opinion*, but rather that such a community has good reason to take those things as facts, and will be likely to have the confidence to act upon them as facts.

Given that we are taking a community's view of what their particular, well-evidenced facts are, we have not been (by and large) concerned with judging whether those community facts are facts according to any meta-standards beyond their own ones. That is, our project *as a whole* did not set out to determine the truth of any particular facts; indeed, we could not do so without sharing that same local field knowledge that would enable us, for example, to recognise the facts in the diagram of a worm's nervous system or in the statistics of death rates. (Of course, individual authors in this book do have field knowledge and may make such judgements.) Nor did we set out to discuss the meaning of facts as a general abstract category – the province of philosophical argument. Rather, we are interested in how facts – bits of knowledge taken by a community to be true – travel, and so our accounts of what makes some facts travel well cannot distinguish between those that are true facts and those that may later turn out to be false facts.

But this does not mean that we totally put aside all the interesting issues of true versus false facts. We recognise that in travelling freely from their original communities over time, space and discipline, there are many chances for facts to be challenged. Some of those facts that travelled well initially later turn out – on the judgement of their relevant communities – to be partly true facts, dubious facts, uncertain facts or even false facts, and this changing judgement of a fact's status can be an important part of what happens to travelling facts. Thus, we include narratives in which travelling facts have been strongly disputed by others in the community, as in the climate science case (Oreskes, this volume), or where facts established and carried along by one community turn out much later after successful travels

---

[14] See also Mansnerus (2009).
[15] As pointed out by Weirzbicka (2006).

to be judged by a later community as false, as in the case of the remarkable longevity of Thomas Parr (Haycock, this volume). We include accounts of facts that turn out to be misleading, such as the date written on a church roof suggesting it was a century older than the wood from which it was made (Valeriani, this volume), along with cases in which fictions declare themselves as facts and vice versa in modern science writing (Adams, this volume). Scientists and humanists themselves have the knowledge to dispute and overthrow particular facts, just as it is they who find, judge and use travelling facts.

Although facts may not be easily described except by their qualities as autonomous, short, specific and reliable pieces of knowledge, nor easily defined except by their contrasts, they are, however, quite recognisable in the field. Their life histories can be traced and documented, examined, explained and understood. Our primary interest in these life stories of facts needs to be remembered. It is not: How did they get to become facts? but rather: What makes some of them travel, and travel well?

3. Travelling Well

Although the travelling facts we study are ones defined and understood within their relevant communities of usage, we, in our commentaries, are responsible for defining the natures of what it means for facts to "travel well." Our answers to the challenging question of our title are first, that facts travel well when they travel with integrity; and second, that facts travel well when they travel fruitfully.

"Travelling with integrity" captures the idea that the content of the fact is maintained more or less intact during its travels. Rather than worry overly much about what this means in abstract terms, it is better to think about it in mundane terms – if a fact changes so much during its travels that it is not recognisable as the same fact or has lost its credibility as a fact, it would be hard to claim that the fact had travelled well. This quality is one of degree, and, once again, is difficult to describe but recognisable in the field.

"Travelling fruitfully" refers to a fecundity in travel. The obvious aspect of this is that facts may travel far and wide in terms of time and of geographical and disciplinary space to find new *users*. More unusual, perhaps, are the ways in which facts find new *uses*: They gain new functions, coalesce in new patterns and make new narratives. Facts may even surprise us *by travelling someway across the space mapped out in those definitional contrasts* (noted earlier), towards fictions, interpretations, theories and so forth.

3.1 Travelling with Integrity

Late-twentieth-century debates about scientific and humanistic knowledge, or its impossibility, were concentrated on the production side and gave us two ways to think about facts. On the one hand, facts are understood to be *found* or *discovered* only after much labour in laboratory, field, archive or museum. An alternative view focussed on the social networks and practical instruments by which – and with equally hard work – facts are *constructed*.[16] Yet, however relevant this dichotomy between *finding* and *constructing* (between an older tradition in sciences and humanities and more recent post-modernist views about both) has been for the emergence of facts, it is not so clear how far these positions prove salient for exploring questions about the subsequent travels of facts. However, they do form the background to various notions, explored within the sociology of science community, of how knowledge circulates, which offer revealing contrasts for our own discussion of the integrity of facts in their travels.

Two strong positions on the travelling possibilities of factual knowledge are due to Bruno Latour and Ludwik Fleck.[17] In Latour's account, the "marks" of science are highly mobile but travel well only if they are immutable, presentable, readable and combinable. From our viewpoint, it is their immutability that will ensure that these marks carry their integrity with them while they travel (though their potential fruitfulness, when used in other contexts, seems dependent on the other qualities he cites). In contrast, in Fleck's account, facts are developed and understood only *within* knowledge communities. As knowledge travels from one community to another, it has to be translated and in the process changes to some degree its meaning and thus, necessarily loses some of its integrity in travelling. Our sense of facts, and of what happens to them as they travel, fits untidily between Latour's marks and Fleck's community facts.

At first sight, our view of facts seems to share more of the qualities of Latour's immutable mobile marks than they do with Fleck's facts, which appear as elements in a clustered and multi-dimensional bundle

---

[16] Although it might seem that in the former view facts would be conceived as hard and objective, while in the latter plastic and subjective, this is not quite so. In the traditional history of science, hard facts are subject to revision over time, while in the latter social studies of science, facts are treated as stabilised pieces of knowledge

[17] See Fleck [1935] 1979 (especially taken in conjunction with his 1936 paper) and Latour 1986 (especially taken in combination with his 1999 paper). There is, of course, a vast literature on knowledge diffusion (not just in science and technology studies); some items are referenced later, while others are discussed elsewhere in the volume.

of knowledge. Some of the things we might think of as small facts seem to be like Latour's marks, such as data observations in economics or the digitised records of plants in bioinformatics. Certainly, facts share with marks the notion of being hardened, stabilised bits of knowledge wrested by humanists and scientists from their investigations. They also share the notion that they are not necessarily linguistic things. Moreover, some little facts, like marks, are often combinable, presentable and readable, but others are not: While the carpenters' marks in St. Paul's Cathedral are obviously immutable marks and are readable to the expert, they are not combinable or mobile in Latour's sense. So our elastic sense of facts seems consistent with understanding some form of facts as marks. But many of the facts we are concerned with seem to be bigger kinds of things, less raw, more produced and with more value added, such as the profitability of an array of companies, the location of particular genetic material or scale diagrams of a specific Greek temple. All of these seem to have more autonomous content than the notion of marks, enough certainly that such facts can separate off, travel well enough on their own and be used elsewhere without having to be combined with other little facts.

On the other hand, our facts, big and small, seem more independent than the fact that Fleck discusses at length, namely that "the so-called Wassermann-reaction is related to syphilis."[18] Fleck suggests that the qualities of definiteness, independence and permanence can only be associated with "well-worn" scientific facts, ones that are already in wide circulation. In contrast, his fact is newly developed and so sits within a whole set of notions about syphilis that includes the disease's symptoms, causes and cures, and the concepts, theories and ideas that hold all those elements together. Facts are found within such integrated clusters of knowledge elements that coalesce together within a local disciplinary "thought collective"; and just as "ideas" and "thoughts" (with which facts sometimes seem interchangeable), they only make full sense within that community and can only be properly expressed in that community. This makes Fleck's community facts vulnerable, for travels of any of these elements beyond that community to another community inevitably involve some transformations in meaning. These transformations are likely to be magnified in travel, for he describes the various processes of communication beyond the community as ones of "propaganda, popularisation and legitimization."[19] So we share Fleck's view

---

[18] See Fleck 1979, p. xxviii.

[19] So perhaps we only recognise or notice that a fact has travelled when it has changed its meaning. For Fleck's account of these processes of circulation, see his 1936/86, pp. 85–7.

that a community defines its own facts, but not his view that the circulation of such knowledge *necessarily* involves its transformation. Nevertheless, Fleck's attention to the ways in which travel may subtly change the nature of the knowledge transferred is very much in tune with our ideas.

When we follow newly made facts, we do find them travelling across space, time and disciplines without major translation as they pass passport control and so without any substantial loss of integrity. Alison Wylie's chapter on archaeology (this volume) offers the materials to distinguish between the facts found in the artefacts versus the facts that we infer or interpret from the artefacts. For example, skeletal remains found in the "eminent mounds" of central North America were interpreted as evidence of violent funerary practices. The facts of the bones and the facts about the practices may both change over time, over space and over disciplinary travels too (for new techniques may reveal new facts previously concealed in the bones, and new comparisons cast interpreted facts in a new light). But we don't expect the facts found in the material artefacts or the facts inferred from the artefacts to change *just because* they have travelled over time or across community boundaries. We also find facts being used in new ways after their travels, where this change of use does not go along with any radical change of meaning for the facts. The ways in which the alpha male character transferred from an account of animal behaviour into evolutionary psychology and thence into romance fiction, where he figured as the Alpha Hero (and perhaps anti-hero) at the same time as being the subject of a feminist critique and rebuttal by such authors, offers a truly surprising story of the steadfast quality of the character. That this fictional avatar was then taken as evidence for his own factual character in his return home to evolutionary psychology is even more surprising (Schell, this volume).[20] The travels of facts are, it seems, sometimes stranger than fiction.

Whereas Latour's little marks are hard enough to remain immutable in their travels beyond their site of production and into other communities, and Fleck's expansive facts are in danger of mutating the moment they step across the community's threshold, our experience of travelling facts lies somewhere in between. Our experience of the bits of knowledge that we understand as facts is that they have sufficient autonomy and separability to be quite mobile without losing their integrity of meaning compared to Fleck's community and multi-connected facts, but that they rarely carry the hard immutability of Latour's marks when they travel. We rather think of travelling facts as rubber balls: They have a certain shape; they can be

---

[20] See also Adams (2006).

carried, rolled, squeezed, bounced, kicked and thrown without harm to them; and they can be used in many different ways and in different situations.[21] This analogy recognises the virtue of steadfastness or sturdiness that we find in Latour's marks, but also of a certain degree of useful mutability around the edges found in Fleck's community-constrained facts. A good example is found in the travels of the classical style of architecture from ancient Greece to nineteenth-century America. Certain details of the style were altered in the process of adaptation, such as the nature of the materials, the addition of windows and the reversal of light and shade in the exterior. While the community of American architects and builders seemed to delight in their ability to adapt the stylistic facts of classical architecture to their own contexts, they also observed certain boundaries. For despite their additions, alterations and subtractions, there is no doubt that such buildings retained a recognisable integrity as examples of "the classical style" in their new domain (Schneider, this volume). This is what we mean when we suggest that facts that travel well exhibit a strong degree of integrity, but they also have a degree of squishiness, a squishiness that may result in them getting their hard edges rubbed off, changing their surface elements or gaining some additional covering as they travel.

Facts may be more separable or modular than most bits of knowledge, but they are not usually born bald. They often have details that we might call "qualifications": circumstantial or contextual elements attached that may be shed on their travels so that they become smoother or rounded off. The nature of this smoothing turns out to be quite field-specific. So, for example, Peter Howlett (2008), in his research on social science debates about the Green Revolution in Asia, found that when economists imported facts from other social sciences, they tended to turn them into generic facts, shearing off the woolly details of time and place that they think hides the important body of the fact beneath. Anthropologists, on the other hand, like to carry over many of the details of content and context when they report case facts from other disciplines; indeed, they would hardly count as facts for their field unless they arrived fully coated. For anthropologists, historians and archaeologists, such circumstantial details are not just the signs of authority that establish the credibility of the author's knowledge, but are part of what it means for something to be a fact in the first place, so their facts tend not to become smoothed out or rounded off by their travels.

Facts may also pick up extra elements on their travels and become covered with additional elements or even sharpened in certain ways. For example, the

---

[21] Thanks to Naomi Oreskes for suggesting this rubber ball metaphor.

set of facts that constitutes the first case report of a new medical syndrome includes both a collection of symptoms and a collection of patient details. As the case moves though the medical system, it gradually emerges which of the original facts about symptoms are related and so become more firmly attached together, which patient facts are really relevant and become more sharply defined, which additional facts are needed to enrich and complete the case and which facts are red herrings to be dropped off as irrelevant details. This process of defining the fact of a disease syndrome entails both smoothing off the edges and sharpening up with additional details at the same time (Ankeny, this volume). A somewhat different process, but comparable in effect, happens when data facts travel around modelling communities: They may be sharpened by specifying their numerical attributes more clearly or have certain details of the process they present clarified to make them operational for simulations, both in ways that maintain the basic integrity of the facts (Mansnerus 2009a and this volume).

A comparison of these two cases with the processes and outcomes of gossip and rumour illuminates this sharpening and enriching of the fact as it travels. Gossip is thought to destroy gradually any initial nuggets of facts that were involved in the first place: The content of the fact is changed each time it is passed along, just as in the game where children repeat a whisper around a circle. Rumour, in contrast, is usually characterised as a process of additions: The claim gains the kind of detail we associate with the factual, but is, in reality, a process of adding falsehoods, as in wartime rumours of invasion.[22] So either the original travelling fact disappears in gossip or it collapses under the weight of additional false details in rumour. Yet, when we made such comparisons, we found many examples of travelling facts that do indeed maintain their reputation for integrity, even while they gained some elements, perhaps becoming sharpened with qualifications or enriched with valid details, in ways that mirror the loss of others in the smoothing process.[23]

These comparisons with other accounts of travelling facts suggest that travelling well can be thought of as the maintenance of a certain basic integrity in the travelling fact, but it does not mean that nothing happens to facts as they travel. As we have seen, some become simpler and lose information while others add information and become more complex as they travel. Facts may not travel entirely intact without some kind of wearing

---

[22] See Knapp (1944) and Buckner (1965).
[23] These processes of enrichment are paralleled in the historical account given by Hasok Chang of the facts about the temperature of boiling water (2007).

down, rounding off or enriching and sharpening: processes that nevertheless enable facts to retain their original content. Facts are relatively hard,
independent things after all, though this does not mean that they are never
subject to different interpretation or to re-formation in their travels. It is
more challenging to understand how facts maintain this integrity when, in
reaching new homes, adapting to their new environments and finding new
uses there they still manage to retain their basic content.

3.2 Travelling Fruitfully

Travelling fruitfully offers another sense of travelling well, a complementary
one, not an alternative one, to that of travelling with integrity. Travelling
with integrity points to stability in the content of the fact. Travelling fruitfully points to the scope of travel where three dimensions prove particularly rewarding to study: the ways that travelling facts find new *users*, find
new *uses* and even begin *to invade the definitional contrasts* that we drew in
defining facts.

The most obvious dimensions for facts to travel to find new *users* are
those over time and space. Just because this notion of travel is the obvious
one, however, does not mean that it should be dismissed lightly, for such
dimensions provide some of the most interesting and unexpected accounts
of fruitfulness in travelling facts, of which the travels of classical architecture are one example. These travels (discussed earlier) were not quite what
they seemed: It was not simply that the facts of ancient Greek architecture
travelled over space and time to nineteenth-century America. Rather, it was
the pictures of eighteenth-century reconstructions of such buildings – the
German Romantics' vision of that architecture, and French and English
scale drawings of those temples – that travelled to America, even while
some of those original buildings were being destroyed. These were picked
up by American master builders who created a nineteenth-century vernac-
ular version of these eighteenth-century European views of the temples of
that old civilization in their new country's official buildings, their banks and
their homes. The fruitfulness of these travelling facts is there for all to see
(Schneider, this volume).

This focus on users points to the importance of the demand for facts in
determining the fruitfulness of their travels. Most facts don't travel very
much because no users find them or choose to use them. Demand is critical.
Here, the recent literature on co-construction of technology is suggestive,
for it treats users, as well as producers, as active agents in both the production and usage of such artefacts, and it assumes that technologies may

find new usages or resist their intended usage, just like travelling facts.[24] Although facts may be paired with artefacts in another definition, that pairing may be one that holds little contrast – for we can understand technologies as embedding facts, much as ancient pottery jars carry facts about the nature of the goods stored within them, be it wine or oil. Certainly, our cases suggest that facts are carried in intimate association with technologies and find new users because of that association. The fertigation project in Tamil Nadu studied by Peter Howlett and Aashish Velkar (this volume) showed not just how the facts of the technology – a precision farming technique that delivered water and fertiliser (fertigation) to scientifically established standards – travelled to farmers (and from those farmers to other new farmers), but that associated facts (both experiential facts about other aspects of farming and institutional facts about marketing processes) were pulled along by the successful travels of the fertigation facts.

An equally fruitful aspect of travels is when facts travel to find new *uses* in other disciplines. Here the case of Calhoun's rat experiments is most informative. How could it be that a set of experimental facts about rats – some with pessimistic and others with optimistic outcomes – set off a chain reaction in the many other fields to which those facts travelled? From the base field of animal behaviour, these facts about how rats behaved when they were crowded in pens travelled to prompt parallel studies of human psychology and the nature of crowding, and thence to the re-design of prisons and college dorms and into science fiction, most notably in the children's book and film *The Rats of NIMH*. No measure of travelling well is needed to be amazed at the disciplinary reach of these travelling facts and the fruitfulness of the new uses to which they were put (Ramsden, this volume).[25]

Travelling fruitfully in such ways might, of course, go so far that the facts might be in danger of losing their integrity; some users recognised this problem and set limits on the way that their travelling facts could be used. In the Indian agricultural scheme, the scientists set rules for the adoption of the technology package they offered in order to ensure that their facts travelled "precisely," that is, with integrity, but they allowed experimentation and potentially fruitful adaptation to other facts that they set travelling on related aspects of cultivation. Thus, the farmers of Tamil Nadu enjoyed experimenting with crop spacing, but were more cautious about taking liberties with the fertigation technology (Howlett and Velkar, this volume). In the animal behaviour studies initiated by Tinbergen and Lorenz, the

[24] For example, see *How Users Matter* (Oudshoorn and Pinch 2003).
[25] See also Ramsden and Adams (2009).

nature of this trade-off between integrity and fruitfulness seemed to have been recognised explicitly in arguments between practitioners in the field about what constituted a legitimate comparison between the behaviour of different species, for example, between chickens and geese, or between other animals and humans. The limits of travelling were fixed by the limits of "meaningful comparison," which ensured that the facts retained their integrity in travelling to other sites. In other words, travelling well – with integrity – was recognised by commentators as a valid boundary to their fruitful use of imported facts (Burkhardt, this volume).

These considerations and case histories point us to another element of the notion of fruitful travel. We can recognise that facts have travelled well not just when they are used somewhere else *by* someone else, but when they are used *for* something else. This entails another user acting upon them or with them in support of another claim, or as inputs or nodes of intervention, or for some new purpose not the one envisaged in their production. At its most fruitful, the use of travelling facts creates a new pattern, a new coherence, a new narrative or fulfils a new role.

A good example where a travelling fact creates a new coherence or pattern is found in the way that a single new fact, when added to a set of apparently disparate facts travelling together, produces an account of a disease syndrome. As we find in Rachel Ankeny's chapter (this volume), for example, it was not until the single, but overlooked, fact that most of the sufferers from a particular condition were menstruating teenage girls that the cause of toxic shock syndrome was located in the new materials in tampons. Once this quite ordinary fact travelled into the case report, it created a new generic fact – about a quite extreme medical condition – that proved powerful enough to make the case travel through medical science. We see from such examples how a particular fact can sometimes pull other unconnected facts into a coherent account.

The ways that individual facts travel into and are thrown out of consideration in the process of medical diagnoses appears to have much in common with the way that detectives are thought to work. Dorothy Sayers suggests that particular "facts are like cows. If you look them in the face hard enough, they generally run away."[26] This claim comes within the genre of detective novels, in which the detective's task is precisely to show how some facts that travel into a case are either misinterpreted or not really facts at all. Although these insouciant heroes generally manage to make such awkward facts run away – either into fictions or into some other facts – they do not do so

---

[26] See Sayers ([1926] 1987).

without using "the little grey cells," smoking opium or other, more physically active, investigations. In this genre, such very particular facts stand inconveniently in the way of a narrative that will fit a series of other particular facts together so as to unveil the means and perpetrator of the crime. But the conventions here are clear: The reader does not know which "facts" are going to turn out to be true, though, of course, all the important "facts" in the narrative, true or false, are understood as fictional.

Such particular facts with the potential to make narratives also inhabit history, where Carlo Ginzburg ([1986] 1989) has famously aligned the mode of micro-histories with the detective novel. Here the historian – like the detective novelist – hopes that his or her particular clues, facts gleaned out of the dust of archives or the corners of buildings, will travel together to make a powerful narrative. These are not metaphors: Simona Valeriani's chapter (this volume) considers the travels of the material facts embedded in buildings. Facts travel, first of all and most obviously, in the timbers and joists of the roofs she studies: facts demonstrated in the materials and in their arrangements. But there is another level here, for when the roof structure or its design drawing is in the "wrong" time or place compared to the local ways of doing things, it sets off alarm bells for the historian to seek answers to how these "foreign" facts became embedded in such artefacts. Making a narrative account here requires not just detective work into the building record, but also an encyclopaedic knowledge of other buildings and of what is typical at that time and place so that the dissonance of a roof joint in a strange place and strange time – a travelling fact out of its normal home – crystallises a new explanation of just why a particular roof is like it is (see Valeriani 2008).

Other travelling facts move their attachment from the well-defined and perhaps small arena of their production to a wider space in which they play a broader function. Sometimes, this is a straightforward process of aggregation of small facts. Facts about deposits and withdrawals collected by banks for their own set of purposes travel to the central bank to be assembled and re-calibrated into bigger facts about the aggregate amount of money in an economy on which monetary policy is made.[27] Archaeological facts about bones held in many museums travel together to become important facts in historical and anthropological debates about the standard of living of different peoples of time past.[28] These bigger facts carry a wider level of generality because they are built up on multiple observations or on multiple cases to

---

[27] See Morgan (2007).
[28] See Steckel and Rose (2002).

describe a regularity or phenomenon, and so are already well connected to a body of other facts. These kinds of facts do not show the same tendency to run away or run together; indeed, they tend to stand their ground, although they, too, can prove vulnerable (e.g., hitherto reliable facts about aggregate money may prove unreliable when individuals change their behaviour).

In other cases, a more unusual move occurs. The facts of Tinbergen and Lorenz's work on ducks, geese and chickens became critical in the arguments over whether certain animal behaviour is, in fact, instinctual or not (Burkhardt, this volume). And we have also seen how the facts of the behaviour of Calhoun's rats in his experimental set-ups came to represent the effects of crowding at a more generic level (Ramsden, this volume). Whereas philosophers have the notions of tokens and types and particulars versus universals as ways to describe and analyse different levels of claims, and statisticians have rules of inference that relate specific findings to generalizations, the interesting problem here is to describe and explain how a fact about particulars turns into a representative fact or a generic fact during the process of travelling from one domain to another. This may occur because someone sees and labels the travelling facts as having that wider scope. But such labelling in itself does not extend the scope without users appreciating the relevance of these travelling facts in the construction of other bigger or broader claims. When particular travelling facts come to stand as evidence for a wider claim, or to stand in for a more general class of findings or to be seen as relevant about a further related field, all in ways that are legitimate and maintain the integrity of the facts within those communities of users, then we can say that such facts have also travelled fruitfully.

A more speculative sense of this notion of travelling well takes us back to our set of defining contrasts. Some of the ways that we have found facts travelling is towards or away from their ostensible contrasts in various ways: for example, facts travelling to fictions, or vice versa, or to and from hypotheses and models and so forth. These kinds of morphing of facts happen quite often and threaten the identity of facts as facts, even where their travels may prove fruitful.

The most obvious, and least problematic, of this morphing is the move between facts and evidence; there is, after all, an intimate relation between them. Some humanistic and scientific fields use a terminology in which facts are produced as evidence for claims, others that evidence is gathered to establish facts: two sides of the same coin. Facts constructed at one level in one domain might serve as evidence at another, as facts established by forensic science become evidence for inferences about a crime. Sabina

Leonelli (2008) discusses how small facts get used beyond their production locality as evidence to support bigger or wider claims and facts in biology. Wylie's chapter (this volume) has these two levels separated off as facts of the record versus facts about the record, where the latter might be termed the inferred facts. This kind of move happens, for example, where little facts (pottery shards) serve not just as inputs for establishing bigger facts (about burial rituals), but for establishing facts about other locations (trade routes in ceramics).

Facts travel into fictions in many guises. Both the sciences and the humanities use folk-facts, factoids and fictions in a variety of ways to make facts travel more easily through narratives. Marcel Boumans has suggested that this fact/fiction ratio varies: Sometimes it seems to be greater than 1 – fictions, lacking the moral fibre of facts, prove less able to travel; and sometimes it seems to be less than 1 – when facts need some extra "colour" to set them travelling.[29] But it is not just a question of the use of fictional devices – such as narrative and character – designed to get facts to travel better. As Jon Adams's chapter (this volume) recounts, the recent explosion in science popularisations has blurred many of our normal distinctions between facts and fictions. We find fictions being constructed or disguised as facts in novels about climate science, and fictional stories being used as facts in polemics about population explosions. This has become such a slippery slope in modern writings that the reader may lose all sense of whether the claims are factual or fictional. This slipperiness is quite different from the way that scientific characters, such as the alpha male, come to be seen as inhabiting fictional domains such as romance fiction (Schell, this volume), not as a popularizing or pedagogical device for science, but simply as a way of defending the quality and integrity of the life presented in novels. By way of contrast, in an earlier period, Swift's *Gulliver's Travels* and Defoe's *Robinson Crusoe* were fictions but were passed off as witnessed, and so factual, accounts.

Facts about the past are equally liable to turn into fictions about the past. The Eyam story of Patrick Wallis (2006) tells how the facts about the plague of 1666 in the Derbyshire village of Eyam were transposed by early nineteenth-century Romantics whose poems both transformed the mining parish of those days into a rural idyll and made a hero out of the rector who corralled the parishioners and quarantined their infection into the village during the plague (and so possibly increased their death rate). The slippage

---

[29] I am indebted to Marcel Boumans for this formulation by which he summarised our two-day project workshop on the *Fact-Fiction Ratio*, held on 12–13th April, 2007.

of fact into poetry and fiction turned a story of nasty disease into one of sacrifice that turned the village into a nineteenth-century tourist site, and thence into a twentieth-century heritage site and the subject of a modern novel before a musical version at least returned the village to its industrial setting.

We can also see how science-based fictions – that is, predictions – become sufficiently fact-like to enable people to act upon them. For example, predictions about the future economy have to become not only sufficiently stable, but also congruent in a particular framework with facts about today and about the past – that is, they have to be brought into a perspectival relation with facts – before they can be acted upon on as facts. These perspectival frameworks in economics were first developed on the basis of conceptual descriptions of the economy and theories about its working behaviour. These important abstract elements have long since disappeared from users' views so that the gross domestic product (GDP) figures that are gathered and collated to measure all economies are now understood to produce a set of neutral accounts that join facts about the past with ones about the future (predictions) in order to provide the basis for economic policy (see Morgan 2008). More broadly, those GDP figures travel freely in the domains of political economy as the facts that align political ambitions with what it means to be a "modern" or "developed" economy in the latter half of the twentieth century (see Speich 2008). Similarly, the climate science account given by Oreskes (this volume) tells a tale of a battle of predictions: The pessimistic predictions that required changes in behaviour had to be sufficiently well grounded with the facts of today and of times past to counteract the optimistic predictions based on other theories and other experiments that were peddled – with the help of market research – to make those "good news" predictions travel better than the "bad news" ones based on sturdier facts.

The more obvious travels of scientific facts towards their contrasts might be found in the intimate practical links between facts and models. The career of Manning's *n*, a factual measure of the roughness of river beds, veers between being a set of numbers learnt through experience by hydraulic engineers, to a number chosen by less-learned practitioners on the basis of photographs of typical rivers, to a number chosen by users to plug in to a computer software programme in flood defence models, to the outcome of a theoretically based equation that estimates the number on the basis of lots of other facts about water in a river system. Its multifaceted career as a fact reflects not just the difficulty of gaining information about a river bed's roughness, but also its practicality to survive well enough from when

Manning developed his formula and measure in the context of nineteenth-century Irish drainage systems into modern flood calculations (Whatmore and Landström, this volume).

These more adventurous travels of facts, and their morphing into their ostensible contrasts, have to be carefully looked out for. Facts and fictions do not necessarily support each other, any more than facts and hypotheses do, in ways that are consistent with our ideas about facts. The lines between facts that are popularised in fictions or the fictions that are used to shore up facts, or between facts that are really masquerading as hypotheses and hypotheses that are really established facts, are ones that have to be enquired into at each particular level and case to see how and where integrity of the travelling fact – that is, its status as a fact – is maintained. Even so, thinking about facts travelling over the bridges between these contrasts does illuminate another way to understand the notion of facts travelling well, for, once again, these travels may be fruitful in both the humanistic and scientific domains, renewing tired genres of fiction as much as enabling governments to take action on climate change.

"Travelling well" has been constructed here in terms of two notions. The first is the idea that travelling well maintains a certain integrity in a fact even while it may have some change in shape, kind or form: that is, it may not be reported or used exactly intact, but it has, in some sense, remained steadfast. Fruitfulness in scope is the idea that travelling well means that facts have been found by users who don't just report them, but act upon them or with them, and so use them to fulfil various other functions than those of their production and intended use. In the process, a well-travelled fact may have travelled far and wide, and may even invade a definitional contrast. These successful travels are what we mean when we suggest that well-travelled facts are ones that have acquired and lived a life of their own.

4. What Makes Facts Travel Well?

Although the networks, community values and instruments of a fact's production may provide it with an initial credibility, they by no means determine the independent, autonomous quality we find in the life of facts. Rather, it is users, in different times, places and disciplines, with different questions and different purposes, who largely determine the uses of facts at various destinations, and thus how well they have travelled to fulfil new purposes. Not all communities attend to each others' facts in a reciprocal way, as Howlett (2008) found when comparing the social scientists' "listening tree" (the papers on the Green Revolution that a writer cites) to their "speaking tree" (the other papers that later cite that writer). Yet it is the tendency of facts to travel relatively independently to other users, without much reference to their producing context, that may perhaps mark facts out from some other forms of travelling knowledge. Indeed, in many ways, facts turn out to be like children: Their parents who found or fashioned them soon lose control of them, they leave home, their product markings become lost as they make their way into all sorts of other unknown communities and fulfil all sorts of unexpected purposes, and sometimes facts remain in limbo for centuries between production and use.

So why then do some facts travel so freely while others do not? It seems that those facts that acquire an independent life of their own depend on a variety of "good companions" and on the "character" that they already have or gain in their travels. These two aspects are not to be understood as necessary or sufficient conditions for facts to travel; rather, they enable us to understand the unexpected and independent travelling lives of facts that form the subject of this book and thus throw light on why some facts travel well and others do not.

4.1 Good Company: Labels, Packaging, Vehicles and Chaperones

The immediate question to be asked and answered is: If facts are things that travel independently, why do they need good companions? Sociologists' studies on travelling knowledge have shown the need to track agents of travel and locations because all too often, what appears to be freely travelling knowledge is, in fact, dependent on tacit or expert community knowledge to make it transfer effectively; that is, such knowledge is not in general separable from its base in techniques and expertise.[30] Historians have focussed on the importance of agents and networks, social and political, for their accounts of what they refer to as "circulating" knowledge. A good example is the circulation of specimens (insects, plants and so forth) during the early modern period, when scientific and commercial interests intersected during the rise of European-based empires. We might even recognise such specimens as facts, as others have done before us: "Every specimen is a permanent fact."[31] Facts are relatively modular pieces of knowledge; they can

---

[30] Collins (1985), on getting lasers to work, offers a classic study.

[31] So says a plaque at the entrance to the Earth History Hall, American Museum of Natural History. Recent literature on the "circulation" of such items treats a variety of other levels of knowledge, as well as facts. Growing out of the literature on the construction of knowledge, and adding political to geographical spaces, it has focussed

be separated off and plugged in elsewhere: They are relatively more able to travel freely from the techniques and expertise that produced them or the people who circulate them. Even facts that are less evidently material things than specimens in jars do not necessarily need identifiable people or communities to carry them (although they may indeed be involved), but they do need carriers of some kind (such as the Internet). So we think of good companions as accompanying the facts rather than being an integrated part of that knowledge system. Good companions here support a fact's travels but are not part of the fact, and can be discarded when the fact reaches a new location.

What do we mean by travelling companions? They range from the mundane level of labels and packaging to the more material vehicles of transportation, as well as to the people involved in chaperoning, and from the various kinds of institutional structures that support travelling knowledge to the technical standards that carry facts with them.[32] Ancient artefacts carry facts embedded in them, as do modern statistical and mathematical models; old-fashioned narratives and fictions help facts travel, as do the witnesses, authorities and celebrities who support them, and even the money that oils the wheels of travelling vehicles – these can all be understood under the idea of companions. When we think of all these different elements explicitly as forms of companionship: chaperones, packaging, processes, carriers and vehicles for travelling facts, we focus not on the agency of the individual producers of facts, but on those who package facts for travel and onto the users or audiences who receive and unpack them, along with the network of people and things via which they travel and the social arrangements within which these travels are embedded.

Labels and packages are certainly one of these several elements that enable facts to travel safely. We can go back to Linnaeus and his innovatory indexing method, designed not just to classify the incoming facts of nature, but to make those facts retrievable again for his own and for others' usage: a classic example of how labelling helps facts to travel (see Müller-Wille and Scharf 2009). Sabina Leonelli's chapter (this volume) on the modern version of this process – bioinformatics – explores how small facts (data from specific experiments reported in one field of biology) are circulated around the broader research communities in biology. This is a travelling facts system in which "curators" label data (so that receivers will see their relevance),

---

attention onto the co-construction of knowledge that comes with such circulation (see, e.g., Raj 2007).

[32] See Velkar (2009) on the way that industrial standards carry facts about the good with them, for example, the sizes of wires.

package the data with care to make sure that they are receivable by others (including the equivalent of a sealed envelope containing an account of their origin) and send them out in digital format by an electronic vehicle. Such packaging serves to ensure that the data travel well in the sense of retaining their integrity as they pass from producer to a holding bay where potential (but unknown) users may locate them. The labelling enables potential users to locate those data that appear fruitful to their own purposes in their own laboratories. Once there, the labels and packaging are discarded, and the curators are often not even recognised when those small facts are used to support larger facts, or facts about other species, in the new domain.

Bioinformatics is a highly organised and specialist version of good companionship, but it is important to notice that although there is a community of possible users in mind when the labelling is done, there are no specific users named on those careful labels so that there are, in effect, no addresses on those labels. Other facts of science are much more lightly packaged and labelled by their producers for presentation to an audience; they, too, are meant to travel within a community of users, and they, too, have no particular addresses on them. Although the Internet has rapidly increased the flow of some kinds of facts in some forms (those that can be read and searched for in Internet terms), the labels identify contents and senders/producers, not users – they tell us where facts came from, not where they are going to. In contrast, the botanical and insect specimens that travelled by boat and cart across the empires of earlier centuries had clearly written addresses on them, even when the identity of the fact inside the box was not known.[33]

A different kind of companionship of packaging and labelling is found in Martina Merz's chapter (this volume) on nano-science. Here, the production technology that reveals the atomic structure also presents, already packaged, the important facts of the structure in the form of visual images. And like many such facts of science, these images are accompanied into the community domain by descriptive labels, while the texts remain secondary. But these imaged facts of nano-science don't like to travel without the company of other images that show various aspects of their character and means of production; indeed these collections of images are *required* to ensure effective travelling of the factual outcomes of the scientific paper. Facts may be found travelling in images in many fields: They cannot be easily portrayed as the linguistic objects treated by modern philosophers, but rather as the outcomes of either a revealing or of a representing technology. The form of these facts seems closer to the earlier notion of deeds or actions

---

[33] See, for example, Terrall (2009).

or the "matters of fact" of legal circles. Here, words and texts form the travelling companions to the facts in the images.

It seems that modern science is full of such vehicles that transport and sometimes transform facts. Facts, in the form of data strings or parameter values, are sent into models, which then calculate other facts from those raw materials. Facts are fed into measuring instruments, and other facts are sent out. From Manning's equation to the models of epidemiology, from the computerised data traces of bioinformatics to the technologies of fertigation, the relations between travelling facts and the practical instruments of science and technology are intimate and varied.

History, too, is a carrier of facts: Facts travel over long gaps, without labels, without much agency or obvious channels of communication, so here the idea of vehicles becomes especially important. The investigative practices of museum curators have long been concerned with all sorts of historical facts that travel in material objects. For example, careful examination with modern x-ray techniques can reveal how a mediaeval sword was made, and thus whether it was made for fighting or as a sign of wealth and rank. The pottery of ancient civilizations carries the traces of contents that can often be analysed to tell of its habitual contents, just as other investigators seek facts from the burnt timbers of a building to understand what was in the building. We can tell much about the techniques of production from the goods themselves even when we don't know their origin (e.g., whether they are hand- or machine-turned), or when their origin labels misleadingly imply a quality such as porcelain or china when those goods were neither porcelain nor came from China.[34] The facts that travel in artefacts often turn out to be multi-layered, rich in both the level of detail about the materials and about the societies in which they were made and survive, as Valeriani shows in her chapter (this volume) on church roofs.

Sometimes the support systems provided by these various kinds of travelling companions are so extensive that they amount to "scaffolding" and when that support is removed, the fact falls down. In David Haycock's chapter (this volume), we have a case of a fact that now seems to us a false fact, a fiction that could hardly be credited, namely the extraordinary longevity of Thomas Parr, who died in 1635 at the supposed age of 152. Yet, during

---

[34] These examples come from wonderful sessions at the project workshop "Facts and Artefacts: What Travels in Material Objects?" (17–18th December, 2007) in which Marta Ajmar-Wollheim (Victoria and Albert Museum), Frances Halahan (Halahan Associates – restorers), Susan La Niece (British Museum) and Andrew Nahum (Science Museum) each talked about various objects from their museums or restoration projects and the facts that travelled in them and with them.

its time it travelled with very good company in the form of associated facts, circumstances, narratives and beliefs about old age. It was "attested" to by no less an authority than the first-class anatomist Dr. William Harvey whose good name chaperoned the fact of Parr's longevity through 250 years or so. This fact lost its status not because of the collapse of trust in the good companion, but because the rest of the supporting scaffolding not only collapsed but was replaced by a new scaffold supporting a new view of what counted as a credible fact. In the nineteenth century, credible accounts of old age came to be reconstructed in statistical terms replacing those earlier facts of particular lives. We can understand this scaffolding as a Fleckian conglomerate community fact about the nature of "longevity," and when the community scaffolding is taken down, the particular fact about Parr that it supported disappeared, though the traces of that well-travelled fact remain in pub names and signs, in a portrait by Rubens and in Parr's burial site in Westminster Abbey.

Chaperones – the people who act as knowing or unknowing companions – come in many guises, as we find in this same old-age case. They might be witnesses (Harvey, witness to Parr's health), general authorities (Bacon, who attested to general claims of longevity) or even celebrities (Lord Arundel, who brought Parr to London). Although witnesses and authorities have particular claims to expertise, the celebrity need have no such expertise. They may all be equally good as companions in getting facts to travel and to travel intact; and sometimes those roles are rolled up into one.

None of this means that facts always travel easily: However much useful packaging and honest labelling they have, nothing guarantees that travelling facts will travel intact. They might, after all, have bad companions, companions who alter the fact to subvert it, re-label it, cast doubt on it and otherwise discredit it as they see it on its way. These facts might meet with hearty resistance, as Naomi Oreskes's chapter (this volume) shows in her account of how commercial and scientific interests worked together to resist the gloomy facts of climate change in the United States and spread an alternative good-news message. Here, the facts of the expert climate science community were heavily contested with arguments over the significance of the facts, and indeed whether they were facts at all, but more significantly for us, by the provision of alternative facts. The public eventually came to accept the facts of the expert community, though they did so believing that the expert community itself was divided. Strangely here, when the facts finally came to travel to the population, they carried with them a picture of bitter division on the part of their parent-producers.

4.2 Terrain and Boundaries

Scenery matters too, of course. The terrain and boundaries of travel are equally important to the possibilities for facts to travel well, particularly when we consider how facts travel not only across centuries in the humanities, but between natural and social sciences and from the scientific to the policy domains. The terms of terrain and boundaries offer a topographical metaphor, which can be interpreted in a number of ways to make sense of travelling facts. We can construct the terrain in *sociological* terms, for example, as a disciplinary landscape in which expertise, trust and power form the features of the terrain and define the barriers to be overcome. Or we can construct it in terms of the *material elements* of the sciences or humanities in which models, instruments and experiments – or archives and previous historical authorities – constitute the terrain. A third possibility is to interpret the terrain and boundaries in *cognitive* and *epistemic* terms, where the requirements for a specific technical understanding, or a knowledge of historical period, limit the range of the travelling facts or their ability to remain intact as they travel.[35] These three spaces may be interconnected in the travels of facts. Peter Howlett and Aashish Velkar's chapter (this volume) investigates how precise technical facts are made to travel over a rather formal boundary between the experts in the university to the farmers working the fields, and how those facts, as facts of experience, are then passed over a more informal boundary from the participating farmers to other farmers. At the same time, the facts transported were integrated – some tightly and some loosely – with a technological process that required scientific and tacit knowledge to work together for the facts to travel well. Here is a case where – by design – money formed the necessary good companion to get the facts over the first border, but the subsequent geographical dispersion of the technical facts relied upon various nodes of transfer as well as the evident cognitive demonstrations in the fat, healthy plants and financial demonstrations in the newfound wealth of participating farmers.

Disciplinary roadways may facilitate the travelling of facts, but at the same time, like rails, they may also limit the range of possibilities for travel.

---

[35] The now-classic genre of science studies that has integrated sociological, historical and philosophical aspects often included elements of travelling facts (even while the topic of travelling knowledge is not their main focus): for example, getting facts about vacuums to travel in early modern Europe (Shapin and Schaffer 1985); the travels of facts in (and about) rocks and fossils in the early nineteenth century (see Rudwick 1985) or the use of "boundary objects" as a focus for facts to traverse communities (see Star and Griesemer 1989).

In Alison Wylie's chapter (this volume), we find attention to how artefactual facts travel along historical and geographical pathways that join communities of archaeologists to particular sites. Here the travelling of facts is inextricably part of their re-production, for successive generations of archaeologists study the same sites again and again, retrieving and recognising different facts, creating a series of historically dated and geographically located chains of reference. In some cases, where the original sites are shut or destroyed, the retrieval becomes an activity of trawling through previous archaeological notebooks and records. The old road map was decided by earlier members of the community, and for archaeologists to make new accounts, they sometimes need to construct new maps with new pathways for their facts to travel on.

For cross-disciplinary travels, the terrain may often look forbidding and the fences rather high. Yet despite this, certain facts seem to show a remarkable ability to travel well across these boundaries. We have already remarked how facts about the behaviour of rats when placed in crowded environments travelled to "speak to" the behaviour of humans in crowded spaces, and thus onto the design of prisons, college dorms and housing estates. Here Ramsden (this volume) shows how the successful travels of the facts about crowded rats depended on some choice words: "the behavioural sink," which captured the imagination of American social commentators who saw the behavioural pattern of urban society as mirroring the behavioural patterns of the rats in their environments. In this case, facts were found capable of passing round or through the boundaries between disciplines and into the various public and popular domains: Those boundaries seemed to be rather porous with respect to these facts, more like tennis nets than prison walls.

A much more unstructured sense of territory is found in history, which offers a repository of facts rather like the Internet: full of facts that are made by producers and that may be found by users, but the links between them are ambiguous to say the least. Although the historian can – by careful research and intuition and imagination – re-create the path by which a fact travelled through time (as the role of Harvey in chaperoning Parr's longevity), this does not mean that the facts of history are displayed ready to be picked up off the historical floor, any more than those facts of biology are plainly seen by a biologist or the facts of society found by a sociologist. The facts that historians use have been addressed at some time past to other people in other places: One of the tasks of the historian is to find and reconstruct them even though those historians are rarely the intended recipient. Whereas a sociologist of science will follow the travels of facts by seeking the community within which facts travel and the human agency by which facts travel, indeed, will even define the travels of facts as being a social event, an historian expects to find and make sense of long-lost and unattended facts by inserting him/herself into a past environment in which the social is not absent, but in which that past society does not purposely make the packaging by which those facts travelled. Some historical facts do travel well over the vast countries of time, but they do so because it is the historian's specialist expertise, craft and aim to find them in the clutter, pick them out and pull them forward.

4.3 Character: Attributes, Characteristics and Functions

This brings us to the character of facts. Here we are concerned with the particular character of specific facts, not with the generic qualities of facts as a category versus other kinds of knowledge. Without pushing the metaphor too far, we want to indicate that facts, like people who have a certain amount of character to start with, will travel well, and those that travel well may tend to acquire more character. That is, we can think of facts as having a certain inherent character, something immutable and unique perhaps, but with a potential to develop their scope, to acquire additional roles or fulfil new functions, to become interpretable at different levels or become more generic. This takes us back to the wellness issue of travelling, since explaining how having character helps certain facts travel well also gives insights into the notion of travelling well, both with integrity and for fruitfulness.

It is often thought that facts – whether they are understood to be found or made – are rather bland kinds of things: either because they are understood to have the transparent quality of objectivity or because they hold the reflective quality of the ever malleable. Yet, whatever their status at production, when we follow the travels of facts, they seem to gain colour. We have found that in explaining why some facts travel and some do not, in accounting for what happens to facts when they travel and in analysing how facts have an impact when they travel, we needed to add adjectives describing the way facts are understood and used by the communities in which they travel. Some of our adjectives were just descriptive, but most reflected characteristics or attributes that suggest why those facts travelled well, for example, they were not just "small" or "big" but "headline" facts; they were "understandable," "surprising," "colourful," "reproducible," "adaptable" and so forth.

In the first place, we found ourselves with a set of adjectives indicating *the character that facts start off with*: These adjectives suggest why certain facts stand out amongst the crowd and so are more likely to be found and demanded by users. For example, a fact that goes against a particular community's assumptions will be found surprising within that community, and this tells us why that fact was noticed and why it was immediately taken up by others and passed along further. This notion of inherent character maybe clarified by the example of two, Internet-found facts (both retrieved on 14 November, 2008) about Taiwan. First, a news report that the Taiwanese president announced the sale of 2,000 tons of oranges from Taiwan to mainland China. This seems a fact of not much interest or importance in the given circumstances: a bland fact of insufficient character to make it travel. The second fact announced itself as "Taiwan's economy is 71% free!" For a social scientist, this is an immediately intriguing fact that asks to be passed on, and then raises questions whenever it is repeated: What could it mean to be only 71% free as opposed to free or not-free? What does it mean for an economy to be free? Who said it? How could such a measure exist or be produced?[36]

The second set of adjectives we found ourselves needing were ones that suggested the roles that facts play in their travels or in their new homes. These alerted us to *the character that facts acquire in their travels*, for example, to become "key" facts or "logo" facts for a much bigger community (and this might be so, even if the fact later turns out to be a false fact, as in David Haycock's case of the longevity of Thomas Parr [this volume]). Potentially the most important aspect of these adjectives is that they point to the functions that facts play in their new environments. We are used to thinking about facts as having considerable causal impact – for example, the social impact of the facts about climate change, or facts about scandals that have a political effect "like dynamite." But travelling facts also have functional importance that might be captured in various ways. Erika Mansnerus's research (this volume) is on the circulation of facts around the various phases and co-operating communities that are jointly responsible for epidemiological modelling in a programme of public health vaccination. She has developed a set of terms for the roles and functions of travelling facts in this context; some are resolutely "stubborn" for they have to be accepted as they are, while others behave as "chameleons" adapting to their new environments. (In effect, although some models have to be adapted to the stubborn facts, in other cases, the facts are adapted to fit the models.) These characteristics are important, for they imply something of the

---

[36] The measure was produced for the "2008 Index of Economic Freedom" and published by The Heritage Foundation and the *Wall Street Journal*.

functional roles facts play in their new homes, where some act as "brokers," creating new possibilities, or "mediators," reconciling items, while others are "containers," carrying items along.

Our experience of proliferating adjectives caught the attention of Susan Hunston, whose analysis of the use of the word "fact" in a ten-year period of *New Scientist*, showed us that scientists, too, liked to attach adjectives to their use of the word "fact," and she grouped these adjectives as indicating disciplinary basis (scientific, historical, technical, etc.), aspects of relative importance/size and quality (crucial, small, obvious, etc.) and something of its affect (illuminating, sad, amazing).[37] Our adjectives tend to fall into Hunston's latter two categories, adjectives that indicate size or qualities and those that denote affective aspects: generic, small, large, surprising, awkward, key – adjectives that were descriptive of character. They prove useful to our purposes of understanding why some facts are taken up and how they are used in their destinations, and so help to explain how facts with character come to travel well.[38]

Facts emerge, develop, mature and pass away, and whether the life of a fact is cut short or lives a good life is as much a matter of character and companionship as it is of context and contingency. Those facts that travel intact and maintain their integrity and extend their scope in fruitful ways can be said to have travelled well. But the plethora of adjectives describing the character of those facts that are taken up and used again signals something deeper, or perhaps more generic, that characterises this combination of integrity and usefulness. These adjectives suggest how facts gain different identities, play different functional roles and create different effects during their travels. They point us to how facts make a difference to other elements in our histories, literatures and sciences during their travels across time and space. And it is this ability to make a difference that lies at the heart of what we mean when we say that a fact has travelled well.

5. Conclusion: How Well Do Facts Travel?

It is usually assumed that the sciences and the humanities have different ways of knowing things, yet our account of travelling facts suggests that we can tell the same sorts of stories about the travels of facts in both domains.

---

[37] "'You can't deny the fact that …': An Application of Corpus Linguistics," Plenary Address, American Association of Corpus Linguistics, Brigham Young University, Utah, March, 2008.

[38] The disciplinary aspect has for us another dimension, for, as noted earlier, different disciplines like different kinds of facts (see Howlett 2008).

Facts can indeed travel well, with integrity, fruitfully and to make a difference in their new homes, though, of course, many do not – some will turn out to be false or be mangled or be of little account. Of the many facts that do not travel at all, we have made no accounts – though we have seen how important demand is for facts to travel. We have demonstrated how various kinds of associates: stories, models, labels and good companions, as well as a good dose of character, will help to set them off and keep them moving to some useful destination. That travel process, however, is quite unpredictable: it is dynamic, extended and interactional.

As a result of this variety of experiences of travelling facts in our studies, we have come to see the nature of facts in a rather different light. Facts are not just a rather useful category of things that scientists and humanists find, produce and fit together to make more interesting narratives, arguments and evidence. Facts are, of course, foundational; they are building blocks for knowledge in the sciences and humanities, but they do not just accumulate usefully and bear interest within a particular discipline or community. Rather, their extraordinary abilities to travel well, and to fly flags of many different colours in the process, show when, how and why they can maintain their integrity and prove sufficiently rocklike to support further facts, ideas and theories in their new domains well beyond and away from their sites of original production and intended use. Facts may just be pieces of separable knowledge, found in many different forms and sizes, but it is in travelling well that they prove how essential they are to our sciences, humanities and society.

Acknowledgements

"How Well Do 'Facts' Travel?" – a research project funded by the Leverhulme Trust and ESRC (F/07004/Z), held at the Department of Economic History, London School of Economics – created this volume: I gratefully acknowledge their funding and the department's support. I owe very considerable thanks to all the members of the project group, to visitors to the project who stayed for periods of several weeks, as well as all those who came in for workshops, for making this research project such a wonderful experience and for their many remarks and arguments that have helped me to create this chapter. Thanks also go to participants at seminars and conferences, and to the referees of this volume, for their searching questions about the ideas in this chapter. A record of the research project can be found at http://www2.lse.ac.uk/economicHistory/Research/facts/AboutTheProject.aspx

Bibliography

Adams, Jon (2006) "How the Mind Worked: Some Obstacles and Developments in the Popularisation of Psychology" Department of Economic History, London School of

Economics: *Working Papers on The Nature of Evidence: How Well Do "Facts" Travel?*,
8/06 http://www2.lse.ac.uk/economicHistory/pdf/FACTSPDF/0806Adams.pdf

Austen, Jane ([1815] 1971) *Emma* (Oxford: Oxford University Press).

Buckner, Taylor H. (1965) "A Theory of Rumor Transmission" *Public Opinion Quarterly*,
29:1, 54–70.

Cerutti, Simona and Gianna Pomata eds. (2001) *Fatti: Storie dell'Evidenza Empirica*,
108:3, *Quaderni Storici*.

Chang, Hasok (2007) *Inventing Temperature* (Oxford: Oxford University Press).

Collins, Harry M. (1985) *Changing Order. Replication and Induction in Scientific Practice*
(London: Sage).

Daston, Lorraine (1991) "Marvelous Facts and Miraculous Evidence in Early Modern
Europe" *Critical Inquiry*, 18:1, 93–124.

(2001) "Perché i fatti sono brevi?" in Cerutti and Pomata, pp. 745–70, 916–7 (English
version: "Why Are Facts Short?" in *A History of Facts*, Max Planck Institute for the
History of Science, Preprint 174, Berlin, pp. 5–21).

Fleck, Ludwik ([1935] 1979 translation) *Genesis and Development of a Scientific Fact*
(translated by F. Bradley and T.J. Trenn; Chicago: University of Chicago Press).

(1936) "The Problem of Epistemology" in *Cognition and Fact: Materials on Ludwik
Fleck*, ed. Robert S. Cohen and Thomas Schnelle, *Boston Studies in the Philosophy
of Science*, 87, 79–112.

Gellner, Ernst (1964) "Facts" in *Dictionary of Social Sciences*, ed. J. Gould and W. Kolb,
(London: Tavistock Publications).

Ginzburg, Carlo ([1986] 1989 translation) *Clues, Myths, and the Historical Method*
(translated by John and Anne C. Tedeschi; Baltimore: The Johns Hopkins University
Press).

Gouk, Penelope (1995) *Wellsprings of Achievement* (Aldershot: Variorum, Ashgate
Publishing).

Hacking, Ian (1983) *Representing and Intervening* (Cambridge, England: Cambridge
University Press).

Howlett, Peter (2008) "Travelling in the Social Science Community: Assessing the
Impact of the Indian Green Revolution Across Disciplines" Department of
Economic History, London School of Economics: *Working Papers on The Nature
of Evidence: How Well Do "Facts" Travel?*, 24/08 http://www2.lse.ac.uk/economic-
icHistory/pdf/FACTSPDF/2408Howlett.pdf

Inkster, Ian (2006) "Potentially Global: A story of Useful and Reliable Knowledge and
Material Progress in Europe, Circa 1474–1914" *International History Review*,
XXVIII: 2, 237–86.

Knapp, Robert H. (1944) "A Psychology of Rumor" *Public Opinion Quarterly*, 8:1, 22–37.

Latour, Bruno (1986) "Visualization and Cognition: Thinking with Eyes and Hands"
Knowledge and Society, 6, 1–40.

(1999) "Circulating Reference" in *Pandora's Hope* (Cambridge, MA: Harvard
University Press).

Leonelli, Sabina (2008) "Circulating Evidence Across Research Contexts: The Locality of
Data and Claims in Model Organism Research" Department of Economic History,
London School of Economics: *Working Papers on The Nature of Evidence: How
Well Do "Facts" Travel?*, 25/08 http://www2.lse.ac.uk/economicHistory/pdf/
FACTSPDF/2508Leonelli.pdf

Mansnerus, Erika (2009a) "Acting with 'Facts' in Order to Re-model Vaccination Policies: The Case of MMR-Vaccine in the UK 1988" Department of Economic History, London School of Economics: *Working Papers on The Nature of Evidence: How Well Do "Facts" Travel?*, 37/09 http://www2.lse.ac.uk/economic History/pdf/FACTSPDF/3709Mansnerus.pdf

  (2009) "The Lives of Facts in Mathematical Models: A Story of Population-level Disease Transmission of *Haemophilus influenzae* Type B Bacteria" *BioSocieties*, 4:2/3, 207–22.

Morgan, Mary S. (2008) "'On a Mission' with Mutable Mobiles" Department of Economic History, London School of Economics: *Working Papers on The Nature of Evidence: How Well Do "Facts" Travel?*, 34/08 http://www2.lse.ac.uk/economic History/pdf/FACTSPDF/3408Morgan.pdf

  (2007) "An Analytical History of Measuring Practices: The Case of Velocities of Money" in *Measurement in Economics: A Handbook*, ed. Marcel Boumans (London: Elsevier), pp. 105–32.

  (Forthcoming) *The World in the Model* (New York: Cambridge University Press).

Müller-Wille, Staffan and Sara Scharf (2009) "Indexing Nature: Carl Linnaeus (1707–1778) and His Fact-Gathering Strategies" Department of Economic History, London School of Economics: *Working Papers on The Nature of Evidence: How Well Do "Facts" Travel?*, 36/09 http://www2.lse.ac.uk/economicHistory/pdf/ FACTSPDF/3909MuellerWilleScharf.pdf

Oreskes, Naomi and Erik M. Conway (2010) *Merchants of Doubt* (London: Bloomsbury Press).

Oudshoorn, Nelly and Trevor Pinch eds. (2003) *How Users Matter: The Co-construction of Users and Technologies* (Cambridge, MA: MIT Press).

Poovey, Mary (1998) *A History of the Modern Fact* (Chicago: University of Chicago Press).

Raj, Kapil (2007) *Relocating Modern Science* (Basingstoke: Palgrave Macmillan).

Ramsden, Edmund and Jon Adams (2009) "Escaping the Laboratory: The Rodent Experiments of John B. Calhoun & Their Cultural Influence" *Journal of Social History*, 42:3, 761–92.

Rosenberg, Nathan (1974) "Science, Invention and Economic Growth" *Economic Journal*, 84, 90–108.

Rudwick, Martin (1985) *The Great Devonian Controversy* (Chicago: University of Chicago Press).

Sayers, Dorothy ([1926] 1987) *Clouds of Witness* (New York: Harper Collins).

Shapin, Steve and Simon Schaffer (1985) *Leviathan and the Air-Pump: Hobbes, Boyle, and the Experimental Life* (Princeton, NJ: Princeton University Press).

Shapiro, Barbara J. (2000) *A Culture of Fact* (Cornell: Cornell University Press).

  (1994) "The Concept 'Fact': Legal Origins and Cultural Diffusion" *Albion: A Quarterly Journal Concerned with British Studies*, 26:2, 227–52.

Speich, Daniel (2008) "Travelling with the GDP Through Early Development Economics' History" Department of Economic History, London School of Economics: *Working Papers on The Nature of Evidence: How Well Do "Facts" Travel?*, 33/08 http://www2. lse.ac.uk/economicHistory/pdf/FACTSPDF/3308Speich.pdf

Star, Susan Leigh, and James R.Griesemer (1989) "Institutional Ecology, 'Translations,' and Boundary Objects: Amateurs and Professionals in Berkeley's Museum of Vertebrate Zoology, 1907 – 1939" *Social Studies of Science*, 19, 387–420.

Steckel, Richard H. and Jerome C. Rose eds. (2002) *The Backbone of History: Health and Nutrition in the Western Hemisphere* (New York: Cambridge University Press).

Swenson, Steven P. (2006) "Mapping Poverty in Agar Town: Economic Conditions Prior to the Development of St. Pancras Station in 1866" Department of Economic History, London School of Economics: *Working Papers on The Nature of Evidence: How Well Do "Facts" Travel?*, 9/06. http://www2.lse.ac.uk/economicHistory/pdf/FACTSPDF/0906Swensen.pdf

Terrall, Mary (2009) "Following Insects Around: Tools and the Techniques of Natural History" (*UCLA paper*).

Valeriani, Simona (2008) "Behind the Façade: Elias Holl and the Italian influence on Building Techniques in Augsburg" *Architectura*, 38:2, 97–108.

Velkar, Aashish (2009) "Transactions, Standardisation and Competition: Establishing Uniform Sizes in the British Wire Industry c.1880" *Business History*, 51:2, 222–47.

Wallis, Patrick (2006) "A Dreadful Heritage: Interpreting Epidemic Disease at Eyam, 1666–2000" *History Workshop Journal* 61:1, 31–56.

Weirzbicka, Anna (2006) *English: Meaning and Culture* (Oxford: Oxford University Press).

PART TWO

MATTERS OF FACT

TWO

FACTS AND BUILDING ARTEFACTS: WHAT TRAVELS IN MATERIAL OBJECTS?

SIMONA VALERIANI

1. Introduction: Travelling Through Cultures, Travelling Through Time

Facts are often expressed as statements or verbal descriptions. But in some cases, facts – particularly facts about technology – are better recorded and transmitted via material objects.[1] Does the nature of the vehicle that contains and expresses (and sometimes almost constitutes) the facts influence how they travel and are received? This chapter contributes to the question of how material objects carry facts from one culture to another, and through the centuries, by considering some examples from the history of construction, while at the same time keeping in mind the praxes of archaeologists, historians, conservators and museum curators. Therefore, we will be dealing with facts travelling on two different levels: facts travelling in a material object to the observer or user, as well as facts from and about buildings travelling in the realm of architectural history.

Material objects can 'carry' facts through different trajectories – across time, of course, but also from one cultural audience to another, which might or might not include a geographical move. Analysing a building or its components – or an archaeological excavation and the artefacts found in it – gives us 'clues' about those historical events and processes that directly involved and physically shaped the objects concerned: their production, the use that

---

[1] In recent years, a growing number of scholars has been concerned with the concept of materiality and 'thingness,' either from a more theoretical perspective (e.g., Baird 2004, Brown 2004, Buchli 2004 and Miller 2005) or from an anthropological-archaeological point of view (e.g., Buchli 2002 and Tilley 2006).

was made of them, possible repairs, changes and so on. The primary vehicle by which these (historical) facts travel is the material object itself, although other means must often be taken into consideration as well to get at the whole story. Following facts through these two trajectories of time and space, we ask which 'facts' travel, via which vehicles and what happens to them when they travel. A specific example may help to make these questions more concrete.

2. Building Structures, Inscriptions and the Shaping of Identity: The Case of St. Cecilia in Trastevere

If you visited the loft space of the church of St. Cecilia in Trastevere in Rome, you would be surprised to notice that both the walls and the roof structure are highly decorated (Figure 2.1). You might then logically conclude that the vaulted ceiling that nowadays covers the nave is a later addition. Looking in detail at the roof, you would be able to describe the kind of structure (a so-called simple palladiana), to see that the beams are painted with both geometric and figurative motives, to notice that most of the beams have marks carved on them near where they connect to other elements of the frame and finally – if you wriggled between the vault and the roof structure and looked really carefully – to discover an inscription underneath beam no. 5 testifying to the construction works on the church as having been carried out in 1472 following instructions given by Cardinal Niccolò Fortiguerra.

Clearly, the material object is telling a story – or, at least, it is providing the interested observer with a host of *facts* and *clues* so they can tell the artefact's story. Most evidently, these travelling facts are about the structure of St. Cecilia and how it holds up, telling us about the know-how of the builders who erected the building. The kind of structure that was used and how the joints work, for example, are directly observable. Intentionally left traces carry facts relating to its production. The carved marks give clues about the assembly method and the organisation of the work on the building site, evidence that the structure was preassembled elsewhere, transported to the building site and then reassembled following these marks – they are a sort of 'incorporated user manual.'

But this will set off an alarm bell in the mind of an historian, who will recognise this 'fact' as being very unusual (Roman Renaissance carpenters did not use marks for such reassembly processes) and start to wonder if the carpenters who worked here might have come from other regions. The facts as to 'why' this uncommon technique was used are not to be found in the object itself, of course: They need to be searched for in the archives. But we

Figure 2.1. St. Cecilia in Trastevere in Rome. Left: longitudinal section of the basilica showing (from the top down) the roof structure, the painted freeze from the sixteenth century, the medieval wall paintings with apostle figures, the profile of the vaults dating to the eighteenth century and the interior of the church. On the right (top): cross-section of the church showing (from the top down) the roof covering, the Annunciation and the freeze painted in the sixteenth century and the mosaics from the ninth century, of which only the apse dome is nowadays visible from the church. On the right (bottom): details of the decoration.

From: Valeriani 2006a, courtesy of the photographical archive of the Soprintendenza per i Beni Architettonici e del Paesaggio di Roma.

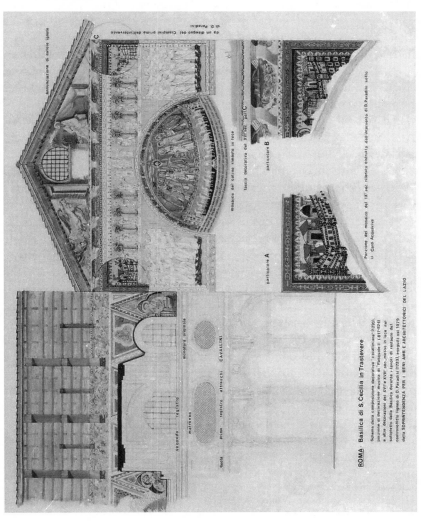

will look into the matter of the archives in more detail in the second case study. The observation that nearly all of the elements composing the roof frames have carved marks that seem to fit in the same system indicates that most of the roof structure originated in the same building phase. This is already a clue toward constructing the chronological sequence of our story. And indeed the artefact carries a number of facts about both its production and its life. The specific relationships between individual parts (A covers B, which covers C; Y cuts Z/is cut by X, etc.) will be key in enabling us to construct a chronological sequence logically. This, in turn, can then be linked to an absolute chronology through laboratory analysis of the elements, or by comparing the object and its characteristics with others known to us from other contexts. For such comparisons to be more effective and significant, the reference materials need to be organised into typologies, which, taken together, build up what (later in the chapter) we call 'the encyclopaedia', which can then be used, for example, for dating the mosaics and frescos or the paintings on the beams in stylistic terms. Here, since the apparent 'facts' (the marks) about its assembly method don't align with what we know (from the encyclopaedia) about contemporary local building techniques, they may be giving us clues about 'other' facts. Following up on such clues, like a detective, may direct us to facts about the import of a specialised workforce from another region, in turn increasing our knowledge about workforce migration patterns.

But, if we return to tie beam no. 5, the story begins to reveal even more complex depths. On the face of it, the material object is being used as a substrate carrying propositional knowledge, in the same way as a piece of paper might. Its inscription asserts the 'fact' that the structure was built in 1472 under the patronage of Cardinal Niccolò Fortiguerra. The problem with this inscription arises when the results of the dendrochronological analysis suggest that the tree from which tie beam no. 5 came was not felled until after 1588 – more than a century after the beam's inscription date.[2] If this is true, the material object is carrying misleading facts.[3] Of course, this peculiarity calls for an historical explanation: To make the next step in our 'story telling' we now need to use our interpretative skills to a greater degree and turn for help to other kinds of sources, namely written ones. Indeed, while the material object carries a wealth of signs of facts about what happened and when (in this case, for

---

[2] In this specific case the data are to be taken with caution because the master chronology has not been finalised; here, for the sake of argument, we will take this as an established fact.

[3] Here (and throughout this chapter) the expression 'misleading' is used in the sense of deliberately misleading. Compare with Haycock, this volume, who provides a discussion of facts as misleading, or false facts in a very different context.

example, about when the tree was felled, which points to the deliberate 'faking' of the inscription), it is often the case that the use of other sources is needed in order to address questions about *why* these things happened.

The available documents, together with stylistic analysis, confirm that the pictorial decorations on the roof frame can all be dated to 1599, which fits well with the dendrochronological results for beam no. 5. Moreover, the coats of arms displayed as part of the decorations of the beams (supported by the age of the wood in beam no. 5) confirm the attribution of these building works to the patronage of Cardinal Emilio Sfondrati (1560–1618). He was a major figure in the counterreformation movement, who put enormous effort into strengthening the image of the Catholic Church by reinforcing its identity as being the authentic heir to the 'church of the origins' dating from the time of the apostles and the first centuries. Cardinal Sfondrati's 'intervention' at St. Cecilia shows how material objects were used in seventeenth-century Rome to carry and communicate to the observer a series of 'facts'. In this case, the material evidence shows that he chose to incorporate the existing roof within a new scheme of interior decoration, whose purpose was to express theological, political and ecclesiastical 'messages' that were particularly close to his heart.

A bit of historical knowledge is needed to support this interpretation. The informed viewer will know that the church of St. Cecilia was first built in early Christian times and has undergone several subsequent renovations. And they will also know that both flat ceilings and visible roofs had been used in basilicas since antiquity and that there was a marked preference for the latter in Rome until the Renaissance, but that during the fifteenth century the preference changed to have vaults above aisles and flat ceilings above naves.[4] Despite that, the visible roof of St. Cecilia was preserved and the very materiality of its trusses was seen as evidence of the church's long and venerable history.[5] The trusses were made more appealing and 'relevant' for the contemporary observer by 'modernisation' – to this end, the old roof structure (at that point in time visible from the nave) was decorated with paintings whose theme (the *Arma Christi*) reflects the counterreformist desire to exalt martyrdom and sanctity.[6] This made the roof an integral part of the spatial,

---

[4] However, this trend was not universal, and even in the seventeenth and eighteenth centuries some restorations preserved the original layout.

[5] Cardinal Sfondrati did consider inserting a new ceiling in line with the architectural fashion of the moment, but then decided otherwise. Part of the reason was structural in nature, as the architect expressed concern about the capability of the walls and columns to carry the additional load (Valeriani 2006a, p. 49 and Valeriani 2006b). Nevertheless, it is remarkable how the patron used the existing building elements to express his political message.

[6] See Valeriani 2006a, pp. 49–52.

Figure 2.2. St. Cecilia in Trastevere in Rome, graphic reconstruction of interior before the insertion of the vaults in the eighteenth century. Decorations from different époques were visible simultaneously.
From: Giovenale, Giovanni Battista. 'Recherches architectoniques sur la Basilique de S. Cecilia,' *Cosmos Catholicus. Le monde Catholique illustré*, 4 (1902), p. 670.

decorative and political concept of the church, whose different elements, in their materiality, were taken to embody facts about the past, but which were also relevant for the present and the future. The mosaics above the arch and the *opus sectile* floor of the St. Cecilia chapel and the remains of the saint were seen as material evidence of the building's paleochristian origin; the frescoes on the sidewalls of the nave as a testimony of the medieval period; the roof structure and the Fortiguerra coats of arms (and the inscription on tie beam no. 5) as further links in the uninterrupted chain of the church's history from its origins to the contemporary period (Figure 2.2).

The very materiality of all these elements underlined their role as a tangible testimony and a powerful symbol of the centuries-old ecclesiastical tradition, and was therefore worth preserving (or even faking if not in a good enough state of conservation). I suggest we can argue that the beam's inscription was a late-sixteenth-century fake (which, in our terms, has endowed the beam with misleading facts) intended to replace a broken link in the chain of meaning stretching back from that time of counterreformation to the church's historic origins.

In the mind of a seventeenth-century patron, the material object was an excellent carrier of facts. In a similar fashion, St. Cecilia is a fascinating building for the historian today, carrying a richness of facts about the past: not only those introduced by Sfondrati's renovation and conservation efforts, but also many other clues of a technical and historical nature, carved (more or less intentionally) into the artefacts by their many makers and users, which can richly repay further detective work.

3. Material Objects as Carriers of Travelling 'Facts'

3.1 Artefacts and the Complexity of Travelling Facts

It is evident that the kind of facts we can 'read out of' an object depends also on the question the observer has in mind – the more complex the material object, the more layered the set of facts that may be travelling with it.[7] If we consider a building, for example, it will often 'carry' a variety of different kinds of facts in different forms (more or less visible, and more or less dependent on our analytical and interpretative skills for their understanding). Buildings (or even town settings) are often conceived as instruments for communicating something about the patron or the way the society is organised and ruled. But they also tell all sorts of other stories about the scientific and technical knowledge needed to realise them, the techniques used to produce them and how the production process was organised, the building materials offered by the local natural environment and so on.

---

[7] Another example of a complex object carrying a layered set of facts can be found in the composite images utilised in scientific journals, which is discussed by Merz in this volume.

This weaving of different kinds of facts transmitted in the material object (within or on its surface) is precisely the core of the case studies illustrated in this chapter.[8] Technical elements (such as roof structures) are taken as the repository of facts about craftsmanship and technical knowledge, but also as vehicles for the transmission of symbolic facts. The cases briefly presented here were chosen to reflect the complexity of the travelling mechanism. They are not, of course, illustrative of all the possible kinds of facts that travel and their ways of travelling, but are used to ground considerations about the relationship between material objects and travelling facts more firmly in their materiality. The focus is on which kind of facts travel through the material objects and what happens to them in doing so. Some attention is given also to which kind of facts *do not* travel in such artefacts, as well as to which elements of the historical narrative emerging from each case study may derive from other sources. The examples concern actual buildings and the attempt to reconstruct their history over the centuries. Embracing the building archaeology approach, the material evidence is understood as the first source for historical enquiry. In both cases presented in this chapter, the analysis starts with the roof structures and the loft spaces, which (since they constitute areas of buildings that are less frequently modernised) may be where original remains, and also subsequent developments, are easier to recognise a posteriori.

The first example has already shown material objects as carrying a whole set of different layers of facts. Let us now look in more detail into different ways in which artefacts more generally carry facts.

3.2 Different Forms in Which Artefacts Carry Facts

We can – at its simplest – imagine a material object as being used as a substrate to carry propositional knowledge, in the same way as a piece of paper. This happens, for example, on gravestones or plaques, where the inscription and decorative elements are meant to tell the observer some particular facts (as the inscription on beam no. 5 in St. Cecilia purported to do). The material object may not always have been intended to carry such information – sometimes this function was attributed to it a posteriori, as in the case, for example, of graffiti (as we will see in the next case study). However, as the materiality of the object concerned plays only an insignificant role, this 'basic' case does not align with the focus of this chapter.

---

[8] Howlett and Velkar (this volume) also discuss how material objects carry technological and technical facts, while (in contrast) Leonelli (also in this volume) shows how digital traces (materials, as opposed to words or texts) carry the facts.

A different manner of fact carrying concerns signs intentionally left on the object by its makers but not intended for an 'end user'. They were often meant to communicate facts relevant for the production process. When observed by the historian (or other practitioner versed in such objects), these signs act as 'clues' that can be used to retrieve a series of facts 'stored' in the object. The carpentry marks observable on wood-framed houses (and wooden structures in general) were symbols or numbers carved in each beam to act as a guide for the carpenter on how to assemble the structure (as at St. Cecilia) – without them it would have been virtually impossible to make sense of the hundreds of timber pieces scattered on a building site. Each single beam carried vital information for the success of the building venture, and – from the historian's point of view – such elements are a precious source of information about both the building process and the technical culture that produced them. Similar considerations can be made when analysing the stonecutter's marks, which are thought to indicate either which artisan made a block or who sourced it, and which therefore 'carry' facts about the division of labour and the accounting mechanisms on a building site, and can also give indirect clues about building sequences.[9] In both cases, the marks allow the retrieval of quite a complex array of facts, far beyond their original scope.

This brings us to the importance of what we can call the 'encyclopaedia' and the 'typified data'.[10] The encyclopaedia is an array of examples, materials and so on, built up from observations from many sources that can be compared with what you find to make sense of it in a broader context. Typified data are more specific, in effect, storing series of data about similar kinds of elements that can then be subdivided into smaller categories, and help analyse or date a specific element in detail. In some areas, this information can be so specific that it is possible, for example, to date a building (approximately) on the basis of the kind of carpentry marks visible on its beams.[11]

To sum up: The characteristics of material objects shaped in a particular way for a specific purpose are also meaningful to us in the pursuit of other, more general questions. There may be many unintentionally left signs that are traces of an array of processes, actions and facts about the production and

---

[9] On the general topic of the use and significance of stone masons' marks, see Friedrich 1932 and Maier 1975, while for a very nice example of how these traces can be carriers of other facts and can help the historian, see Rogacki-Thiemann 2007, pp. 47–55.

[10] Pucci 1994, see later in this chapter.

[11] Valeriani 2006a and Valeriani 1999 (with references). Obviously, the problem of reuse and the co-existence of relicts from different building phases need be taken into consideration.

the life of the object: the facts' technical nature and, indirectly, their social nature.

Some clues are easily seen without the help of any sophisticated devices – but instrumentation and technical know-how may be required to make others visible. The texture of a wood or stone artefact, for example, can tell us much about the tools used for its production – and looking at a tool and the way it has aged can tell the schooled eye what it was used for. But if we, for example, want to retrieve from an ancient vessel facts about what materials it used to contain – which may in turn give us clues as to our ancestors' diets – we will often need to employ more sophisticated techniques. Even facts that are actually visible to the naked eye can remain unrecognised for a period, either because familiarity with the object has decreased over the course of time – and with it the ability to 'read' such facts – or because the object has come into a geographical environment where it is unknown. The 'failure' of some technical facts to travel well is exposed in cases such as those where attempts to replicate an artefact fall short because the new environment fails to appreciate which of its characteristics are the essential ones. Where a new recipient lacks some key item of knowledge that would allow them to make a 'correct' interpretation of the object, the fact's 'journey' can be seen to have been interrupted.

Nevertheless the object still 'contains' the facts – and another user may be able to discover them. It is just a matter of learning how to 'read' the artefact,[12] or of deciding which questions need to be asked, which layer of facts brought to the surface. This complex relationship between the object, the marks and the facts they can 'reveal' will be addressed later in the chapter – but, clearly, making sense of the marks and linking them to the facts – which we could say are 'embedded' in the material object – requires a certain degree of expertise.[13]

Linking the physical characteristics of objects to facts about their history or their production and so on, can, in some cases, be quite a complex process. Sometimes the material qualities are enough in themselves – but sometimes they only provide us with clues, and extensive 'detective' work needs to be carried out in order to get to the facts. In some cases, the object

---

[12] Tarule 2004 presents an interesting investigation of which facts can be read out of an artefact.

[13] The literature is rich in contributions that discuss whether artefacts are social constructed entities highly dependent upon our perception or, in contrast, are to be seen as the most 'objective' and sturdy of all the entities we can study. Taking a pragmatic attitude, many authors are trying to find a balance between the two ideas (see, for example, discussions in Baird 2004, Brown 2004 and Miller 2005; see also the overview offered by Daston 2004 in the introduction [pp. 7–24]. This also represents this chapter's take on the matter.

may carry 'qualities' but without revealing much about the techniques used to produce them. Thus, in preparation for the five-hundred-year anniversary of Columbus's discovery of America, extensive work was undertaken in Genoa's historic harbour, where great celebrations were planned. As historians and archaeologists had warned, excavations for new tunnels brought the medieval quays to light. Apart from questions of how appropriate it was to dismantle these historical structures, a very practical problem also arose: The quay walls, and specifically the mortar in them, were incredibly hard and the work proved much more challenging than expected (Mannoni 1988). Even the best contemporary materials and production processes could not match the extraordinary quality of the mortar, and this discovery kicked off a line of enquiry to try to determine how this had been achieved.[14] The physical and chemical composition of the mortar was relatively easy to establish, but was not necessarily enough to explain the superior quality of the end product – younger mortars made with similar ingredients failed to perform as well. Further analysis carried out by the Department of Material Sciences at the Engineering faculty in Genoa revealed that the mortar's hardness was linked to particular microstructures, which had evidently been formed in the material through specific processes about which no knowledge seemed to be available. A group of experts have since been working on the topic, interrogating historical sources and experimenting, but while some forward steps have been made in the last twenty years, we are still struggling to match the quality of the medieval mortar exactly. The information stored in the object has only partially helped us solve the mystery – the rest of what we now know has been achieved through study of the production sites and processes (including archaeological investigation, analysis of historical documents and oral history: Vecchiattini 1998, Mannoni et al. 2004) and, starting from those clues, through experiments trying to reproduce the mortar using different settings (Pesce and Ricci forthcoming). In this case, the artefact didn't carry all the facts about the methods used for its production – but it did demonstrate that it was possible to make such high-quality mortars. Similar stories have been told for many other – more famous – materials, such as porcelain.[15]

---

[14] Mortars with similar qualities had occasionally been observed previously in different parts of the historical harbour, as well as (for example) in the foundations of Palazzo Ducale, but no systematic enquiry into their composition and historic production had been launched.

[15] It is interesting to note that the hydraulic component of these mortars was known under the name of 'porcellana,' the same word used for 'porcelain.' And indeed this material (kaolin) was an important ingredient in the production of porcelain. It was probably imported by Genoese merchants alongside Turkish alum for which Genoa had long a monopoly.

3.3 Little Clues for a Big Question: Which Facts Are to Be Found in Material Objects?

So far we have discussed the ways in which facts can travel in (or through) a material object via its material characteristics and (more or less) intentional marks. But what kind of facts travel in these artefacts – and which kind of questions can they help us to address?

Facts 'carried' in material objects are of very different kinds and concern the materials and technologies used to make the object as well as how it was used and so on. And these 'material' facts, in turn, will carry facts about the practitioner and the society that produced them: their knowledge, their skills, their taste and maybe even their aspirations, so that this 'fact-carrying' can be seen, in a sense, as closing the gap between the object and its society We could even argue that the *artefact is a fact* or *a collection of facts* about material qualities, production techniques, design principles and strategies, trends, customs and social circumstances.

This becomes evident if we take, for example, an object where materials of inferior quality are being used to imitate a product of higher status, as often happens with pottery, porcelain or precious metals. The object can clearly be seen as carrying misleading facts, and it is more likely to do so successfully if it travels from one environment or geography to another. But even if the 'facts' it carries about its material qualities are misleading from the perspective of a social historian, it nevertheless carries 'valid' facts about consumption patterns, taste, fashion and so on. Taking on a false identity often goes beyond the formal similarities expressed in an object's materiality, and is also transmitted through its 'naming' (thus, there is a certain kind of ceramic called 'porcelletta', a name that – falsely, but deliberately – suggests similarities with porcelain). The names attached to material objects can also carry other kinds of misleading facts – history is rich with material objects known under names that relate them strongly with places or personalities to which they are, in reality, only loosely or 'secondarily' connected.[16] The so-called Genoese pottery and the 'Serlian window' provide good examples. The first was actually produced in southern Spain by Islamic potters, but was named Genoese because it was exported to England via Southampton by Genoese merchants.[17] The latter is a particularly shaped window (of three parts, with a large, arched central section flanked by two narrower,

---

[16] Adams, this volume, provides a discussion of when facts are misleadingly displayed as fiction and vice versa.

[17] Marta Ajmar, presentation at the workshop *Facts and Artefacts*, LSE, December, 2007, paper forthcoming.

shorter, square-topped sections) that, in fact, is not at all the invention of Sebastiano Serlio (c. 1475–1555), the famous Italian Renaissance architect, but somehow became attached to his name, and is still regularly identified with his work by the general public – indeed, it has become almost a symbolic element of Renaissance architecture.

Another type of fact that can be seen as adhering to an object without really being part of its materiality is the symbolic fact. It can sometimes be 'readable' from the artefact itself, but in other cases is just attached to it because of its history, or has become known to us through oral or written tradition.[18] A decorated sword can be seen as an example where symbolic facts are stored visibly in a material object: The elaborate features of the blade and handle are a clear sign that the object was not exclusively, or even primarily, intended as a fighting or cutting instrument. If we analyse which kind of metal was used and look in detail at the traces of the production process, we will probably be able to tell if it is to be understood primarily as a functional object or as a status symbol (e.g., assessing how effective it would have been as a cutting tool vs. its decorative elements).[19] If the sword was found, for example, in a burial site, this set of material traces will tell us quite a lot about the status of the owner. The object carries symbolic facts in different ways: At one level, it works quite plainly as a symbol of the status of the owner, through the unmistakable richness and elegance of the manufacture alone. And beyond that, it carries other, more specifically symbolic facts – expressed, for example, through shape and decoration – that are more complex to decode. In studying this artefact, an anthropologist will be more interested in the social implications of the possession of a decorated sword (looking chiefly at the symbolic facts carried by the object), while the historian of technology (seeing mainly the object's technical 'facts') will use it to understand the development of production techniques or warfare.

Symbolic facts can also be loosely attached to an object and not relate to its materiality at all. Examples of this could be the pictures and objects that Freud used in his famous practice, which are particularly interesting to us because they symbolise his preferences and his 'worldview'. But – to be relevant to us – someone must tell us their story and explain the context

---

[18] Interesting contributions to this topic were made during the *Facts and Artefacts'* workshop organised by the author at LSE in December, 2007; for details see http://www.lse.ac.uk/collections/economicHistory/Research/facts/Workshop-Simona.htm

[19] Here, my thanks go to Susan La Niece (British Museum) who shared her knowledge on the matter with me.

they come from.[20] Our museums are full of objects whose material qualities have no particular value but which we preserve and admire because of
their symbolic value (and a similar process of attaching 'symbolic facts' to
an artefact relatively independently from its materiality happens, of course,
with sacred objects).

These are extreme examples, where a whole range of additional information – other than the artefact itself – is needed if the facts are to travel. An
object itself can often tell only part of its own story – evidence about 'how
and why' a particular object happened to be built/produced/used at a particular time and place often has to be searched for in other places, 'outside' the
artefact. In this case, the artefact itself will not be the only vehicle via which
facts travel – its materiality also signals the existence of interesting facts and
encourages the informed observers to go and seek explanations for them.

Therefore, material objects may serve the function of being a sort of
'springboard for enquiry,' perhaps by carrying facts which signal anomalies
that call for an historical explanation. An example could be an artefact of
unusual shape found in a burial, where its atypical shape, material or production technique sets off an alarm in the mind of the scholar, who then
looks for other clues and, after further consideration and cross-referencing
(using, for instance, the encyclopaedia and the typified data), interprets
this (perhaps) as evidence of the dead person being a migrant from a distant country. In this sort of case the object itself carries the fact of its 'difference': It tells the observer (both then and now) that (for example) this
building design, this structure, this concept of how its plan is organised,
how people circulate, which functions are performed where and so on,
exists and does work – even though it is alien to the local tradition (the case
study presented next will allow us to consider this in more detail).

In order to get from the 'little clues' to the 'big questions,' the archaeologist uses a two-part process: He starts with an 'objective' analysis of the
artefact (often exploiting data-gathering and analytic methods from the
natural sciences) and of its spatial relationship with the surrounding layers.[21] After this he makes a more interpretative effort to answer questions of
the 'why' and 'how' kind. As Andrea Carandini notes about the archaeological method and stratigraphy:

> The stratigraphic units are the result of actions, but many analytically identi-
> fied actions still cannot give the sense of an activity; in the same way that a

---

[20] This point is also made in a different context by Merz (this volume) in discussing the use
of nanotechnology to create the IBM logo.
[21] The notion of 'mediating facts' is discussed by Alison Wylie in this volume.

single observation carried out on the crime scene doesn't automatically explain to Sherlock Holmes the reasons behind the criminal's actions. In archaeology, the sense of an activity is decided by the archaeologist's interpretation of a group of minimal actions ... In the process of getting from the single actions to the activities, and from those to the activity groups, and finally to the events/periods, the synthesis is increased, and with it the degree of subjectivity of the interpretation.[22]

4. History, Archaeology and the Evidence-Facts Question – Archaeologists Versus Historians; Bricks Versus Paper

We have seen the particular ways in which material objects carry facts and how the information stored in the objects sometimes needs to be complemented via data derived from other sources to be able to fully make sense of the object. If the materiality of the sources of our facts does matter – if they carry different information/facts than do texts (or carry them differently from how texts would carry them) – then we should expect to find differences in the work by scholars dealing with different sources. Given my background, I focus here on the different disciplines dealing with the study of the past and look at 'historianship' (in the narrower, traditional sense of the word) and archaeology, the kind of historical study most concerned with material objects.

In this context, the positions taken by some archaeologists in the methodological debate that heated the community of historians in the late 1970s and early 1980s (and since) seem particularly revealing. The discussion was prompted by Carlo Ginzburg's brilliant and influential description of how the historian uses evidence in his essay 'Spie. Radici di un paradigma indiziario'[23] (and followed up in other well-known contributions to the topic, chiefly in his book *Clues, Myth, and the Historical Method*).[24] Ginzburg's proposition was that the historian works with a method based on clues, which is fundamentally different from the Galilean paradigm that forms the basis of natural science reasoning. The historian doesn't aim at reconstructing general rules, but particular histories, employing a 'clue-following' methodology resembling that of our hunting ancestors. This *'paradigma indiziario'* (conjectural paradigm), he suggests, became more widespread in the nineteenth century, when many disciplines that use this methodology

---

[22] Carandini [1991] 2000, p. 67, translated by author.
[23] Ginzburg 1978; see also Ginzburg 1983. Much of the discussion concentrated on the question of the 'two cultures,' which I will not take up here in detail.
[24] Ginzburg 1989.

were codified. These include (arguably most famously) Freud's psychoanalysis, with its interpretation of dreams and their details, as well as Morelli's method of attribution of paintings based on apparently insignificant features (such as fingernails, toes, lobes, etc.). Medical semiotics and criminology (with the development of fingerprint recognition) can be added to the list, as well as (albeit in a fictional setting) the 'methods' of Sherlock Holmes in the new detective novel genre developed in the same years by Arthur Conan Doyle.[25]

It seems particularly relevant for our question of 'What Travels in Material Objects?' to consider how Ginzburg's ideas were taken up by archaeologists. An interesting input was given by Giuseppe Pucci, in the themed issue titled 'La Prova' (The Proof) published by the journal *Quaderni storici* in 1996. Following a long tradition, Pucci asserts that the terms 'evidence' and 'proof' assume a specific meaning in history and archaeology because the historian and the archaeologist normally don't demonstrate (as might the mathematician or physicist) – rather they persuade by adducing proofs, aimed at convincing their audience through rhetoric:

> The archaeologist, more often than he assumes, does not demonstrate; rather, he argues. Demonstration and argumentation both aim to have a conclusion accepted by means of adducing proofs, but the meaning of proof changes with the change of the field of use ... In reality, we can say that archaeological discourse belongs rightfully to the field of rhetoric.[26]

But, Pucci argues, this doesn't affect the rationality and validity of the archaeologist's findings – it simply means they employ a different form of knowledge to the conventional 'scientific' one.

Another interesting reflection on this topic was offered by Andrea Carandini, who asserted the scientific character of the historian's work, contending that continuity can be observed between the Galilean and conjectural paradigms. The soundness of the archaeologist's findings (as of the detective's) is assured by the use of pieces of knowledge derived both from the experimental sciences and from different kinds of typological knowledge.[27] Like Holmes, the archaeologist uses a kind of abduction that aligns with known rules and codices. Pucci makes a similar observation in

---

[25] See also Eco 1983, which deals with the topic of logic, reasoning and Sherlock Holmes' method and the logic of abduction in different disciplines (although not including archaeology). Clue-orientated puzzle solving is also discussed in the context of medical cases by Ankeny (this volume).

[26] Pucci 1994, p. 60, translated by author. See also Ginzburg 1999.

[27] Carandini [1991] 2000, p. 256.

pointing to the use of typified data and of the 'encyclopaedia' as a fundamental element of archaeological argumentation:

> But if it is true that many clues together don't formally constitute a proof, the idea
> that many convergent clues strengthen the line of reasoning can be maintained,
> particularly when they agree with the encyclopaedia. We could say that if the
> symptom/clue is the brick with which the archaeologist builds his building; and
> if the conjecture represents the project, the mortar is given by the encyclopaedia.
> When we say encyclopaedia, it is not simply about taking in account *quod ple-*
> *rumque accidit.* The comparisons, essential in the archaeological argumentation,
> are more persuasive – we could say that they have a greater proving strength –
> when they descend from typified data and, in general, when they are quantita-
> tively significant.[28]

The availability of the encyclopaedia, the collection of examples or even
specimens of certain kinds of objects, or buildings or architectural features,
is essential if the archaeologist is to be able to interpret clues and construct
a narrative. As for the natural scientist, the comparison with only one other
instance of the same phenomenon is only relatively significant. The description and cataloguing criteria for the objects also need to be coherent for the
comparison to be significant. Therefore, the availability of banks of typified
data is an important condition for the development of the interpretation of
archaeological materials (such as the carpentry marks noted earlier in this
chapter).

It will be argued here that the lack of agreement between Ginzburg and
his colleagues depends to a great extent on the different kinds of sources
they use for their enquiry as carrier of facts: material objects and texts. The
discussion in the original publications deals with history and the historical method, rather than concentrating specifically on the kinds of sources
used. But it should be stressed that archaeologists' 'objections' obviously
reflect and elaborate upon their own practice. Therefore, we can argue that
the dispute should be seen not as the clash of two views about the historical method, but as echoing the differences between historical disciplines
mainly concerned with different sources. In other words, historians and
archaeologists use different methods to make their 'facts of the past' travel[29]
because these facts come in different forms. The nature of the vehicle 'containing' the facts influences the travelling process in a complex manner, as
we have seen. But, it must be underlined, the difference in the relationship
between the historian or the archaeologist and the evidence they use seems

---

[28] Pucci 1994, p. 69, translated by author.
[29] The expression 'facts of the past' is introduced in this volume by Alison Wylie.

to be *one of degree rather than one of kind* (in fact, the historian also needs to consider the text as a material object in some cases – for example, to prove if a manuscript is authentic, or to judge the significance of the way in which the text is arranged on the page, etc.).

If we look more closely at the archaeologists' responses, we notice two recurring elements: one is the reference to the use of methods derived from the experimental sciences (C14 [carbon-14], dendrochronology, chemical/physical analysis, etc.) and the other has to do with the availability of large numbers of samples and of typified data. The possibility of conducting various types of analyses from the natural sciences is clearly connected to the material nature of the sources the archaeologist is dealing with. It also reflects the potential of the material object to store, carry and express a layered plurality of facts in a way that is not given to a text.

In terms of the use of typified data, it can be said that although the process of comparing with what is understood as 'typical' and the use of series of data are not alien to 'traditional' historians, this method is generally more distinctively characteristic of the archaeologist: Again, it can be argued that this difference is a consequence of their more extensive use of material objects as evidence. It is in this light that we should interpret Pucci's claim that the archaeologist is more similar to the medical doctor and the detective than the traditional historian is

> The traditional kind of historian uses, in fact, methods from the humanist tradition more than those derived from the medical semiotic, which is fundamentally alien to his culture. The archaeologist working in the field is, on the contrary, forced to use both; he therefore is among the first to have the right to a seat at the same anatomical table as Morelli, Freud and Conan Doyle.[30]

5. The Travelling of Technical Facts: The Case of Christopher Wren and St. Paul's Cathedral

As we have seen, the relationship between the facts stored in the object and the encyclopaedia is central to the development of archaeological reasoning.[31] This next example examines which facts are directly readable in the object and what kind of further interpretative steps need to be taken to tell the artefact's story in its entirety, including understanding why a phenomenon

---

[30] Carandini [1991] 2000, p. 256, translated by author.
[31] Here the term 'archaeological' is used in a broad sense to indicate also, for example, building archaeology. The accent is not on excavation but on the kind of history that is chiefly made through the study of material remains.

Figure 2.3. St Paul's Cathedral, roof truss over the nave.
From: Photo by the author.

we can still observe took place: why the building was built in terms of its techniques, materials, styles, shape, functional concept and so on.

5.1 Plaster, Timber and Iron: Facts and Artefacts in St Paul's Lofts

Climbing up to the whispering gallery in St Paul's Cathedral (1677–1708) you will notice a small, dark door at the top of the impressive spiral staircase. If you are lucky enough to get hold of a special permit (and the right key), you will be able to open it and enter a fascinating space spanning above the nave. You will be walking on a floor interrupted by the spherical extradoses of the nave's vaults, with the roof structure just above your head. You will notice it is composed of king post trusses (Figure 2.3) presenting slightly different shapes depending on the portion of the building they are used in – for example, featuring risen tie beams over particularly high vaults, such as the one at the entrance side of the nave.

If you have a closer look at the shape of the elements composing the trusses (particularly the rafters and king posts), you will see that they are often carved out of bigger logs in a way that must have involved wasting wood. The 'head' of the king post is indeed wider than the body, meaning that the sides have had to be subtracted, leaving timber pieces too small to be used for any structural parts. A similar detail also can be observed on the rafters: At their bottom end they are characterized by a 'step' of just a couple of centimetres that leaves the foot thicker than the rest of the beam. This may all appear particularly surprising when we remember that there was a dramatic shortage of timber in England at the time, particularly in London due to intensive building activities after the Great Fire (1666) and to the contemporary demand for new fighting ships after the Anglo-Dutch wars of 1665–1667 and 1672–1674. But now we are already using our historical knowledge to 'let the objects speak' – we will return to this topic later and use an array of sources to interpret these clues. Here, the artefacts are acting, as mentioned previously, as a sort of 'springboard for enquiry,' giving signs of anomalies that call for historical explanations.

Other noteworthy details in the roof are the metal elements of different shapes used for different purposes throughout the structure. Metal straps reinforce the connections between posts and tie beams as well as between rafters and tie beams, and other metal bars are used to stabilize the wall-roof structural system above the 'transept.' Looking closely at these elements we will notice that they have features that make it unlikely that they were produced and fitted in the early eighteenth century, suggesting that structural troubles arose after the original construction.[32]

Changes to the structure are suggested also by other clues. The beams composing the trusses are generally marked where they meet with other elements (e.g., at junctions between rafters and tie beams or posts and tie beams and so on). As mentioned before, this is to ensure that the pieces prefabricated off-site are reassembled correctly on-site. In the case of St. Paul's Cathedral, the marks not only indicate which pieces belong together to form which specific truss, but, at least originally, also gave instructions as to the order in which the trusses needed to be erected (e.g., number one is next to the dome, and the numbers progress towards the entrance side). Charting the marks enables us to understand clearly how the assembly process was organised. But it also makes apparent any irregularities in the system that might act as clues toward the fact that the structure was partially mended and reassembled, probably following some kind of damage, in the process of which some original elements were misplaced (the numbers on the trusses are not in the right order). Checking against the historical record, we can see that, in fact, the cathedral was hit by a (luckily) unexploded

---

[32] Confirmed by the written records (Burns 2004, pp. 98–101).

Figure 2.4. St Paul's Cathedral, loft space above nave, graffito.
From: Photo by the author.

bomb during World War II, which came through the roof above the quire
and shattered the High Altar.

If we then turn our attention to the walls and look carefully in a side-
light, we can make out a graffito carved in the plaster: '*G. Reeve May Y3
1705*' (Figure 2.4). Again we have here an example of a material object (in
this case, a plastered wall) used as a carrier of facts in the same way as a
page, and here it gives us a *terminus ante quem* this portion of the church
was roofed, suggesting that works were at a final stage in 1705. Looking
at the cathedral's building accounts we can, in fact, trace back a certain
'*Geo. Reeves*,' who, between October 1704 and May 1706, is employed by
the master bricklayer Richard Billinghurst. Specifically, in October 1704,
the bricklayers are '*Laying Bricks in Spandrils over ye Mid Isle Westward*';
in November 1704, they are '*Laying bricks in the Vaulting over the Middle
Isle W of the Dome*'; and in May 1706, '*Cleansing the joynts of Brickwork
in Dome and Mending the Paving in Church.*' Moreover, at the same date,
Richard Billinghurst gets paid '*For Bricklayers, & Labourers to serve them,
employed at the Dome from 16th April 1705 to 24th December following*'
(Wren Society 1938, XV, p. 135).

5.2 Facts from Artefacts and Facts from the Page: Putting Evidence into Context

If we reach back to what Pucci calls the 'enciclopedia' and to our typological knowledge, we will discover that St. Paul's roof structures are far more significant than we have so far implied. In fact, they are a novelty in the context of seventeenth-century English building culture; before Wren, hammer-beam roof structures were the most common (Figure 2.5).[33] Therefore, the presence of king post trusses and the use of metal elements to complement the timber structure in Wren's masterpiece are evidence of the introduction of new building techniques that arrived on the back of the new architectural style, which was derived from continental Renaissance architecture.[34] In order to understand the path followed by this set of innovations, it is important to note that – as the sources make clear – they were introduced by the architects and not (as one might think) by the master carpenters (Campbell 1999, 2002).

Probable sources of inspiration for Wren's designs are to be seen in the Italian and French architecture of the time. In fact, some of the brilliant technical solutions Wren adopted, although new to the English architectural world, had been already widely discussed and established abroad. (Thinking of king post trusses, for instance, we look particularly at the Italian carpentry tradition.) But how did these facts about building techniques actually travel? Wren wasn't a travelled man – the only journey he made to the continent was a relatively brief one to France, where the kind of trusses used in St. Paul's were not particularly common.[35] Therefore, we must assume that if the facts about carpentry techniques travelled to England through Wren, it was thanks to writings and drawings and descriptions, and not via his direct observation. We know for certain that Wren's library contained some important texts addressing these problems from a 'scientific' point of view. The most prominent one is Bernardino Baldi's *Exercitationes*, which was also mediated through Wotton's *Elements*.[36] Baldi's work discusses in detail the shape and structural behaviour of a king post truss, but, being a book on mechanics, the sketches and descriptions presented are more explicative of general principles then useful for reproducing the structures in detail.

---

[33] King post trusses had already been occasionally used by Inigo Jones and, before him, Robert Smithson, but they were still unusual (Yeomans 1992, p. 26).

[34] The new style, as well as some innovations in the building techniques, had already been partially introduced by Inigo Jones, although the civil war kept him from building much (Yeomans 1992).

[35] It has been suggested that Wren might have also travelled to the Netherlands (Kuyper 1980).

[36] Baldi 1621, see Italian translation and commentary in Becchi 2004. Wotton (1624).

Figure 2.5. Westminster Hall, hammer-beam roof structure (c.1399).
From: http://www.essential-architecture.com/LO/047-Westminster_Hall_edited.jpg

If we look for the sources that might have enabled Wren to import not just a general structural model, but also detailed technical solutions (which were obviously partially adapted to the local building tradition), we have to look further. King post trusses with struts – as those employed in St. Paul's – are a very *old* type of structure. Material evidence of their use goes back at least to the early Christian period, but they were probably already used well before that (Valeriani 2006a, pp. 107–27 and 2008). Nevertheless, there are various versions of king post trusses, differing from each other in how

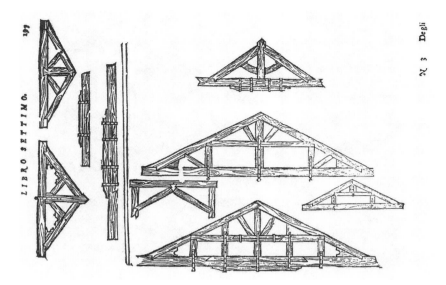

Figure 2.6. Roof structures after Sebastiano Serlio (Serlio 1575).
From: Serlio, 1575, p. 197.

the elements are shaped, for example, or how they are connected to each other. The details of the structure adopted by Wren (which later became common in England) resemble very closely those shown in Serlio's treatise and its chapters on roofs (Figure 2.6).[37] Evidence substantiating the thesis that Serlio was an important source for Wren's carpentry designs is given, amongst other things, by the king posts with joggled head described previously.[38] This solution – not previously used in England – is unlikely to have been developed by a carpenter, being relatively impractical and difficult to

[37] Wren owned an edition of Serlio published in Venice in 1663 (probably *Architettura di Sebastian Serlio bolognese, in sei libri diuisa, … Nuouamente impressi in beneficio vniuersale in lingua latina, & volgare, con alcune aggiunte. Sebastiani Serlij Bononiensis, De architectura libri sex. …*, In Venetia: per Combi, & La Nou, 1663). As yet, it has not been possible to clarify if he knew the seventh book, where the most interesting notes and drawings as regard roof structures are found, although it is probable that he had seen it, as Jones, for example, owned a complete edition of Serlio's works (*Tutte l'opere d'architettura, et prospetiua, di Sebastiano Serlio bolognese, …*, in Venetia: appresso Giacomo de' Franceschi, 1619).

[38] Another important point is made by the striking analogies between Wren's solution for the production of the long tie beam at the Sheldonian Theatre and the technique clearly depicted in Serlio's seventh book. An in-depth analysis of Wren's carpentry for the Sheldonian, as well as of the links with French building practices of the time, would exceed the scope of this chapter. A first, provisional report on this subject is to be found in Valeriani 2006c, and an updated version is forthcoming.

make compared to other variants, as well as uneconomical. Therefore, it seems unreasonable to assume it was a 'faulty' interpretation by English carpenters of a generic instruction from the architect – rather, it can be advanced as evidence of what we could call 'faulty transmission' of facts on the part of the designer(s). Actually, this somewhat inconvenient solution is not found in any extant Italian roof, but is the result of a 'thought experiment' by Serlio (treatise writer and practicing architect, but not builder). Nevertheless, the technique was picked up by Wren (the scientist-architect, but not practical carpenter), who introduced it to English carpentry.[39] It then became a basic feature of the new structures, which came into English architecture on the back of the new architectural style. It is reasonable to assume that the differences in detail between the English and the Italian solutions are due to the inputs of the local carpenters, who recognised some weaknesses of the Serlian design and reinterpreted it.

5.3 Reprise

The material objects in the loft spaces of St. Paul's allowed us to retrieve – more or less directly – facts concerning the history of the building (such as construction date, structural problems arising after construction, etc.). But in other cases, the evidence 'stored' in the timber, plaster and iron could only act as interesting clues whose significance would only be understood via what we have been calling 'the encyclopaedia.' Comparison with typified data at the local level highlighted anomalies that called for historical explanations, but the use of a wider, international database has enabled us to recognise elements of continuity and to grasp broader historical phenomena, such as the introduction of new building techniques into England in the seventeenth century. Here, the material object gave us clues about an anomaly – the use of a new kind of structure – but the historian needed many more sources to understand why and how this specific carpentry technique came to be used. Once a possible source for these new technical facts had been identified (in Serlio's book), we looked at the material object again in detail to see if it fitted with the suggested source or not. Any 'discrepancies' led to further work to assess whether the presumed source was indeed the right one and, if so, how/why the original object/design/technical fact had changed while travelling.

---

[39] It must be noted that Inigo Jones had already designed posts with joggled heads (see, for example, Stoke Bruerne, Northamptonshire), but in contrast to Wren, also used some posts without enlarged heads, more in line with the Italian carpentry tradition (see, for example, Queen's Chapel, St James's Palace).

In the case of St. Paul's, while discussing the most probable 'line' along which the new roof structures characterising English Renaissance architecture could have been imported on the back of the general architectural style, we have highlighted how the vehicles of travel in this case were books and drawings, rather than the architect's direct 'experience' of foreign structures. The contemporary thirst for renewal of the English architectural tradition ensured that such technical facts travelled well, remained quite intact and took root in good 'soil', which allowed incoming techniques to succeed in replacing existing ones, albeit adapted at times to local traditions. While travelling, these facts were transformed along two main lines: Some changes were due to 'faulty interpretation' of original information, and others to the influence of the local craftsmen's existing know-how.

6. 'Everyday Things Can Tell Secrets to Those Who Can Look and Listen'[40]

As with all kinds of sources, the ability of facts to travel depends on an observer being able to 'read' the artefact and 'recognise' the facts. A lack of observer expertise can result in facts travelling badly – or even not at all. This is not peculiar only to material objects: The same applies to any kind of information carried by whatever medium. For example, if you don't know they are written characters, you could easily misinterpret hieroglyphics as being purely decorative elements. And even if you know the signs are building words and phrases, you first need to figure out how to interpret them. Still – even if you don't recognise them as writing, or if you can't interpret them – the marks will stay on the object, together with the embedded facts they are carrying, waiting to be recognised or understood. This raises the issue of stability of facts' meanings – and material objects don't seem to differ significantly from other media in this regard: Despite the relative immutability of their materiality (in terms of, for example, chemical composition), the interpretation of facts that travel via artefacts still seems to change over time, and can depend greatly both on the observers' abilities and so on, and on the community/culture observing the object.

Taking for granted the importance of the social constructed identity of material objects and how the observation of those objects is socially determined, this chapter has aimed to analyse – with examples – how artefacts carry facts. Material objects have facts to tell or, as Lorraine Daston put it,

---

[40] *'Le cose di ogni giorno raccontano segreti, a chi le sa guardare ed ascoltare'* Gianni Rodari, text of the song *'Per fare un tavolo.'*

'Things talk and are not merely repeating or playing back the human voice' (Daston 2004, p. 11).

Looking at methodological differences between historical disciplines – archaeology and history – dealing with material objects in different ways, it has been argued that the material nature of the element storing or transporting facts influences how they travel and the methodologies needed to 'retrieve' them. Artefacts – material objects – carry a complex and layered collection of facts of different natures in a fundamentally different way from how facts are transported by the written page. This seems to be because material objects are a particular kind of vehicle that do not 'carry' facts as in an empty box – rather, the facts are embedded in the material of the object. Accent has been placed on the propensity of artefacts to carry facts about what happened, 'how' and 'when' rather than 'why.' (In like manner, written texts lend themselves to easily communicate intentions and reasons, but are not the best or most reliable vehicle to carry information about physical qualities and techniques.) Moreover, even though a text can inform about these aspects, it cannot normally offer the same richness of unintentionally expressed facts as can material objects. Although this is obviously a difference of degree rather than of kind, it still underlines a peculiarity in the ability of artefacts to carry facts, as against the written page or other 'vehicles.'

Acknowledgments

I would like to thank the many people who helped me thinking about these issues and gave important feedback on earlier versions of this chapter, particularly Mary Morgan and Peter Howlett, the members of the facts group, and its guests, for conceptual suggestions; David Yeomans for our discussions on English and Italian carpentry and Jon Morgan for his editorial help. My thanks goes also to the Leverhulme Trust and ESRC that generously financed the 'How well do Facts Travel?' Project (grant F/07004/Z) and gave me the opportunity to conduct this piece of research.

Bibliography

Baird, Davis. *Thing Knowledge: A Philosophy of Scientific Instruments*, Berkeley, California: University of California Press, 2004.

Baldi, Bernardino. *In mechanica Aristotelis problemata exercitationes*, Moguntiae: Typis et Sumptibus Viduae Ioannis Albini, 1621.

Becchi, Antonio. *QXVI. Leonardo, Galileo e il caso Baldi: Magonza, 26 Marzo 1621*, Venezia: Marsilio, 2004.

Brown, Bill. *Things*, Chicago, Ill., London: University of Chicago Press, 2004.

Buchli, Victor. *Material Culture: Critical Concepts in the Social Sciences*, London: Routledge, 2004.

The Material Culture Reader, Oxford: Berg, 2002.

Burns, Arthur. 'From 1830 to the Present', in *St Paul's. The Cathedral Church of London 604–2004*, edited by Derek Keene, Arthur Burns and Andrew Saint, London: Yale University Press, 2004, pp. 84–112.

Campbell, James W. P. *Sir Christopher Wren, the Royal Society, and the Development of Structural Carpentry 1660–1710*, PhD Thesis, Department of Architecture, University of Cambridge, Trinity College, 1999.

'Wren and the Development of Structural Carpentry 1660–1710', *ARQ: Architectural Research Quarterly*, 6(1), March 2002, pp. 49–61.

Carandini, Andrea. *Storie dalla terra*, Torino: Einaudi, [1991] 2000.

Daston, Lorraine (editor). *Things That Talk: Object Lessons from Art and Science*, New York: Zone, 2004.

Eco, Umberto ; Sebeok, Thomas A. *The Sign of Three*, Bloomington: Indiana University Press, 1983.

Friedrich, Karl. *Die steinbearbeitung in ihrer entwicklung vom 11. bis zum 18.* Jahrhundert, Augsburg 1932.

Ginzburg, Carlo. 'Spie. Radici di un paradigma scientifico', *Rivista di storia contemporanea*, 1, 1978, pp. 1–14.

'Morelli, Freud, and Sherlock Holmes: Clues and Scientific Method', in *Eco* 1983, pp. 81–118.

Clues, Myth, and the Historical Method, Baltimore: Johns Hopkins University Press, 1989.

History, Rhetoric, and Proof, Hanover: University Press of New England, 1999.

Kuyper, Wouter. *Dutch Classicist Architecture: A Survey of Dutch Architecture, Gardens and Anglo-Dutch Architectural Relations from 1625–1700*, Delft: Delft University Press, 1980.

Maier, Konrad, 'Mittelalterliche Steinbearbeitung und Mauertechnik als Datierungsmittel', *Zeitschrift für Archäologie des Mittelalters*, 3, 1975, pp. 209–16.

Mannoni, Tiziano. 'Ricerche sulle malte genovesi alla "porcellana"', in *Le scienze, le istituzioni, gli operatori alla soglia degli anni '90*, Atti del IV convegno di scienza e beni culturali, Bressanone, 1988, pp. 137–42.

Mannoni, Tiziano Pesce, Giovanni and Vecchiattini, Rita. 'Rapporti tra archeologia, archeometria e cultura materiale, nello studio dei materiali impiegati nelle opere portuali', in *Anciennes Routes Maritimes Méditerranéennes (ANSER): Le strutture dei porti e degli approdi antichi*, edited by Anna Gallina Zevi and Rita Turchetti, Roma: Rubettino, 2004, pp. 113–26.

Miller, Daniel. *Materiality*, Durham: Duke University Press, 2005.

Pesce, Giovanni and Ricci, Roberto. 'The Use of Metakaolinite as Hydraulic Agent of Aerial Lime Plasters and Mortars. The Case Study of Genoa (Italy)', in *Proceedings of the HMC08 Historical Mortars Conference*, Lisbon September 2008, forthcoming.

Pucci, Giuseppe. 'La prova in archeologia', *Quaderni storici*, 85 (1), 1994, pp. 59–74.

Rogacki-Thiemann, Birte. *Der Magdeburger Dom St. Mauritius et St. Katharina – Beiträge zu seiner Baugeschichte 1207 bis 1567*, Petersberg: Michael Imhof Verlag, 2007.

Serlio, Sebastiano. *Il settimo libro d'architettura di Sebastiano Serlio Bolognese*, Frankfurt: Officina typografica Andreae Wecheli, 1575.

Tarule, Robert. *The Artisan of Ipswich*, Baltimore: John Hopkins University Press, 2004.

Tilley, Christopher Y. et al. *Handbook of Material Culture*, London: SAGE, 2006.

'La Bauforschung a Lubecca: metodi di datazione', *Archeologia dell'architettura*, 4, 1999, pp. 83–92.

Kirchendächer in Rom – Zimmermannskunst und Kirchenbau von der Spätantike bis zur Barockzeit. Capriate ecclesiae – Contributi di archeologia dell'architettura per lo studio delle chiese di Roma, Berliner Beiträge zur Bauforschung und Denkmalpflege III, Petersberg: Imhof Verlag, 2006a.

'I Metodi dell'archeologia dell'architettura applicati allo studio delle coperture lignee di alcune basiliche a Roma', in *Archeologie. Studi in onore di Tiziano Mannoni*, edited by Nicola Cucuzza and Maura Medri, Bari: Edipuglia, 2006b, pp. 519–22.

'The Roofs of Wren and Jones: A Seventeenth-Century Migration of Technical Knowledge from Italy to England', Working papers on The Nature of Evidence: How Well Do 'Facts' Travel?, 14/06, LSE, Department of Economic History 2006c (http://www2.lse.ac.uk/economicHistory/pdf/FACTSPDF/1406Valeriani.pdf)

Valeriani, Simona. "'In the Ancient Forme". The Reception and "Invention" of Ancient Building Techniques in Early Modern Times,' *Hephaistos. New Approaches in Classical Archaeology and Related Fields*, 26, 2008, pp.169–88.

Vecchiattini, Rita. 'Unità produttive perfettamente organizzate: le calcinare di Sestri Ponente – Genova', *Archeologia dell'Architettura*, III, 1998, pp. 141–152.

Wotton, Henry. *The Elements of Architecture*, London, 1624.

Wren Society. *Photographic Supplement of St. Paul's Cathedral and Part III of the Building Accounts from October 1st, 1695 to June 24th, 1713. Also the Chapter House accounts, 1712-14 and Outline of Cathedral Accounts, 1714-25*, vol. 15, Oxford: Oxford University Press, 1938.

Yeomans, David. *The Architect and the Carpenter*, London: RIBA Heinz Gallery, 1992.

THREE

A JOURNEY THROUGH TIMES AND CULTURES? ANCIENT GREEK FORMS IN AMERICAN NINETEENTH-CENTURY ARCHITECTURE

LAMBERT SCHNEIDER

1. Introduction

Archaeology constantly deals with so-called "facts." Public opinion clearly associates the field with demonstrable facts. Since the object of archaeology is investigating the past by analysing material phenomena, the discipline is expected to have something substantial to say about the "travel" – meaning in this case the *historical continuity* – of such "facts."[1] The existence of ancient civilizations with their apparent immutability has generated confidence in the existence of cultural and artistic continuity, or at least of a gradual development that transmits facts through time. The numerous modern revivals of ancient forms and ideas, both in scholarship as well as in the broader context, have seemed evidence for the existence of a "cultural memory"[2] within which facts might comfortably travel through time.

This article examines this widely held popular assumption. I suggest that the answer to the question of what travels and how largely depends on the interest and focus of the beholder rather than on the phenomena beheld. Seen in this light, both classical revivals in art and architecture and the academic investigation of ancient Greek culture turn out to be creative undertakings that mould and even invent the shape and meaning of

---

[1] Other chapters in this volume that take different cuts through the issues of travelling facts in material objects, including construction, architecture, archaeology and history, are those by Valeriani and Wylie.

[2] Assmann 1999; Schneider 1999a; Schneider 1999b; Assmann 2006.

the past. The material with which I will illustrate this is Greek-inspired American architecture of the nineteenth century and the public response to this phenomenon.

2. Architecture as Sculpture: Winckelmann and the European Classicism

When the sculptures that Lord Elgin had taken away from the Athenian Acropolis arrived in England in 1809 and were subsequently exhibited in the British Museum[3] (Figure 3.1), they became almost immediately world famous. In particular, the pedimental sculptures of the Parthenon, despite their fragmentary condition, rose to celebrity status. At that time, classical Greek sculpture such as the Parthenon pedimentals was considered a symbol of freedom, an embodiment of a freer, unfettered lifestyle than was possible in most European countries. Looking back into the past was linked to hopes for a better future, and therefore had utopian overtones. To the early nineteenth-century European beholders, the Parthenon sculptures (Figure 3.1) represented freedom from restrictive etiquette of court dress, from wasp waist and corset, from stifling ties and measured steps, but also freedom of thought and of political action.[4] Even nakedness was approved of in this case, with so-called "wet drapery" supporting the illusion of powerful, flowing movement. Whether casually stretching or in vigorous action, the gods proudly present their bodies to the beholder. Might not all people at one time have been able to behave as such, freed from traditional restrictions? These were the dreams of intellectuals of the time.[5]

In a remarkable double equation, classical Greek sculpture, like architecture, was understood as a symbol of naturalness, even as a perfection of nature; and nature, in turn, as a metaphor of freedom. So it was not only the fact that one now possessed fragments of Greek sculpture of the epoch that was considered the cradle of democracy – it was the specific *form* of these sculptures that met with an interpretation that had been awakened at that time but was soon eclipsed by other readings.

Europe's enlightened public was well prepared to view these works in the way described here. It had been Johann Joachim Winckelmann – in a sense, the founder both of classical archaeology and of stylistic-

---

[3] King 2006, Cook 2007, Hazlitt 2008; Schneider 2010.
[4] Forster 1996; Schneider and Höcker 2001; Schneider 2003.
[5] Lessing 1769a; Lessing 1769b; Herder 1778; Schiller 1793a; Schiller 1793b; Goethe 1772; Goethe *Laokoon* 1798; Goethe *Propyläen* 1798; Goethe 1805.

Figure 3.1. Parthenon, East Pediment goddesses.
From: A. Michaelis, Der Parthenon (1871).

orientated art history – who decades before had formulated the daring analogy between classical Greek sculpture, nature and freedom,[6] thereby initiating a pattern of thought that was met with widespread interest and enthusiasm all over Europe. Winckelmann imagined classical art to be so natural, so unspoiled by luxury and over-sophistication, that he even compared it with the supposed innocence, simplicity and grace of the American Indian. During the first half of the nineteenth century this notion was occasionally adopted in American art. Like the Dionysus in the East pediment of the Parthenon, Henry Kirke Brown's figure of an Indian of 1850 in Philadelphia[7] (Figure 3.2) reclines in a most relaxed manner and is clad in the "costume" of ancient Greek nudeness. Similarly, Shobal Clevenger's rendering of an "Indian Chief" of 1843[8] (Figure 3.3), which by its rigidity appears naïve to modern eyes, impressively demonstrates how highly autopoetic and unfounded on observation these equations were, while at the same time very effective.

By shifting classical art into a lofty realm of superiority, the material products of Greek society of a specific historic situation mutated into something timeless and even transcultural.

Selected forms of ancient statues found in Rome crept into the minds of modern beholders only on the basis of the belief that they were classical Greek rather than Roman – they were delineated in engravings, thus

---

[6] Winckelmann 1756. Similarly: Herder 1778.
[7] Vance 1989, p. 302 following.
[8] From: *United States Magazine and Democratic Review*, February 1844. Quoted by Vance 1989, p. 304.

Figure 3.2. Philadelphia, PA. Fountain with allegories of the continents. Henry Kirk. 1850. *From*: Photo Lambert Schneider.

reducing their sculptural character to a dry contour, and in this form redistributed geographically. In a further step, they were then reactivated and reinterpreted as models of man-in-the-state-of-nature by applying them to the rendering of American Indians, who were thus transitively allocated similar "natural" nobility as the ancient Greeks. So even in this provisional and superficial first overview of the process, "travelling facts" seem to vanish almost completely. Or do they?

The beholders of the late eighteenth and early nineteenth centuries ascribed this outstanding quality of naturalness particularly to works of the fifth and fourth centuries BC, which they called the only true classical ones. The previous broad definition of the classical was thus narrowed. Within antiquity it was only the *Greek* that was to be awarded the elitist honorific of "classical." Within this, Athenian culture of the fifth and fourth centuries BC was privileged most of all, with art and architecture of the period considered in the same terms as sculpture. Consequently, the corpus of ancient Greek relics was viewed as a kind of plastic art, a view that would have far-reaching consequences.

This new way of looking at architecture and at art in general *as if it were sculpture* was largely based not upon the observation of objects or processes from the past but made up "at home," created by an inner process. For

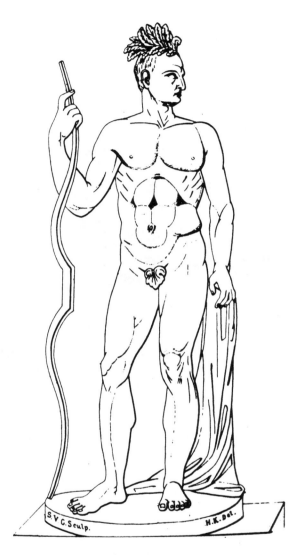

Figure 3.3. "Indian Chief," Shobal Clevenger, 1843.
From: *United States Magazine*, February 1844.

instance, Winckelmann – that daring prophet of the message of Greek art to modern times – was during his early years in Germany unable to see many Greek originals, and the few he physically encountered apparently made no great impression on him. He managed to write his famous and influential work of 1756, *Thoughts on Imitating the Works of Greek Painting and*

Sculpture, before he had ever seen and thoroughly studied original works of Greek art. And even later in his life, when he resided in Rome as kind of a pope in the field of scholarship in ancient art and was at least *economically* able to visit Greece (where he was invited to go to by friends more than once), he refused to do so, turning down the opportunity to see classical Athens. "I am already in firm mental possession of this Greek ideal. I am not at all convinced to discover anything new there," he annotated in a letter to his friend Johann Hermann Riedesel. The episode illustrates well the degree to which this new and sparkling classicism was not a *reconstruction* of an ancient past but instead a creative act of modelling a vague dream into a firm and detailed picture of classical Greece, which subsequently gained physical existence both in sculptural art and architecture. From time to time, this creative act made use of archaeological observation, even minute observation, but it was never really derived from archaeological observation as it is often believed to be.

Winckelmann's conception of classicism incorporated social and political implications, yet was romantic from the start – unreal yet uplifting. Winckelmann and the following generations of intellectuals in Continental Europe like Johann Gottfried Herder or Wolfgang Goethe had no means of enacting or even effectively promoting democracy in their home countries, not to speak of establishing radical democratic practices as had arisen in Athens in what had been (according to Winckelmann's classification) the most classical epoch.[9]

It was this idealised conception of classical antiquity that was enthusiastically welcomed all over Europe – first in England, but soon on the continent as well. Here it fed into the desires of the enlightened public, and yet must have appeared utterly harmless to any established powers, even the most reactionary. In the first half of the nineteenth century, the ardently Greek-minded rulers of Bavaria and Prussia rivalled each other in turning their capitals (still backwaters in comparison to metropoles like London or Paris) into a new Athens. Even politicians like Count Metternich or Tsar Alexander III of Russia seemed enchanted by this dream. So it was not only that the original social and political message of these thoughts was soon discarded, but rather, from the beginning, this concept of classicism never actually interfered with even the most (as Winckelmann had it) "unnatural," and therefore "un-Greek," attitudes and practices.

Digging for classical remains, conserving and reconstructing ancient buildings, as well as erecting new ones in the classical style in an astutely

[9] Marchand 1996.

archaeological manner: All this fit perfectly well not only with democratic ideas but also with monarchic rule. Meanwhile, the Greek order – in the sense of the architectural order with all its metaphorical connotations – soon became the language of the establishment all over Europe, of stately or private authority, in milder or (more often) severe form, especially so in German-speaking countries and in Greece itself.

The original meaning of the Latin word *classicus* already implied association with an upper class, but as the nineteenth century wore on, this more social definition acquired an added depth and severity previously absent. In particular, the Greek Doric order and the slightly less severe and more elegant Greek Ionic order were now interpreted as physical embodiments of what Sigmund Freud would later term the "super-ego." Winckelmann's original viewing of Greek sculpture and architecture as symbolic of naturalness and freedom had given way to a new definition: a manifestation of class-conscious order, of externally enforced discipline and of internalized self-discipline through education.[10] Classical Greek art had, in a most problematic way, become symbolic of human culture.

Classical archaeology became a tool for attaining the classicistic goal. In Greece itself, archaeological activities did not seek to disclose the ancient world as it had been, but only to confirm the ideals of the so-called classical period through the excavation of monumental relics. All that did not accord with these ideals was deconstructed, cleared aside and annihilated with a terrible rigour. The few remaining skeletons of ruins of the classical period were then heavily restored to form a view fitting the ideology. The Acropolis at Athens, for example, came to resemble more and more places like Munich or Berlin. Archaeologists thoroughly adjusted the physical reality of the ancient sites to their idealistic vision. They created sculptural architectonic ensembles of a kind that had never existed in antiquity. Seen in this light, classical archaeology appears as a structural complement to other endeavours within the whole bundle of undertakings of modern classicism. The ensuing disintegration and destruction of historical traces happened not *in spite of* but *because* of classical archaeology.

No wonder that parallel to this archaeological strictness, rigorous conformance to the classical and an almost obedient devotion supported by archaeology were the dominating principles in contemporary domestic building. In reality, these constructions were rarely real buildings in the traditional sense. Rather, they functioned as plastic monuments, signifiers in

---

[10] Marchand 1996; Schneider 1996; Schneider 2003, pp. 148–50.

stone – Walhallas,[11] grave-monuments or gate-monuments, such as that in Munich by Leo von Klenze, or that of Wassili Petrovich Stassow of 1838 at St. Petersburg.[12] These were not integrated into daily life but instead placed on a pedestal for veneration. Thus, most of the archaeologically astute uses of the classical Doric and Ionic order no longer functioned as true architecture, but rather as symbols of a given law and of internalized order.

The classicistic "sculptural dream", as described earlier, was realized in a physical form by contemporary building activities as well as by archaeological excavation, restoration, presentation in museums and publication. Winckelmann and other writers of the eighteenth century had provided the ideals that stirred the imagination largely based on an "armchair ideology" as far as Greek architecture is concerned; Le Roy and, above all, Stuart and Revett furnished the public with true and detailed visual material based on autopsy and applicable to actual contemporary building.[13]

3. Democracy, Entrepreneurial Pride and the Classical Notion in the New World

In comparison to this, how does the re-use of the same classical models manifest itself in a country that for so long lacked any foundation in classical archaeology as a scholarly discipline and educational pursuit?[14] America actually offers the richest variety of Greek-inspired architecture in the world,[15] in both a quantitative and qualitative sense. American classicistic architecture is often closely associated with the idea of democracy. Hence, the title of Henry-Russell Hitchcock and William Seale's book on state capitols erected in Doric, Ionic and Corinthian order: *Temples of Democracy*.[16]

---

[11] Traeger 1987; Schneider and Höcker 2001, pp. 32–4; Nerdinger 2002.

[12] The gate, executed in iron technique, is a free adaptation of the Propylaea on the Athenian Acropolis. It commemorated Russia's successful war against Turkey and Poland in 1834–38. Schneider and Höcker 2001, pp. 34–6.

[13] Merz, this volume, is also concerned with the reading of images that are indicative of ideals (such as the IBM logo rendered by nanotechnology), as opposed to images that are accurate in conveying the phenomena in the travels of facts.

[14] Yeguel 1991; Dyson 1998; Meckler 2006.

[15] Downey 1946; Newton 1952; Scully 1973; Waddel and Liscombe 1981; Crook 1987; Ackermann 1990; McCormick 1990; Curl 1991; Peck 1992; Höcker 1997; Reed 2005. Most of these architectures in the United States were not designed by professional architects but rather by *builder-carpenters* (Minard Lafever calls them *operative workmen*). Written records are rare. The following literature mainly stems from or deals with renowned and literally well-documented architects: Benjamin [1833] 1972; Lafever 1839; Downing 1850; Lafever 1852, Lafever 1856. Gallagher 1935; Downing 1988; Bryan 1989; Lane 1993; Lane and Martin 1996; Seale 1996.

[16] Hitchcock and Seale 1976.

And in a sense they are that. Nevertheless, the title is a misnomer, for it suggests that Greek-inspired forms were primarily understood as an expression of democratic principles. This was not the case.

In the first place, it does not fit chronologically. Greek-inspired architecture swept across the states from New England, through the Midwest, and out into the most remote locations. This wave started no earlier than the second decade of the nineteenth century[17] – more than a generation after the "fathers" of American democracy.

These fathers, the signatories of the Constitution, had also adamantly associated themselves with antiquity, as evidenced by written sources. But it was not classical Athens with its undesirable fate that they chose for a model, but rather the Roman Republic. Roughly speaking, their attitude seemed to be *Antiquity, yes; Greece, no.* Therefore, they never compared themselves with Pericles, but always with figures such as Cato or the legendary Cincinnatus: Roman politicians who were in antiquity, as well as in modern reception, representatives of a hard-working and austere lifestyle – not unlike American farmers and ranchers at the time, people who would in literature be portrayed standing behind a plough but at the same time were concerned with the community and the state.[18] Later, as Greek elements became fashionable in architecture, decoration and sculpture, this attitude persisted. So visualisation of democracy was not primarily the impetus of this wave, and even later Greek forms were generally not interpreted in this way.

Admittedly, Thomas Jefferson was well acquainted with French revolutionary classicistic architects and intellectuals, who introduced him to Winckelmann's thoughts.[19] So one finds various speculations in scholarly texts that these connections strongly influenced the American artistic and architectural scene at the turn of the eighteenth century into the nineteenth.[20] However, this alleged impact is just not based in reality. Neither George Washington's residence, Mount Vernon (1743 and later), nor Jefferson's Monticello are characterized by anything that could be called Greek revival. The same applies to Washington's governmental architecture during this time period. Both the White House and the Capitol[21]

---

[17] Meyer Reinhold 1984.
[18] Kennedy 1989, pp. 7–103.
[19] Höcker 1997; Bernstein 2005.
[20] Hitchcock and Seale 1976, Höcker 1997; Bernstein 2005.
[21] 1793–1863: The Doric columns in the room under the Old Senate. Architects: William Thornton, B. H. Latrobe, Charles Bulfinch, Robert Mills. Allen 2005; Reed and Day 2005.

are overwhelmingly Roman. Truly *Greek* forms were introduced no earlier than 1818, by Charles Bulfinch. And it is only in the basement of the Capitol where you find archaic-looking Doric columns copied not from a building in Greece but from an early temple at Paestum,[22] which here supports a cap vault. However, this Greek element remained isolated within the architectural complex and remained isolated historically in the sense that it inspired no successors in the United States.

Rather, it was the new self-confidence of the next two generations, fuelled by Andrew Jackson's victory over the British troops in 1812, a new pride following years of depression, that was visualized by this fashion. So it was not so much "temples of democracy" in a strictly political sense as it was an expression of the new economic prosperity and the new trend towards conspicuous consumption while at the same time signalling diffusion of civilisation.

It is revealing that it was not so much the old founding families who followed this fashion, but rather the young entrepreneurs. I think this is one explanation for the fact that – although you find some examples of Greek revival in places like Boston – there are, by far, more and more impressive examples found further west, in newly developed areas: in Troy[23] or Geneva[24] (both in upstate New York).

This new class of entrepreneurs neither saw in classical Greece a democratic model, nor did they in any way reverentially look back to a distant past. For them, Greek forms were something akin to a garment suitable for their social status and new-found wealth. A telling example of this attitude is Whale Oil Row at New London, CT, aligned by houses with truly Greek Porticos in Ionic order (Figure 3.4), all copying a tiny temple at Athens that has meanwhile completely vanished but was drawn and published in printings by James Stuart and Nicholas Revett in their famous work of 1762–94, *The Antiquities of Athens Measured and Delineated.* (Figure 3.5). "Whale Oil Row," indeed! The clients and owners of these wooden buildings of c. 1850[25] definitely were not classical philologists or any other ardent admirers of the ancient past, nor were they civil servants or politicians schooled in and devoted to ancient democracy. Instead, they were more like Melville's Ahab.[26]

---

[22] Major 1768, Table XII.
[23] Scully 1973, pp. 23–6; Smith 1976, p. 189; Smith 1981, pp. 468–9; Schneider 2003, pp. 154–6.
[24] Smith 1976, p. 188; Smith 1981, pp. 434–5; Schneider 2003, pp. 154–6.
[25] J. R. Ruddy: New London, Connecticut (1998).
[26] Melville, 1851.

Figure 3.4. New London, CT. "Whale Oil Row". c. 1850.
From: Photo Lambert Schneider.

Figure 3.5. Ionic temple at the Ilissos River at Athens, no more extant.
From: Drawn by Stuart and Revett (1762).

4. Transformations of the Classical Models in American Architecture and their Public Reception

It is not just the circumstances in which this architecture was built that speak against a tight linking of democracy to these Greek-inspired forms. It is also the contemporary assessment of the phenomenon that points to another direction, and, as we will soon see, the buildings themselves.

Some of the buildings, especially the earliest ones, look, at first sight, very much like those you would find in England and Continental Europe: close copies of ancient classical architecture – for instance, William Strickland's remake of the Parthenon (1819–24, Philadelphia, Figure 3.6).[27] Even these strict copies, however, were seen in a different light by contemporary beholders: light not only in a metaphorical sense, but also in its literal meaning. It is Philip Hone, a typical entrepreneur of the time, politician and amateur in the field of architecture and the arts, who gives us an assessment of this building on 14 February 1838: [28]

> The portico of this glorious edifice, the sight of which always repays me for coming to Philadelphia, appeared more beautiful to me this evening than usual, from the effect of the gas-light. Each of the fluted columns had a jet of light from the inner side so placed as not to be seen from the street, but casting a strong light upon the front of the building, the softness of which, with its flickering from the wind, produced an effect strikingly beautiful.

Hone's view is a contemporary one, but these lights still exist and give "physical" proof of his impression. The basic concept of Greek temple-building is totally inverted by this. Whereas the massive walls of the cella of ancient Greek temples appeared as something compact and dark behind the shining columns, here the core building shines like a jewel behind the darker fence of the columns. The columns still appear important, but more dominant is the actual building itself, which, after all, in this case was "The Second Bank of the United States," so not an empty monument but a building intended for actual use.[29]

---

[27] Kennedy 1989, pp. 114–5, 194–5; Tournikiotis 1994, p. 213; Schneider and Höcker 2001, pp. 29–32; Schneider 2003, pp. 158–61.

[28] Quoted by Hamlin 1942, p. 78 n. 19.

[29] The fruitfulness of travels in which some elements of the facts hold their integrity and others are qualified or even subverted can be found also in Ramsden's account of Calhoun's rat behaviours and in Schell's account of the alpha-male facts in romance fiction, both this volume.

Figure 3.6. Philadelphia, PA. Second Bank of the United States. View with authentic gas lights.
From: Old postcard, source unknown.

This radical inversion of an otherwise minutely copied ancient model is not an isolated case. Similar lighting is reported of the Old Custom House at Erie, PA,[30] of 1839 and can still be seen at Bethel United Methodist Church at Charleston, SC,[31] (1852–53). Also once a noble bank – even with living quarters to house the president of the United States when he visited this place – was the now First Church of Christ Scientist at Natchez, MS,[32] erected in 1833

---

[30] Now Erie Art Museum: Muller 1997.
[31] Regarding the front elevation, an astute copy of the Athena and Hephaistos temple, the so-called Theseion at the Agora of Athens: Schneider 2003, pp. 159–60; Foster 2005.
[32] Kennedy 1989, pp. 116–7; Schneider 2003, pp. 160–1.

Figure 3.7. Louisville, KY. Actor's Theatre, formerly a bank. Architect: James H. Dakin. 1835/37.
From: L. Schneider 2003, p. 161 fig. 18.

as a fine copy of the already mentioned Ilissos Temple at Athens (Figure 3.5); the light behind the columns is again authentic. No less impressive is the appearance of the Actor's Theatre at Louisville, KY,[33] of 1835–37 (Figure 3.7), again originally a bank designed by James H. Dakin. These examples reveal

[33] Scully 1973, pp. 26–40; Smith 1976, p. 350; Smith 1981, pp. 284–5; Kennedy 1989, p. 372; Schneider 2003, p. 161. Even neoclassical buildings of the twentieth century, such as Henry Bacon's Lincoln Memorial in Washington, D.C. (1913–22), continue this American tradition.

not only a new attitude towards architecture quite contrary to that found in Greek antiquity, but also a new characteristic of Greek-inspired American architecture itself, through which the new and truly sovereign American way of dealing with the phenomenon "Greek classical" manifests itself.

Not just by gas lights, but by whole rows of large windows – often double-storey and complemented by spacious doors – American architects converted Greek temple architecture into something completely new: into buildings that seem to wear Greek orders like clothing. The core building never hides behind the columns. The attitude we found reflected by the artificial lighting is also reflected in the buildings themselves.

This, of course, is not just a question of aesthetics. It is the vital functions of the buildings, their uses in life that were proudly shown to a public: proudly with respect to any beholder and proudly with respect to the ancient models. Like jewels, the inner cores of the buildings glow behind rows of Greek columns. It is not just this view from the outside-in that is important and underscores the proud display of function, but also the view from the inside-out. The beholder looks through the ancient columns to the present beyond.

Most popular in American Greek revival was the design of the Ionic front of a classical fifth-century temple near the Ilissos River at Athens, as drawn by Stuart and Revett (Figure 3.5) and reprinted in various nineteenth-century American books on architecture. The columns of this since-destroyed little temple were employed for instance as a model for a villa of around 1850 at Eutaw, AL,[34] (Figure 3.8) to "clad" the core building and to support not only the roof but also a surrounding balcony, attached in a most un-classical manner directly to the shafts of the columns; the roof again is here crowned by a little belvedere that imitates the main structure on a smaller scale. The same applies to Neill-Cochran House at Austin, TX,[35] (Figure 3.9), where again the rows of beautiful large windows on both storeys behind the classical Ionic order are to be noticed. Another fine example is Wilcox-Cutts House of 1843 in Orwell, VT,[36] (Figure 3.10); this time, there are five columns – gently abandoning classical rules of Ionic order in favour of attaining a colonnade that does not obscure the view from the large windows behind.

A similar care for the building itself and pride in what is going on inside is shown in public American architecture. Nashville's Tennessee State Capitol[37]

---

[34] Kennedy 1989, p. 241; Schneider 2003, pp. 162–3.

[35] Erected during the same years: Smith 1981, pp. 635–6; Kennedy 1989, p. 242; Schneider 2003 p. 164.

[36] Pierson 1976, p. 449; Kennedy 1989, p. 35; Schneider 2003, p. 162.

[37] Hitchcock and Seale 1976, p. 119; Hudson and Ballard 1989, pp. 314–5; Schneider and Höcker 2001, p. 27; Schneider 2003, p. 165.

Figure 3.8. Eutaw, AL. Kirkwood or H.A. Kirksey House. c. 1850.
From: Photo Lambert Schneider.

of 1845–59 (Figure 3.11), designed by William Strickland – the architect of
the Second Bank of the United States noted previously – presents an enlarged
version of the main front of the Erechtheion on the Acropolis with its characteristic capitals (Figure 3.12), but now with eight instead of six columns,
and once more there appear blinking rows of windows behind the colonnade. On top of the roof – above a dome not visible from the exterior – is
placed a minute copy of Lysicrates Monument at Athens (again taken from
Stuart and Revett's book).

The same inverted use of the classical Erechtheion is found also in
many private buildings of the time.[38] Even in cases where the classical
order was accurately copied in toto, the same fundamental inversion of
the classical concept is to be noticed[39] – all the more so when the classical

[38] Madewood Plantation House at Bayou Lafourche near Napoleonville, LA, erected in
1846–8; architect, Henry Howard. Smith 1981, pp. 313–4; Kennedy 1989, p. 185. Avery
Downer House at Granville, OH, 1842, by Minard Lafever. Kennedy 1989, p. 323.
[39] Judge Robert Wilson House of 1843 at Ann Arbor, MI: Kennedy 1989 p. 235; Schneider
2003, pp. 162–3, 166.

Figure 3.9. Austin, TX. Neill-Cochran House (governor's mansion). Architect: Abner Cook. 1853/55.
From: R. G. Kennedy, *Greek Revival America* (Stuart Tabori & Chung, New York 1989), p. 242. © Jack Kotz.

models were changed in form and proportion (Figure 3.13)[40]: That the builders and architects of the antebellum time were not afraid to frivolously install five (!) columns when otherwise copying their model quite accurately is due to the same new and distinctly American approach to the classical (Figure 3.14).[41] Exactly the same features are to be found in the American use of the Doric order: houses with temple porticoes but proudly presenting the inner building itself with spacious doors and windows (Figure 3.15).[42]

Once aware of this phenomenon, it is worth having a second look at Strickland's Parthenon remake, whereupon one is able to see it with somewhat different eyes. This building originally had not one, but five large doors

[40] The front of the Erechtheion on the Acropolis, for instance, reappearing at Clifton Place, Mount Pleasant, TE (1839). Kennedy 1989, p. 50.
[41] Fitch-Gorham-Brooks House in Marshall, MI, of 1840. Kennedy 1989, p. 48.
[42] House in Central Massachusetts, designed by Elias Carter. Kennedy 1989, p. 245. William Risley House of c. 1837 at Fredonia, NY. Kennedy 1989, pp. 238–9; Schneider 2003, pp. 162–3.

Figure 3.10. Orwell, VT. Wilcock Cutts House. Architect: Thomas Dake. Alteration to present appearance 1843.
From: R. G. Kennedy, *Greek Revival America* (Stuart Tabori & Chung, New York 1989), p. 35 upper side. © John M. Hall Photographs.

and as many square windows above, which were only later closed for reasons of structural security. So even this rather astute copy of the Parthenon front decidedly remodels the original concept.

One of the most ingenious and daring nineteenth-century American re-uses of classical models was the enlarging of the design of the small Athenian theatre-monument of Thrasyllos[43] (Figure 3.16) to a structure that could be adapted not only to private villas but also to large-scale structures, such as warehouses and hotels. See the simple ancient Greek model in the drawing of Stuart and Revett, and what American architects like James Gallier,[44] James Dakin[45] and, above all, Alexander Jackson Davis[46] made of it: structures with pillars that can be repeated endlessly, always exposing a wide open core building, again with the effect of light described before.[47]

---

[43] Travlos 1971, pp. 562–5.
[44] Gallier 1833; Gallier 1973.
[45] Scully 1973.
[46] Peck 1992.
[47] Ashland-Belle-Helene near Napoleonville, LA, of 1841 by James Gallier. Pierson 1976, p. 456; Smith 1981, pp. 312–3; Kennedy 1989, p. 159; Schneider 2003, pp. 169–70. Belle Meade at Nashville, TE, of 1853–4; architect, William Giles Harding. Hudson and Ballard 1989,

Figure 3.11. Nashville, TN. State Capitol. Architect: William Strickland. 1845/59.
From: Photo Lambert Schneider.

The most lavish specimens of this Pillar Order derived from the Thrasyllos Monument are to be found in A. L. Davis's studies for various projects:[48] the huge Astor Hotel in New York of c. 1830 or the New York Commercial Exchange, projected in 1862.

pp. 317–19; Schneider 2003, p. 170. Bocage Plantation (Houmas House) near Burnside, LA: again with doors, windows and a balcony directly attached to the pillars, a concept similarly applied to temple-type houses too. Smith 1976, p. 570; Schneider 2003, pp. 169–70.

[48] Peck 1992; Schneider 2003, pp. 169–70.

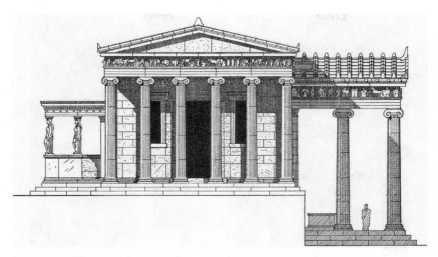

Figure 3.12. The Erechtheion on the Acropolis of Athens. Late fifth century BC. Front elevation.
From: L. D. Caskey et al., *The Erechtheum* (Cambridge Mass. 1927).

Figure 3.13. Madewood-Plantation-House at Bayou Lafourche near Napoleonville, LA. Architect: Henry Howard. 1846–48.
From: Photo Lambert Schneider.

Figure 3.14. Marshall, MI. Fitch-Gorham-Brooks House. c.1840.
From: Photo Lambert Schneider.

It is again Philip Hone who has left us a vivid portrayal of the aesthetic and practical functioning of this peculiar type of classical adaptation. In his diary of September 1, 1835,[49] he writes about such a building:

> We had last night at the pavilion a farewell hop in the dining room, at which the girls enjoyed themselves very much. At eleven o'clock, I retired to my room, lighted a cigar, and seated myself at the front window. The view was unspeakably grand. The broad red moon ... threw a solemn light over the unruffled face of the ocean, and the lofty pillars of the noble ... building, breaking the silver streams of light into dark gloomy shadows, gave the edifice the appearance of some relic of classic antiquity.

This it did not quite do, but "some relic" is quite to the point.

[49] In fact, he is referring to the Rockaway pavilion designed by Town & Davis and Dakin: Nevins 1927, p. 74.

Figure 3.15. House in central Massachusetts. Architect: Elias Carter.
From: Photo Lambert Schneider.

The attitude toward classical models expressed in this architecture and its evaluations sometimes included connoisseurship but did not at all require scientific archaeology, which might have guaranteed a safe travel of ancient facts into modern times. In fact, American builders and architects did not travel to Greece, and with rare exceptions the same applied to their patrons. They simply copied from the same few books – most often Stuart and Revett – reproduced and altered these examples in their own books and then just built: usually in wood and executed not by trained and learned architects but by carpenter builders.[50] Another phenomenon, in its own way quite convincing, is a kind of grafting of different pieces onto others, resulting in a new creature – and one that might even have enchanted the ancient Greeks were they not so constrained by traditional building doctrines: Minard Lafever's leaf-capital (Figure 3.17), which was very popular, especially in the Southern states, is one such example. His publication of 1839,[51] in which he presented this creation, has the

---

[50] Kennedy 1989. For exceptions see: Meyer Reinhold 1984, pp. 256–79.
[51] Lafever 1839. Lafever's anti-Roman and pro-Greek attitude is well documented by his statements in Lafever 1852.

Figure 3.16. Theatre monument on the south slope of Athenian Acropolis. 320/19 BC. As depicted by Stuart and Revett (1762).
From: Stuart & Revett 1762.

telling title: "The Beauties of Modern Architecture…". He proceeded as if following instructions in a cookbook: Take from Stuart and Revett the lower half of the Corinthian capital of Lysicrates Monument at Athens[52]

---

[52] Stuart and Revett 1762–1794, vol. I, chap. IV pl. VI.

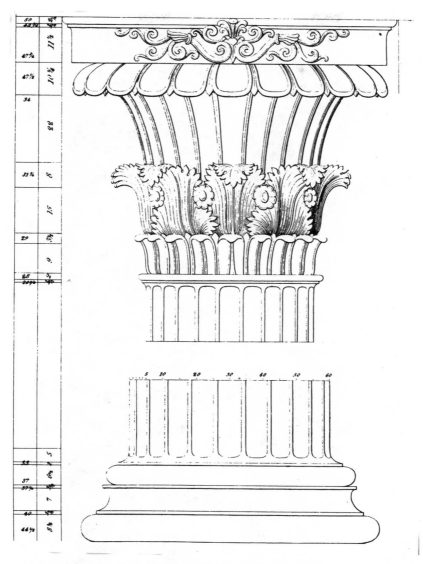

Figure 3.17. American "Greek" leaf-capital after Lafever (1839).
From: Lafever 1839.

with all its characteristic leaves and blossoms (Figure 3.18), then without
hesitation add the upper half of a capital from another monument, the
"Tower of the Winds,"[53] found in the same neighbourhood in Athens and

---

[53] Horologium of Andronikos: Travlos 1971, pp. 281–8.

Figure 3.18. Corinthian capital of Lysicrates Monument at Athens (335/34 BC) by Stuart and Revett.
From: Stuart & Revett 1762.

Figure 3.19. Leaf-capital of the "Tower of the Winds" at Athens. First century BC. *From*: Stuart & Revett 1762.

also published by Stuart and Revett[54] (Figure 3.19). This second element, however, was not strictly copied but infused with life, its leaves becoming more juicy and plant-like. Playfully dealing with historical models, these

---

[54] Stuart and Revett 1762–1794, vol. I (1762), chap. III, pl. VII.

variations were compatible with other newly created capital forms, such as the American Tobacco and the American Corn Order.[55]

In accordance, Lafever explained his capital in the following words[56]: "This is a design composed of antique specimens, and reduced to accurate proportions; with a view to render it acceptable in many places, instead of the standard orders. ...In many situations this design will be preferable to those generally in use." And, as a comment on his Erechtheion-capital variation shown in the same book, he wrote: "This example has neither the proportions nor general features of the antique Ionic order, nor is it pretended that it is in general equal to it; but it is hoped that it may not be ... inferior."[57]

5. Odyssean Travels of Classical Forms and Ideas

I am well aware that I have completely omitted an important stop on the Odyssean voyage of forms from ancient Greece to nineteenth-century America: the English Greek Revival, which started a good twenty years earlier than in the United States. It had been England, after all, that through publications such as those of Stuart and Revett, had furnished pattern-books for American classicism and that showed an abundance of Greek-inspired buildings that could have served as models for American architecture. It was likewise England, in contrast to continental Europe, where Greek orders were applied to buildings of actual use – such as churches and residences. Did these so-called predecessors really inspire their American followers, as is often implied?

A close comparison would show the contrary to be true. Since the beginning of this new style, Greek-revival architecture in England looks markedly different. Normally complete temple-fronts were applied as facades of mansions and churches in an appropriate archaeological manner. So these English examples, even with their variety and relative freedom, remain rather devout of the ancient models in comparison to their American counterparts. Almost all English architecture of that time not only looks very Greek in general, but – at least seen from the front – also comes in the disguise of temples. And, what is most important, the basic concept of ancient Greek temple-building described previously remains largely untouched in these

---

[55] Pierson 1976, p. 403
[56] Lafever 1839, p. 102.
[57] Lafever 1839, p. 142, pl. 31.

Figure 3.20. Belle Helene near Napoleonville, LA. Architect: James Gallier. 1840/41. *From*: Photo Lambert Schneider.

cases. Telling examples are Henry Holland's Sculpture Gallery at Woburn Abbey in Bedfordshire[58] of 1787–89 and 1801–03, an accurate copy of the late-fifth-century temple at Ilissos River in Athens drawn and published by Stuart and Revett (Figure 3.5) and St. Pancras church[59] in London, designed by the Inwood brothers and built in the years 1819–22 reusing various parts of the Erechtheion.

In England – much more than on the European continent – the classical temple concept had already been applied to buildings with ordinary life functions such as mansions, churches and commercial buildings, and this application undoubtedly remained not unknown in America. But the Greek temple-front as a representative model on the one hand, and the contemporary use of the buildings on the other, always remained in an irresolvable state of conflict, necessitating varying degrees of compromise in every case, as may be seen from Grange Park,[60] Hampshire, executed in 1804–9 as an astute copy of the Athena and Hephaistos temple at Athens. Here the core of the building almost hides behind the fence of the Doric temple-front.

[58] Angelicoussis 1992; Crook 1995, fig. 71.
[59] Schneider and Höcker 2001, pp. 26–7; Höcker 2008, pp. 159–60.
[60] Crook 1994, pp. 97–134; Forster 1996, pp. 628–83.

Not so the more playful and relaxed American treatment of these precedents (Figure 3.20). Why not arbitrarily stretch columns, even Doric ones, to support the roof of a house? And in addition frivolously fasten a spacious balcony to them in a most un-tectonic way, without capitals or even impost blocks? For the first time in history, the ancient Greek order was, without compromise, adapted to meet both the needs of using a space for living or working and for social representation. American classicists were certainly fond of the classical ideal created in Europe, but they rarely fell to their knees to worship a remote classical past. With their decidedly non-archaeological and non-devotional approach, they have produced a rich and sometimes wild variety of Greek-inspired artefacts.

It was both North and South in the antebellum era and the New England states as well as the Midwest that adopted this style and became a harbour for travelling Greek architectural elements. It was state architecture and private buildings. It was profane architectures as banks and churches. And within this last category, it was, astonishingly enough, all congregations that used this style: Jews, Freemasons, Methodists, Lutherans, Presbyterians and so forth. Why so? What did these ancient Greek orders mean to them? More research has to be done to answer this question in detail. But one factor seems to me apparent: The Greek order was considered perfectly suited for ennobling one's own titles and demands while at the same time being a kind of empty vessel, void of specific ideological ties. To the Americans of the early nineteenth century, Greek was not specifically English (as was the colonial style that had dominated American architecture before); it had not been the mainstream style in England during the clashes between English and American troops. Nor was it French, neither in the sense of the *ancien régime*, nor in the revolutionary one. Greek was neither decidedly democratic nor did it serve as a symbol of the Southern states with their slavery. It was this openness that was welcomed by the various religious denominations and ethnic and social groups, who had found a relatively safe home in this new country. *Greek* architecture functioned as a uniting social and cultural tie. This concept remained successful for two generations, until the Civil War fractioned the American society deeply and made a style like this obsolete, leading to more specialized and thus fractionized forms of self-representation.[61]

We have encountered various attitudes towards the classical: rigidly devout ones, and paired with these, scientific archaeology. And – in nineteenth-century America – a more upright, unfettered and relaxed attitude to fifth-century antiquity (Figure 3.21). However, classical archaeology did not belong to this.

---

[61] Kennedy 1989.

Figure 3.21. "Classical Greek" elements in American nineteenth-century architecture and ornament.
From: Photos Lambert Schneider.

From the beginning, it had never been pure curiosity but devotion that led people to look back to that far-distant past. What was taken as a fact of antiquity, and what was deemed to be worth incorporating into the present, was determined by contemporary interests and conceptions. The devotion to the distant past was always partial: aimed at only a small fraction both of time and of material.

If we look at these processes from a distant viewpoint, facts indeed seem to have travelled (Figure 3.22). We even know the routes. The forms of ancient Greek Doric and Greek Ionic columns and capitals – these specific

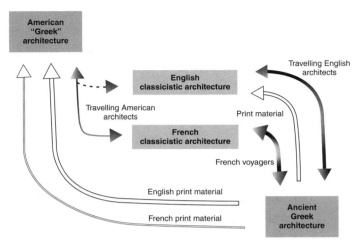

Figure 3.22. On the way from ancient Greece to modern America. Transmissions of architectural elements and thoughts.
From: Lambert Schneider/Maria Witek (Hamburg).

architectural orders – started out on their voyage from Greece by means of drawings and measurements: first to France and then, with more impact, directly to England. They were published there in books, and they were, of course, also popularized through true modern Greek-inspired architecture in England and all over the European continent. But the main vehicle of their further travel to the United States was not so much detailed knowledge of European classical architecture. Instead, it was graphic reproduction that functioned as a vehicle for this transport – above all the engravings of Stuart and Revett, which made their way to the New World.

At the time, only a handful of publications[62] – mostly French –existed at all that were able to convey some imagination or idea of the appearance of ancient architecture at Athens and other places in Greece. The views printed in these publications, however, widely appear as kind of ideal landscapes with ensembles both of ancient-looking ruins and complete buildings. The aim of most of these early literary descriptions and pictorial representations was not so much to transmit precise data, but rather to fire the public with general enthusiasm for ancient Greek architecture. Lacking accurate drawings concerning the proportions of the buildings, they sufficed for motivating classicistic building ambitions at home, only as long as "true" quotations were not asked for.

---

[62] Spon 1687; Pococke 1745; Le Roy 1758; Barthélémy 1788; Barthélémy 1789; Dalton 1791.

A prominent example of this kind of transfer is Julien David Le Roy's *"Ruines de plus beaux monuments de la Grèce"* of 1758, which was widely distributed in France and Italy, and came out only a year later in London.[63] Only the original French edition of 1758 contains a second, smaller, volume with true elevations and metrical information by measurements of classical Greek buildings such as the unfinished temple of Apollo at Delos[64] and the archaic temple at Corinth[65]: Neither architectures, however, were taken as models in American revival architecture. Only Le Roy's single rendering of the Parthenon entablature[66] and his more detailed drawings of the Erechtheion columns and entablature may have influenced modern building on the continent and abroad. All later editions of Le Roy lack even this limited architectural information. Here one is confronted with a picturesque mixture of reality and fantasy interwoven in a way that could by no means be recognized and separated by the beholder who had not seen the originals: Extant parts of architecture and uncertain reconstruction are rarely discernable; even different buildings are mixed into one, or coherent building structures split into separate "independent" architectural units.

Le Roy's publication is known to have circulated also in the United States,[67] but during the 1820s it was superimposed by James Stuart and Nicholas Revett's ambitious publication (see Figures 3.5, 3.16, 3.18, 3.19),[68] which by the quantity of the monuments taken into account, and even more so by the quality of precise data via drawing and measurements, was a step into a new dimension of transporting ancient Greek architectural forms through time and space. Stuart and Revett's three volumes were significantly more expensive than Le Roy's work,[69] but they did find their way to American architects and were thoroughly studied and copied there – as is documented by American architecture books and even more so by the extant buildings themselves. In

---

[63] Wiebenson 1969; Höcker 1997; Middleton 2004; Steiner 2005, reviewed by Christoph Höcker in: H-Soz-u-Kult, 15.05.2006, <http://hsozkult.geschichte.hu-berlin.de/rezensionen/2006-2-108>

[64] Figures 3.2 and 3.4.

[65] Figure 3.6.

[66] Figure 3.13.

[67] Hitchcock 1976; Sutton 1992; Höcker 1997, Hafertepe and O'Gorman 2001. The degree to which this fact is granted impact on American architecture goes along with the general evaluation on French influence on the American Greek revival. To my knowledge, this was rather minimal, aside from Thomas Jefferson's rather "Roman-based" introduction of ancient architectural design into American building. For this aspect, see Schneider 2003.

[68] Stuart and Revett 1762–1794.

[69] Höcker 1997; Höcker in a lecture at Hamburg 2008, forthcoming in 2010.

the end, Stuart and Revett's publication influenced American building more thoroughly than any other work. This applies not only to the rather accurate copies of Greek architecture (as, for instance, Strickland's Second Bank of the United States, dealt with earlier), but also to the majority of free variations on Greek models. Even for these, the use of Stuart and Revett's documentation based on fundamental research on the spot seemed more apt than the French predecessors. This is proven by the fact that various important models for the American Greek revival do not come up in Le Roy's work at all and could have only been taken from the concurrent English publication – as, for example, the capitals of the "Tower of the Winds,"[70] the temple at Ilissos River at Athens[71] or the roof crowning of Lysicrates Monument,[72] which is given by Le Roy only in minimal scale and with false details.[73] Even the famous Ionic capitals of the Erechtheion on the Athenian Acropolis depicted in both publications were clearly taken as models for American buildings – not via Le Roy's drawing but from Stuart and Revett's second volume of 1768.[74]

All this seemingly safe travelling of facts from Greece to America and from ancient times to the recent past, however, makes up only half of the process, at best. First, this was by no means a continuous flow of transmission. On the contrary, the specific forms of classical Greek Doric and Ionic columns and capitals had been totally forgotten for eighteen centuries! They had vanished in architecture already during the Roman era; they do not reappear in Byzantium and in the Middle Ages of Western Europe, and were even not revived in the Renaissance.[75] It was not before the middle of the eighteenth century that they suddenly began their voyage from Greece to the west and north. Truly, this was a geographical journey they made, but through time? "Facts" of the fifth and fourth centuries certainly reached modern times, but it was only after a long sleep that they set out for their journey through time.

Even taking into account this gap of time, the description of the process remains far from complete. How can "facts" sleep and then wake up? "Facts," even in their metaphoric sense, are unable to act in that way, or in any way. In the beginning, it was contemporary interests, ideas and preferences that

---

[70] Stuart and Revett 1762 (vol. I), chap. III pl. VII.
[71] Stuart and Revett 1762 (vol. I), chap. II pl. III. An accurate copy of Stuart's front elevation is found as a proposal for a country house in Lafever 1852, pl. 75.
[72] Stuart and Revett 1762 (vol. I), chap. IV pl. IX: copied at William Strickland's Tennessee State Capitol (p. xx with n. 37 and fig. 9).
[73] Le Roy 1758, pl. 25, 26.
[74] A telling example is Lafever 1839, pl. 32.
[75] The Doric order appears in the Renaissance only in its Roman form of Tuscan, and the Ionic order also is only derived from Roman models.

led members of the English Society of Dilettanti[76] and other voyagers to far-away Greece, and there let them select what they selected as worth copying or taking back to their home countries.

The same applies to the further travel of our "facts" from Western Europe to America. Here again it was contemporary – now specifically American – attitudes and preferences that led entrepreneurs, architects and craftsmen to draw freely on the forms found in architectural books, creating a wild and decidedly non-archaeological "Greek revival." And never mind the subsequent reinterpretations of these forms, which in ancient Greece had never been seen as symbols of nature or of freedom, and certainly not as symbols of a supposedly transcultural and all-embracing humanity. But it was not just a new meaning and new functions that were superimposed on old forms; it was the decision of what is a "fact" at all – and what is a "fact" being worthy of transport and transplantation – that enacted and determined the further travel of our architectural forms. So all this was provoked and determined by American modern ambitions and attitudes.

Both the distant view and the scrutinizing close view appear worth undertaking. The distant look lets us perceive "facts" travelling from one region to another and through times: sometimes comfortably and continuously; sometimes erratically – more like an Odyssean voyage. As soon as you look at the same processes in detail, however, "facts" that before seemed to travel to us from distant regions and distant times will eventually vanish. The acts or movements in our case then turn out to have been rather directed backwards, from a respective present to a far-away past. Furthermore, it was not one continuous move; it was a whole bundle of acts of retrieving and reactivating these distant "facts." True enough, the eventual modern revival of ancient forms that resulted from these activities was vigorous and thus has induced historians to see true tradition here. But this impact never depended on authenticity. Aside from the fact that the venerated Greek past was a highly selective one, even the single formal elements were not required to be truly Greek at all. They had simply to be *considered* Greek, and they had to fit the needs and ambitions of a particular nineteenth-century society.

The past always has been and will remain, to a large extent, an invention of the present. Within this creative act of reconstructing a past, single elements of times ago may well be transported to the present. However, they cannot act themselves. Present actors decide which of them is given access

---

[76] Redford 2008.

to the boat, and present actors decide what is done to them on their long and unsafe travel. When we refer to a voyager, we admit that the person at the end of a journey is not exactly the same as that at the beginning, but still somewhat the same. This also applies to travelling "facts." Within the frame of my case study, they do seem to exist, although their appearance has changed in subtle ways during their journey. Classical Greek *column architecture* had in its original context been restricted to religious and state buildings and was understood as a symbol of social pride shared by free citizens of a democratic state. This symbolic value was not unknown in the modern era. So, together with formal elements, some of the meaning seems to have been transported though time. All this, however, does not form something like a coherent "tradition." Rather, the "facts" in this case are to be considered as elements of a "memoria."[77] a contemporary set of paradigmatic models created by drawing back on selected past entities.

Acknowledgements

The work on which this paper is based was carried out with the support of the J.P. Getty Research Institute for the History of Art and the Humanities in Santa Monica, where I was active as Scholar in 1996. Further reflection on the materials took place in the following years in conversation with many colleagues in Hamburg and elsewhere. This paper was undertaken as a contribution to the Leverhulme/ESRC Trust project, "The Nature of Evidence: How Well Do Facts Travel?" (grant F/07004/Z) at the Department of Economic History, LSE. I would like to thank the members of the "Facts" research group and the authors of the other contributions to this volume for their comments and advice. My thanks goes particularly to the editors for a fruitful and interesting intellectual exchange. I would also like to extend my gratitude to Simona Valeriani who helped with the editing.

Bibliography

Ackermann, G. M. *A Chink in the Wall of Neoclassicism*. In: June Hargrove (ed.), The French Academy. Classicism and Its Antagonists (Congress Del/London, 1990), pp. 168–95.

Allen, W. C. *History of the United States Capitol: A Chronicle of Design, Construction, and Politics* (Washington, D.C., 2005).

Angelicoussis, E. *The Woburn Abbey Collection of Antiquities*. Monumenta Artis Romanae 20 (Mainz, 1992), pp. 11–43.

---

[77] Assmann 1999; Schneider 1999a; Schneider 1999b; Assmann 2006.

Assmann, A. *Erinnerungsräume. Formen und Wandlungen des kulturellen Gedächtnisses* (Munich, 1999).

Assmann, J. *Religion and Cultural Memory* (Stanford, CA, 2006).

Barthélemy, J-J. *Recueil de cartes géographiques, plans, vues et médailles de l'ancienne Grèce, relatifs au voyage du jeune Anacharsis. Vol. 1* (Paris, 1788).

Recueil de cartes géographiques, plans, vues et médailles de l'ancienne Grèce, relatifs au voyage du jeune Anacharsis. Vol. 2 (Paris, 1789).

Benjamin, A. *Practice of Architecture* (New York, 1833, reprint [= *The Works of Asher Benjamin vol.* 5] New York, 1972).

Bernstein, R. B. *Thomas Jefferson* (Oxford, 2005).

Bryan, J. M. (ed.) *Robert Mills, Architect* (Washington, D.C., 1989).

Caskey, L. D. et al. *The Erechtheum* (Cambridge Mass. 1927).

Cook, B. F. *The Elgin Marbles* (London, 2007).

Crook, J. M. *The Dilemma of Style. Architectural Ideas from the Picturesque to the Post-Modern* (London, 1987).

The Greek Revival. Neo-Classical Attitudes in British Architecture (London, 1995).

Curl, J. S. *The Art and Architecture of Freemasonry* (Woodstock and New York, 1991).

Dalton, R. *Antiquities and Views in Greece and Egypt with the Manners and Customs of the Inhabitants, from Drawings Made on the Spot, A.D. 1749* (London, 1791).

Downey, G. *On Some Post-Classical Architectural Terms.* In: Transactions of the American Philological Association 77 (1946), pp. 22–34.

Downing, A. J. *The Architecture of Country Houses* (1850; reprint New York, 1969).

Pleasure Grounds. Andrew Jackson Downing and Montgomery Place, with Illustrations by AJD, ed. by J. M. Haley (New York, 1988).

Dyson, S. L. *Ancient Marbles to American Shores. Classical Archaeology in the United States* (Philadelphia, PA, 1998).

Forster, K. W. *L'ordine dorico come diapason dell'architettura moderna.* In: I greci 1.1. Noi e i Greci (S. Settis, ed.) (Torino, 1996), pp. 707–42.

Foster, M.P. *Charleston, South Carolina. A Historic Walking* (2005).

Gallagher, H. M. P. *Robert Mills, Architect of the Washington Monument 1781–1855* (New York, 1935).

Gallier, J. *Popular Lectures on Architecture* (Brooklyn, NY, 1833).

Autobiography (1973).

Goethe, J. W. *Von deutscher Baukunst* (Frankfurt, 1795).

Einleitung in die Propyläen (Tübingen, 1798a).

Über Laokoon (Tübingen, 1798b).

Winckelmann und sein Jahrhundert (Tübingen, 1805).

Hafertepe, K. O'Gorman J. F. *American Architects and their Books to 1848* (Amherst, MA, 2001).

Hamlin, T. *The Greek Revival in America and Some of Its Critics*, The Art Bulletin 1942, pp. 244–58.

Hammet, R.W. *Architecture of the United States* (New York, 1976).

Hazlitt, W. *On the Elgin Marbles* (London, 2008).

Herder, J. G. *Plastik* (Riga, 1778).

Hitchcock, H-R. *American Architectural Books and Writers* (Minneapolis, MN, 1976).

and W. Seale. *Temples of Democracy: The State Capitols of the USA* (New York and London, 1976).

Höcker, Ch. "Greek Revival America? Reflections on Uses and Functions of Antique Architectural Patterns in American Architecture between 1760 and 1860." *Hephaistos* 15 (1997), pp. 197–240.

London (Stuttgart, 2008).

Hudson, P. L. and Ballard, S. L. *The Smithsonian Guide to Historic America. The Carolinas and the Appalachian States* (New York, 1989).

Kennedy, R. G. *Greek Revival America* (New York, 1989).

Kennworthy-Browne, J. *The Sculpture Gallery at Woburn Abbey and the Architecture of the Temple of the Graces*. In T. Clifford and H. Honour, The Three Graces. Antonio Canova. Exhibion National Gallery of Scotland (Edinburgh, 1995), p. 61–72.

Kidder Smith, G.E. *The Architecture of the United States* (Garden City, NY, 1981).

Kidney, W. C. *The Architecture of Choice* (New York, 1974).

King, D. *The Elgin Marbles. The Story of the Parthenon and Archaeology's Greatest Controversy* (London, 2006).

Lafever, M. *The Beauties of Modern Architecture, Illustrated by 48 Original Plates Designed Expressly for this Work* (New York, 1839).

The Modern Builder's Guide (New York, 1852).

The Architectural Instructor, Containing a History of Architecture from the Earliest Ages to the Present Time… (New York, 1856).

Lane, M. *Architecture of the Old South* (New York, 1993).

Lane, M. and Martin, V. J. *Architecture of the Old South: Greek Revival & Romantic* (Abbeville, GA, 1996).

Le Roy, J. D. *Les ruines des plus beaux monuments de la Grèce* (Paris, 1758).

Lessing, G. E. *Laokoon oder über die Grenzen der Malerei und Poesie* (Berlin, 1769a).

Wie die Alten den Tod gebildet (Berlin, 1769b).

Major, Th. *Members and Measures of the Hexastyle Ipetral Temple*. In: Ruins of Paestum, Otherwise Posidonia (London, 1768).

Marchand, S. L. *Down from Olympus. Archaeology and Philhellenism in Germany, 1750–1970* (Princeton, NJ, 1996).

McCormick, Th.J. *Charles-Louis Clerisseau and the Genesis of Neo-Classicism* (Cambridge, MA, 1990).

Meckler, M. *Classical Antiquity and the Politics of America: From George Washington to George Bush* (Baylor, TX, 2006).

Melville, H. *Moby Dick or, the White Whale* (London/New York, 1851).

Meyer Reinhold, L. *Classica Americana. The Greek and Roman Heritage in the United States* (Detroit, MI, 1984).

Middleton, R. (ed.) *Julien-David LeRoy, The Ruins of the Most Beautiful Monuments of Greece* (Los Angeles, CA, 2004).

Muller, M. M. *A Town at Presque Isle. A History of Pennsylvania* (1997).

Nerdinger, W. *Leo von Klenze. Architekt zwischen Kunst und Hof 1784–1864* (Munich, 2002).

Newton, R. H. *Town and Davis. Pioneers in American Architecture 1812–1870* (New York, 1952).

Nevins, A. (ed.) *The Diary of Philip Hone. Vol. 1* (New York, 1927).

Peck, A. *Alexander Jackson Davis, American Architect 1803–1892* (New York, 1992).

Pierson, W. H., Jr. *American Buildings and Their Architects. The Colonial and Neo-Classical Styles* (Garden City, NY, 1976).

Pococke, R. *A Description of the East and Some Other Countries, vol. 2* (London, 1745).

Redford, B. *Dilettanti. The Antic and the Antique in Eighteenth-Century England.* J. Paul Getty Museum and the Getty Research Center (Los Angeles, CA, 2008).

Reed, H. H. and Day, A. *The United States Capitol: Its Architecture and Decoration* (Washington, D.C., 2005).

Schiller, F. *Über Anmut und Würde* (Leipzig, 1793a).

Über das Pathetische. (Leipzig, 1793b).

Schneider, L. "Das archäologische Denkmal in der Gegenwart., In: Die Antike in der europäischen Gegenwart" W. Ludwig (ed.), *Symposium der Joachim Jungius-Gesellschaft der Wissenschaften Hamburg 1992* (Göttingen, 1993), pp. 31–42; 171–8.

Il classico nella cultura postmoderna. In: Salvatore Settis (ed.), I Greci I: Noi e i Greci (Torino, 1996), pp. 707–41.

Das Pathos der Dinge. Vom archäologischen Blick in Wissenschaft und Kunst. In: B. Jussen (ed.), Archäologie zwischen Imagination und Wissenschaft. Anne und Patrick Poirier (Göttingen, 1999a), pp. 51–76.

Postmodernes Vergessen und schmerzfreie Erinnerung. Gedanken zur Akropolis von Athen. In: U. Borsdorf and H. Th. Grütter (eds.) (Frankfurt/M., 1999b) pp. 245–66.

"Klassik ohne Devotion. Ein Blick auf Amerikas griechisch inspirierte Architektur des 19. Jahrhunderts," in *Aktualisierung von Antike und Epochenbewusstsein. Erstes Bruno Snell-Symposion der Universität Hamburg am Europa-Kolleg* (Munich/Leipzig, 2003) pp. 143–78.

"Einem Chamäleon auf der Spur – Der Klassische Körper in Kunst und Leben. Winckelmanns problematisches Erbe in der deutschen Kultur," review of Esther Sophia Sünderhauf, *Griechensehnsucht und Kulturkritik. Die deutsche Rezeption von Winckelmanns Antikenidal 1840–1945* (Berlin, 2004), In: *Hephaistos* 23 (2005), pp. 245–55.

"Der Parthenonfries. Selbstbewusstsein und kollektive Identität." In: Stein-Hölkeskamp, E. (ed.), *Erinnerungsorte: Griechenland* (Forthcoming, Munich, 2010).

Schneider, L. and Chr. Höcker, *Die Akropolis von Athen* (Darmstadt, 2001).

Scott, P. and A. J. Lee, *Buildings of the District of Columbia* (Oxford, 1993).

Scully jr., A. *James Dakin, Architect. His Career in New York and the South* (Baton Rouge, LA, 1973).

Seale, W. *The Old Georgia Governor's Mansion* (Milledgeville, GA, 1996).

Severens, K. *Charleston Antebellum Architecture and Civic Destin* (Knoxville, TN, 1988).

Smith, G.E.K. *Architecture in America* (New York, 1976).

The Architecture of the United States. An Illustrated Guide to Notable Buildings Open to the Public (Garden City, NY, 1981).

Spon, J. *Voyage d'Italie, de Dalmatie et du Levant* (Lyon, 1687).

St.Julien Ravenel, B. *Architects of Charleston* (Charleston, 1945).

Steiner, U. *Die Anfänge der Archäologie in Folio und Oktav. Fremdsprachige Antikenpublikationen und Reiseberichte in deutschen Ausgaben.* Stendaler Winckelmannforschungen 5 (Ruhpolding, 2005).

J. Stuart, J. and Revett, N. *The Antiquities of Athens Measured and Delineated.* 3 vols. (London, 1762–1794).

Sutton, R. K. *Americans Interpret the Parthenon: The Progression of Greek Revival Architecture from the East Coast to Oregon 1800–1860* (Niwot, CO, 1992).

Tournikiotis, P. *The Parthenon and Its Impact in Modern Times* (Athens, 1994).

Traeger, J. *Der Weg nach Walhalla. Denkmallandschaft und Bildungsreisen im 19. Jahrhundert* (Regensburg, 1987).

Travlos, J. *Bildlexikon zur Topographie Athens* (Tübingen, 1971).

Trowbridge, B. Ch. *Old Houses of Connecticut* (New Haven, CT, 1923).

Vance, W. L. *America's Rome* (New Haven/London, 1989).

Waddell, G. and Liscombe, R. W. *Mills's Courthouses & Jails* (Easley, SC, 1981).

Wiebenson, D. *Sources of Greek Revival Architecture* (London, 1969).

Winckelmann, J. J. *Gedanken über die Nachahmung der griechischen Werke in der Malerey und Bildhauerkunst* (Dresden, 1756).

Winterer, C. *The Culture of Classicism: Ancient Greece and Rome in American Intellectual Life. 1780–1910* (Baltimore/London, 2002).

Yeguel, F. *Gentlemen of Instinct and Breeding. Architecture at the American Academy in Rome 1894–1940* (Oxford/New York, 1991).

FOUR

MANNING'S N – PUTTING ROUGHNESS TO WORK

SARAH J. WHATMORE AND CATHARINA LANDSTRÖM

1. Introduction

In a research project on the science and politics of flood risk,[1] we found our-selves fascinated by the ubiquity of a small, italicised symbol – 'n' – in the working practices of hydraulic modellers. On closer examination, it became clear just how densely packed this symbol is as a factual statement about the world claiming that hydraulic roughness is a property of rivers that can be approximated and represented by a single numerical value.

n is a parameter that does crucial work in a commonly used equation for calculating discharge for uniform water flow in open channels:

$$Q = \frac{AR^{2/3} S_f^{1/2}}{n}$$

where Q = discharge, A = channel cross-sectional area, R = hydraulic radius, S_f = energy slope and n = Manning's roughness coefficient (Fisher and Dawson 2003). Despite recent academic challenges to the validity of n, it remains undisturbed as a cornerstone of the working practices of engineering consultants that inform the policy and management of flood risk in the UK.[2]

---

[1] This chapter, and the presentation on which it is based, was written under the auspices of a research project funded under the Rural Economy and Land Use Programme (www.relu.ac.uk) on 'Environmental knowledge controversies: the case of flood risk management' (www.knowledge-controversies.ox.ac.uk). We are grateful to our collaborators, particularly Stuart Lane and Nick Odoni, for enlightening discussions on Manning's n.
[2] Flood risk policy and management in the UK is divided between Defra (the Department of Environment, Food and Rural Affairs), which is responsible for policy development, and the EA (Environment Agency), which is responsible for implementation (with organisational variations) in England and Wales.

Our fascination deepened when our pursuit of *n* drew us into the work of its originator, Robert Manning, an Irish drainage engineer practising in the second half of the nineteenth century who presented it first, in a still-cited paper, to the Institution of Civil Engineers of Ireland in 1889 as a proxy for roughness, or the effects of friction on the movement of water, that can be derived from a visual assessment of the shape and character of a river channel. How and why has this parameterisation proved so durable in the changing practices of hydraulic science and engineering?

We do not address these questions through a chronological account of the travels of Manning's *n* but rather through one that reflects our process of investigation, which began with the demands of flood risk science and politics today. This genealogical device works against the deceptive production of a singular 'trajectory' and the historical determinism that this would imply, insisting instead on an enfolding of past and present, not least through the framing interests of those embarking on any historical investigation. By focussing on the work Manning's *n* does at different moments in time, we aim to capture the combination of stability and elasticity that enable this conceptualisation of hydraulic roughness to travel through time and between communities of practice. Our account draws on historical archives and documentary records, interviews with flood modellers in academic and consultancy practice and our own first-hand experience of one-dimensional (1D) modelling software through participation in professional training courses.

Our interrogation works through three specific moments in the career of Manning's *n*. We begin in the early twenty-first century, a moment witnessing a surge in scientific critiques of the *n*-value and Manning's equation (invented to calculate flow velocity) as a formula that over-simplifies the complex dynamics of energy loss in water flow. These critiques reach beyond the pages of scientific journals, and we analyse a concerted attempt to develop and institutionalise an alternative calculus – the Conveyance Estimation System (CES), sponsored by the policy agencies responsible for flood risk management in the UK. Its limited success in breaking the hold of Manning's *n* as an industry standard provides an important lens through which to examine the extraordinary durability of Manning's formula. In the second of our analytical moments, nineteenth-century Ireland, we examine Manning's work in the land drainage regime that underpinned the programme of public works of the British colonial administration. Our analysis focuses on the interwoven influences of Manning's day job as a water engineer and the mathematical calculations that occupied his spare time in the development of a 'general equation' for calculating discharge that was simple and effective enough to appeal to his contemporary practitioners over established and competing methods. The last of the three moments examined here moves us forward in time again to the re-packaging of *n* in the twentieth century that underpins hydraulic modelling to this day. In this intervening period, we focus on the ways in which Manning's formula becomes incorporated as a standard element in hydraulic modelling software and the visual estimation of *n*-values for rivers becomes regularised through photographic reference handbooks compiled for engineers.[3]

2. Moment 1: Manning's *N* under Fire

2.1 Too Simple for the Twenty-First Century

We begin this first moment of investigation with journals in the geosciences in which the ubiquitous use of Manning's equation and the *n*-value in hydraulic engineering practice has recently become a target of critique. The critics question the idea that hydraulic roughness is a phenomenon that can be represented as a single numerical value. In other words, it is a critique of what we might call the 'fact' packaged as Manning's *n*. Its indictment as a formulation that over-simplifies roughness, both conceptually and empirically, is illustrated here by reference to two papers.

In a paper in *Earth Surface Processes and Landforms*, Lane (2005) exhorted his fellow water scientists to re-evaluate the hydraulic variable of roughness because of its lack of conceptual clarity. His primary concern is the habitual treatment of roughness as a singular independent variable, arguing that it is a more complex feature of an already complex physical system and ought to be treated as such. His critique is directed at those who routinely elevate Manning's formula – today mainly used to estimate the impact of roughness on water levels – to the status of a law, thereby effectively taking its assumptions for granted, rendering it immune to interrogation (examples cited include Govindaraju and Erickson 1995 and Zhang and Savenije 2005).

Lane goes on to argue that because roughness, formulated as Manning's *n*, has become an automated calibration parameter in flood modelling on which production of 'the correct relationship between flow and water level' (2005, p. 251) is reliant, the concept of roughness has become even further

---

[3] For parallel cases in which man-made facts (as opposed to these 'facts of nature') travel in artefacts, and for pictorial representations of them, see Valeriani and Schneider, both in this volume.

distanced from the dynamics of friction in any actual physical system. For example, he points out that roughness is physically both laminar- and scale-dependent. It is laminar in that 'provided the surface topographic variability extends beyond a thin layer of fluid (the laminar sub-layer) close to the bed, the bed is hydraulically rough, and friction between the bed and the flow will depend upon surface topographic characteristics (e.g., grain size)' (op cit). It is scale-dependent in that 'as the spatial scale of consideration is changed, we change the amount of topography that must be dealt with implicitly, that is, parameterised as frictional resistance' (2005, p. 252). Lane's paper challenges his fellow scientists to find better ways of articulating current scientific knowledge about this complex physical phenomenon and, thereby, of improving the calculability of roughness.

Just such an alternative way of articulating roughness is suggested in a paper two years later by Smith, Cox and Bracken (2007). Appreciating the entrenchment of Manning's n in the flood science community, Smith and his colleagues begin their paper by identifying and challenging the assumptions that underpin the study of overland flow hydraulics. They develop a detailed argument for an alternative formulation, the most interesting aspect of which, for our purposes here, is their discussion of the enhanced technical capacity for measuring roughness (as resistance to overland flow) since Manning developed his original formula. Reviewing a large number of research publications, Smith and his colleagues profile a range of experimental methods that have been used in attempts to measure resistance more accurately but which they consider deficient because such laboratory-based studies ignore 'real-world' processes such as 'changing soil surface configurations with distance downslope' (2007, p. 382). They go on to argue that such deficiencies can be overcome 'by embracing new technologies available to assist the acquirement of accurate measurements of flow depth and velocity' (op cit) over different surfaces. They champion terrestrial laser scanning as a technique likely to enable much better measurement of resistance to flow than methods used to date. On this account, new techniques for measuring hydraulic roughness are rendering its parameterisation as n redundant; when the phenomenon can be empirically described and measured, there is no need for estimating it for use in an equation.

These critics of Manning's n do not, it appears, take issue with its ability to capture relationships between water levels and energy loss due to friction in pipes or artificial channels. It is the routine application of Manning's n to flow in natural rivers and over floodplains that is in dispute. For these academic scientists, Manning's formula used for estimating the energy loss due to hydraulic roughness is too simple, even simplistic, an approach to a complex phenomenon that is now amenable to much more effective conceptualisation and empirical analysis. It may be too soon to judge whether the critique will effect the changes in practice for which these authors call. However, it seems doubtful that lack of attention to conceptualisation or methodological innovation suffices to explain the persistent use of Manning's *n*, given that just these issues have been the dedicated focus of a practitioner-led research programme on 'Reducing uncertainty in river flood conveyance,' which concluded in 2004, before either of the critical papers cited previously had been published.

2.2 The Conveyance Estimation System

'Reducing uncertainty' was a research programme initiated by the major governmental sponsors and users of flood risk science in the UK – the Department of Food, Environment and Rural Affairs (Defra) and the Environment Agency (EA). The programme enrolled a number of flood scientists and engineers in the quest for new ways of working with roughness in computer modelling. Led by the water engineering consultancy HR Wallingford (Ltd.), the programme brought together scientific experts from the academic, public and commercial sectors in a review of current practice. This included a concerted effort to replace Manning's *n* as the standard parameter for roughness in the estimation of conveyance in the flood models on which Defra and the EA base their policy and management activities. Documents archived on-line provide some insights into this programme, including why it was considered important enough to fund at the time. [4]

> The Environment Agency for England and Wales identified the need to reduce the uncertainty associated with flood level prediction through incorporating the recent research advances in estimating river and floodplain conveyance. Existing methods for conveyance estimation that are available within 1D Hydrodynamic modelling software, e.g., ISIS, MIKE11, HECRAS, HYDRO-1D, are based on some form of the Manning Equation, first published in 1890. With the substantial improvement in knowledge and understanding of channel conveyance that has taken place over the past twenty years, there is a need to make these more advanced techniques available for general use in river modelling. (Defra/EA, 2004, p. 1)

The premise of this initiative was that Manning's equation was dated in its approach to roughness and surpassed by improved scientific understandings of the physical process of conveyance. As one of the senior scientists

---

[4] See: www.river-conveyance.net/index.html

working on this programme told us in an interview, their efforts centred on treating roughness as a more complex phenomenon.[5] This involved subdividing roughness into three friction types or zones: skin friction (energy loss from movement over a surface or bed, a factor similar to Manning's n), secondary occurrence (energy loss from the movement of water around a river bend) and turbulent shearing (energy loss from turbulence within the water itself). This threefold re-conceptualisation of roughness was tested using a combination of experimental flume studies, field data and computer models. The initiative both addressed the physical complexity that was later emphasised by Lane and allowed for empirical testing not unlike that subsequently proposed by Smith and his colleagues. The rationale for replacing Manning's n at work here is epistemic, to improve the representation of a phenomenon occurring in nature. The ambition of the programme, according to another scientist who took part, was to take n apart, to approach the different aspects of energy loss with an empirically derived equation and then re-conceptualise n as a value relating solely to the friction of the surface over which water moves.[6]

The 'Reducing uncertainty' programme produced a new 'Conveyance Estimation System' that treats roughness as one component of a complex physical phenomenon and accounts for uncertainties in the relationship between energy loss, velocity and water levels more comprehensively. Programme records claim that considering 'the substantial improvement in knowledge and understanding of channel conveyance that has taken place over the past twenty years, there is a need to make these more advanced techniques available for general use in river modelling' (Defra/ EA, 2004, p. 1).

Interrogating the programme records made us aware that Manning's n is rarely encountered by those engaged in the modelling of flood events as an element in an equation to be solved by assigning a numerical value to a variable. Rather, in the everyday working practices of modellers working on flooding, it is more usually encountered as an embedded feature of routinely used software packages. This helps to explain the programme's investment in creating a new software product – the CES – which, as the programme literature describes it, is

> a software tool that enables the user to estimate the conveyance or carrying capacity of a channel. /.../ The CES includes a component termed the 'Roughness Advisor', which provides advice on this surface friction or 'roughness', and a

---

[5] Interview by S. W. 2007.
[6] Interview by S. W. 2008.

component termed the 'Conveyance Generator', which determines the channel capacity based on both this roughness and the channel morphology. In addition, the CES includes a third component, the 'Uncertainty Estimator', which provides some indication of the uncertainty associated with the conveyance calculation. (Defra/EA 2004, p. 1)

In this, the consortium of academic scientists and engineering researchers involved in the 'Reducing uncertainty' initiative can be seen to be attempting to package their understanding of roughness in a way that would make it travel as readily as Manning's *n*. In so doing, they draw attention to the ways in which the effectiveness of *n*, which they hoped CES could emulate, rested less in the mobilisation of roughness as an accepted fact than as a working tool in the production of knowledge about flood risk. As a tool rather than as a fact, the success of this re-packaging would be reliant on flood modellers and river engineers changing the ways in which they worked. The three components of the CES software require the modeller to undertake three different activities to estimate the energy loss previously parameterised as Manning's *n*. The 'Roughness Advisor' requires input of measurement data in order to provide output values for surface friction in units that are then used to compute values for 'roughness zones', which provide numbers that are then input in the cross-sections as 'n_1' values.[7] Next, the modeller needs to use the 'Conveyance Generator' to compute energy losses due to other factors – for example, sinuosity. The third step is to use the 'Uncertainty Estimator' to generate upper and lower bands of values within which modelled water levels from a given flow may vary.

This new way of modelling roughness as a discrete three-step activity that feeds into the normal model-building process is presented as a change for the better.

> This task is now modularised and mimics the model building activities. /.../ Modellers are provided with a flexible interactive tool with a great deal of freedom. /.../ As a result of this new freedom, defensibility of the results becomes a more important issue than before. /.../ The key difference to previous modelling is that an insight is gained into the role of conveyance in the overall hydraulic performance of the system, in an uncertain background. (Defra/EA 2004, p. 4)

The CES has been included in the ISIS modelling software package as a separate application that a user may choose to use or not to use. On the evidence of the ongoing scientific critique, as well as our interviews and ethnographic work with flood-modelling practitioners, few users appear to

---

[7] Mansnerus, this volume, also discusses how facts are used as inputs to create other, 'model-produced' facts.

choose to employ the CES. To begin to understand why these efforts to replace Manning's *n* have made so little headway, we must go back in time, first to look more closely at Robert Manning's achievement in the nineteenth century, and then to examine some of the devices through which it has taken hold in twentieth-century engineering practice.

3. Moment 2: Making Roughness Estimation Practicable

3.1 Drainage Engineering and Public Works in Nineteenth-Century Ireland

Robert Manning was elected to membership of the British Institute of Civil Engineers in London in 1858 and rose to become president of the Institution of Civil Engineers of Ireland in the year in which it received its royal charter – 1877. This was the audience to which he presented his still-famous paper: 'On the flow of water in open channels and pipes,' first in December 1889 (Manning 1891) and later, in a refined and copiously annotated version, in June 1895 (Manning 1895). His career as an engineer had been rather more precarious than these impressive credentials of professional standing suggest. Born in British-occupied France and schooled in Ireland, Manning's eminence as an engineer was an achievement born of practical learning rather than university education – an approach to knowledge that he came to advocate at the height of his career. In his presidential address to the institution in 1877, he observed that

> [w]hen I entered the profession more than thirty years ago I found that it was considered a greater disgrace not to know the workmen's name for a tool or a particular kind of work than to be ignorant of the very elements of mathematical and mechanical science. /.../ But things have changed since then. The knowledge that was then looked upon as ridiculous and impractical theory is now viewed as the merest elementary smattering. /.../ I trust that while our younger members will not fail to acquire a competent knowledge of mathematics there are none of them who are so immersed in the integration of circular functions, or other applications of the calculus, as not to learn how a dozen men are to be set profitably to work with a pick, shovel and barrow .... (Manning 1878, p. 80)

His working life began in estate management for his uncle in County Wexford. In the late 1840s, his skills found their place in the Public Works regime at the office of the Drainage Engineer in Louth, initially in a clerical post, and two years later as district engineer in the drainage districts of Meath and Louth and subsequently Ardee and Glyde. After an interlude working as estate engineer to Lord Downshire (1855–67), Manning returned to the ranks of the Board of Works (now Office of Public Works) in 1868, first as Second Engineer and then as Principal Engineer, a post to which he was promoted in 1874 (Dooge 1989). Manning's career in public administration coincided with a sustained investment in arterial drainage as the lynchpin of a colonial project to raise the productivity of land and the profitability of agriculture in Ireland. Public administration in Ireland was directed by the British Parliament via a system of grand juries, the jurisdiction of which extended from the administration of law to fiscal and then civil government at national and local levels via a system of boards and agencies operating in the 34 counties. In 1817, the Board of Works took responsibility for coordinating the activities of county surveyors whose appointment became subject to a system of public examination introduced in the Grand Jury (Ireland) Act of 1833. Pay was poor, and most county surveyors were Irish nationals whose activities were regulated by the allocation of grants and loans for major infrastructural investments (from bridges to drainage) by the Board of Works. As a district engineer employed by the Board, Robert Manning would have been directly responsible for overseeing the work of county surveyors, whose job description was more accurately that of county engineers (McCabe 2006).Throughout his varied career, Manning made and recorded extensive observations of aspects of rainfall, river volume and water runoff, publishing papers on his methods and findings – for example, 'on the flow of water off the ground,' describing rainfall-runoff measurements in connection with a new water supply system in Belfast (Manning 1866),[8] and on 'triangulation for survey of the Downshire estates' (Manning 1882).

One of the most challenging drainage schemes on which Manning worked was that concerning the River Glyde in County Louth, which flows into the sea in confluence with the River Dee at Annagassen in the northeast of Ireland. He was personally responsible for much of the surveying carried out in the mid-1840s in his capacity as assistant to the District Engineer Samuel Roberts. In Manning's annual report to the commissioners in 1851, he records the employment of some 76,122 men in drainage work in the county, with a maximum number in any one day of 656, commenting that he had work (but not funds) for at least double that number (Commissioners of Public Works 1851). Along with the clearance of some fifteen miles of waterway, his report provides details of the excavation of the River Glyde, deepening its seventy-foot-wide channel by some five

[8] This paper was awarded the Telford Gold medal by the Institute of Civil Engineers in London.

feet above its confluence with the River Dee. The purpose of these labours, he notes, is to relieve the 'lands between these points' from floods and, in the process, to increase the land values (and rents) by an estimated 7sh 6d per acre. He pays particular attention to the construction of two mill-races at the junction with the Dee, diverting water to the mill industries at Annagassen. Manning's extensive experience here and elsewhere in practical river hydraulics, from surveying to engineering, generated one of the two main sources of data that informed his efforts to render roughness a calculable dimension of the 'mean forward velocity' of water (discharge) and its management. The other was his extensive private reading of the theoretical works and empirical observations of his civil engineering contemporaries, particularly those in France and the United States, who were at the forefront of their profession in his day. In an early paper on 'the flow of water off the ground' (1866) reporting on the results of a series of observations in Woodburn District for a twelve-month period between 1864 and 1865, he traces the dependence of 'all formulae for the discharge of water... upon the principle that the velocity is proportional to the square root of the head' to the work of Torricelli, a student of Galileo, published in 1643, which derives this 'settled principle accepted by all hydraulicians [from] the laws of the fall of heavy bodies' (1866, appendix, p. 467). Notwithstanding sustained endeavours in the science of hydraulics, Manning defines the outstanding problem thus:

> Although the science of hydraulics is now nearly 250 years old, it is less than half that time since anyone could calculate even approximately the velocity or surface inclination of water flowing in an open channel of given dimensions. /.../ Anyone who has carefully studied the subject must have come to the conclusion that it is almost hopeless to obtain a strictly mathematical solution of the problem, and that even to observe and record correctly the physical data required is a matter of extreme difficulty, not to say impossibility. (1891, p. 161–2)

3.2 *N* Makes Discharge Calculation More Reliable

Given the precariousness of employment and heavy workload that the under-resourced regime of public works afforded him, it should come as no surprise that Manning valued his practical experience as a working engineer as highly as the published work of leading 'hydraulicians' over three centuries in Europe and North America. The work he most admires is that of those who, like himself, base their scientific formulations of the laws governing the motion of water in channels on first-hand observations and experiments. Characterising this approach as one concerned with 'empirical formulae' (i.e., formulae deduced from experimental observations), Manning is insistently circumspect about their 'generalisability' and about the balance to be struck between the ambition to formulate a 'rationale theory' with ever-greater demands in terms of mathematical complexity and the exigencies of practical engineering, which 'force the profession [into] the habit of rough generalisation and what is called "rapid approximation"' (1866, p. 466). Thus, for example, he later refers approvingly to Cunningham's observation in a paper to the Institution of Civil Engineers in 1882[9] that for all the impressive increase in mathematical sophistication over more than a century between the hydraulic formula of de Chezy (1775) and that of Kutter (1876), 'practical hydraulicians /.../ should determine to abide by the [de Chezy's] simple formula that has stood the test of so many years, which most of them had verified for themselves, and which they know was practically accurate within the limits they had occasion to use it' (1891, p. 169). This also goes some way to explaining the modesty with which he presents his own formulation as a furtherance, or supplement, to those of some of his predecessors rather than as a superior replacement.

Manning's working method in both the 1891 and 1895 versions of his seminal work on the 'flow of water in open channels and pipes' is to survey the empirical formulae produced by a selection of earlier 'hydraulicians' and compound the experimental observations (and varying measurements) on which they are based through a series of tabular composites, thereby magnifying, so to speak, their deductive power (see Figure 4.1).

He is careful to stress that while the 'close agreement between the observed and calculated velocities [across such an] extended range of data /.../ must to a certain extent give confidence in its [the equation's] use as a general formula' (1891, p. 164), such an agreement 'is not an absolute proof of the correctness of such formulae' (ibid). Rather, it is the 'great difficulty (if not impossibility) of establishing a strictly mathematical theory of the motion of water in canals [that] excuses, if it does not justify, their adoption' (ibid). The formulae of de Chezy (1775) and Du Buat (1786, 2nd edition) in the eighteenth century and of Bazin (1865) and Ganguillet and Kutter (1889) in the nineteenth are particularly influential in framing the contribution he sets out to make to the 'science of hydraulics.' This he defines as finding a 'general equation [for the uniform motion of water in open channels] which will hold good for all measures without the necessity of changing coefficients' (1891, p. 162) and be 'sufficiently accurate for practical purposes and calculations by which are easy' (1891, p. 167). Where

---

[9] Cunningham (1883) was presented orally to ICE in 1882.

TABLE. I.

No.	Authority	Description of Channel	Units	R.	S.	Velocity		C.
						Observed	Calculated	
1	Revy – – –	Parana –	E.F.	36·45	·000000745	1·804	1·819	11
2	Humphreys & Abbott –	Mississipi –	„	64·52	·00004365	6.825	6·993	„
3	Du Buat – –	River Rayne –	F.I.	55·347	·0001654	27.620	25·763	„
4	Same – – –	Canal du Jard –	„	40·428	·0001121	17.420	17·430	„
5	Ganguillet & Kutter –	Linth Canal –	S.F.	5·200	·00029	3.470	3·450	12
6	Same – – –	Same –	„	9·800	·00037	5.620	5·663	„
7	Bazin – – –	Burgundy Canal –	M.	·0980	·10100	3.747	4·030	13
8	Same – – –	Same –	„	·1424	·10100	4.931	4·908	„
9	Same – – –	Same –	„	·1967	·10100	5.694	5·670	„
10	Same – – –	Same –	„	·2017	·10100	6.429	6·202	„
11	Darcy – – –	Old Cast-iron Pipe –	„	·0090	·00025	0.051	0·051	15
12	Same – – –	Same –	„	·0608	·04105	2.073	2·229	„
13	Same – – –	Same –	„	·0608	·18981	3.833	4·115	„
14	Same – – –	New Cast-iron Pipe –	„	·0205	·00232	0.358	0·364	20
15	Same – – –	Same –	„	·1250	·00045	0.449	0·476	„
16	Ftely & Stoains	Sudbury Conduit {	E.F.	1·0709	·00018929	1.844	1·856	21
17	Same – –	lined with Brick	„	2·3297	·00018886	2.937	2·915	„
18	Bazin – –	Puro Coment –	M.	·1116	·00150	0.921	0·935	23
19	Darcy – – –	Now Lead –	„	·0035	·00336	0.165	0·166	29
20	Hamilton Smith, jun	Glass –	E.F.	·01045	·2309	4.439	4·356	31

UNITS { Time = Second. Length, E.F. = English Feet; F.I. = French Inches, ancien systeme; S.F. = Swiss Feet; M. = Metres.

IN OPEN CHANNELS AND PIPES.

Figure 4.1. Table from Manning's 1891 paper explained in the original article as a list of the experiments he recalculated to show the reliability of his formula which numbers agree with the flow velocities attained by other scientists. The first column names the scientists whose data he used, the second identifies the river channel studied and the third provides the units in which the original calculations were made. Columns 'R' and 'S' note the hydraulic radius and the slope for each channel studied; these are followed by the observed velocities and those calculated by his own formula in inches per second and, finally, 'C' the 'coefficient which varies with the nature of the bed,' later to be known as *n*.

these earlier formulations had made important advances in calculating the 'four functions of velocity – "gravity," "surface inclination," "mean hydraulic depth" and "viscosity" [friction between water molecules]' – they had failed to provide satisfactory or reliable answers to key questions, like whether or not the nature of the bed surface affected the velocity of flow, or whether the surface velocity of the water was different to that at the bed.

Taking his lead from the parallel but separate efforts of Du Buat and de Chezy,[10] Manning directs his efforts to the 'key of hydraulics' – the taxing issue of 'the equation between the accelerating and retarding forces in uniform motion' (1891, p. 184) and the unresolved question of the effects of the 'rugosity' or roughness of the bed. As contemporary hydraulicians like Bazin had already recognised, this was highly consequential because it meant that 'it was certainly not accurate to represent all [channel] radii by

[10] Du Buat and de Chezy's accounts are more or less contemporaneous and very similar, but differ critically on the question of the influence of rugosity or roughness, such that du Buat's equation states the resistances to be in a less ratio than the square of the velocity or, in other words, that the velocity increases in a greater ratio than that given in de Chezy's formula (Manning 1891, p. 183).

the symbol R whether the bed is rough or smooth' (1891, p. 182). Where others had conducted experiments that permitted the qualification of R by coefficients for 'rugosity' derived for a variety of different bed conditions in specific contexts, the problem, as Cunningham (1883) had already noted, was that 'the truth of any such equations must altogether depend on that of the observations themselves, and it cannot in strictness be applied to a single case outside them' (Manning 1891, p. 191). This is the problem to which Manning directed his efforts, to produce a formula that accounted for roughness but which avoided these objections and provided an 'equation [that] is homogeneous /.../, consistent with such natural laws as we are acquainted with and [corresponds] very closely [with] experimental velocities' (op cit).

For all the regard in which his work as a drainage engineer was held by his fellow engineers and colonial administrators in Ireland[11] and his own modesty about his scientific abilities, Manning's enduring reputation is freighted by a single algebraic letter, '*n*', the coefficient for roughness in the equation he devised for calculating the 'forward velocity of water flowing in an open channel.' He first presented this formulation in a paper read to the Institution of Civil Engineers of Ireland in December 1889 (published 1891). In it, he compares the velocities calculated by seven of the leading hydraulics authorities in circulation among his contemporaries (including calculations from the United States, France and Germany, as well as his own), which, in turn, derive from some 160 experiments and 210 observed cases (see Figure 4.1). Having standardised their diverse units of measurement into metres (length) and seconds (time), he takes the mean results of all seven to arrive at 'an approximation to the truth' (1891, p. 172). His formulation was founded on five principles 'upon which there is little, if any, disagreement among hydraulicians' (1891, p. 191). These are the 'laws of gravity' (the accelerating force); the 'retarding forces which balance the acceleration' (principally friction between the bed and water molecules and, to a lesser degree, between water molecules themselves and between the surface of the water and the air); the 'resistance of the bed,' which 'increases directly as the length of the perimeter in contact with the fluid, and inversely as the area of the transverse section'; and 'the resistances increase in a less ratio than the square of the velocities' (1891, p. 191).

[11] For example, Manning's accounts of designing channel dimensions for mill-power and navigation, such as a catchment involving 250 falls per mile increasing in depth by a tenth of a foot, with 8,200 different sections and 11 different side slope conditions, were published by the Board of Works and became the norm for engineering irregular channels under the Drainage Acts textbook cases for standard practice.

After having undergone some further work (including a review of 643 experiments and observed cases), his formula was represented to the Institution of Civil Engineers of Ireland in a paper read in June 1895 (published that same year). What became known as Manning's equation took the form that remains a staple element in any calculation of discharge (forward velocity) underpinning the science and management of flood risk to this day:

$$V = \frac{k_n}{n} R^{\frac{2}{3}} S^{\frac{1}{2}}$$

where V is the cross-sectional average velocity, R is the hydraulic radius, S is the energy slope, $k_n = 1.486$ (English units) and n is the Manning resistance coefficient. The 'n-value' is an estimation of roughness, or the effects of friction on the movement of water generated by the shape and character of the channel through which it is flowing. Where abstract theory and empirical measurement failed, this pragmatic proxy rendered friction amenable to calculation in a reliable way. We may be no nearer to formulating the 'exact theory,' which his interlocutors at the Institution of Civil Engineers of Ireland in 1889 held dear, but his pragmatic approach to producing a 'simple formula as easily remembered as Chezy's' has proved sound in that engineers for more than a century since have been

> ... satisfied to consider the velocity sought as that which, multiplied by the area of the transverse section, will give the discharge (which has been well called 'the mean forward' velocity), [such that] a general equation may be found which will hold good for all measures without the necessity of change to the coefficients. (Manning 1891, p. 162)

4. Moment 3: Automating Manning's N

4.1 Twentieth-Century Software Embedding

Present-day hydraulic engineers do not need to worry about working out equations in the field; they can use computers to perform extremely complicated calculations. However, this growth in computational power and complexity seem to have intensified the importance of the n-value that Manning invented to create a 'simple formula' rather than supersede it. What has made it such a durable and ubiquitous component of flood-modelling practice? In the first instance, we would point to the incorporation of Manning's equation and n-values into the software packages that established 1D computer models as the standard technology of hydraulic modelling. Over the course of the twentieth century, the assessment of flood risk and appraisal of management options on which public policy agencies in the UK rely came to fall increasingly to commercial engineering consultancies. The hydraulic modelling practices of these consultants have become more standardised, coming to rely on three widely used software packages: HEC-RAS, a free download developed by the U.S. Army Corps of Engineers; ISIS, proprietary software jointly developed by Wallingford Software and Halcrow in the UK; and MIKE 11, another commercial product developed by the Danish Hydrological Institute (DHI).[12]

Participating in courses introducing new users to these packages, we quickly realised that a university degree in engineering or hydrology was not necessary to work with them, since, like most software products today, they are designed to be user-friendly. Training to use HEC-RAS to simulate flow in a river, we followed instructions to begin by constructing a 'geometry' of the river to be modelled, using survey data from an actual river. These first steps are undertaken in a graphic interface that enables us to draw a line representing the river by 'clicking' on the symbol of a pencil. The next step in creating the virtual river geometry is to construct a series of cross-sections, river stations at which the profile of the river bed and sides are defined on two axes. In the training session we were shown how to bring up an on-screen table in which to enter the measurements for the first cross-section with a roughness value (see Figure 4.2). Our instructor told us to enter a Manning's n-value of 0.03. A brief lecture on what this action amounted to led us to understand that for workaday flood modellers, Manning's n is a number that influences how fast the model lets the virtual water move down the virtual river. If the virtual channel is rougher (i.e., has a higher n-value), the loss of energy in the flow of water will be greater and, hence, the forward movement of the water will be slowed, whereupon the level of water in the river will rise and eventually spill over the banks.

We learned that the way to find the appropriate value for Manning's n was to use a photographic reference guide showing values for different types of river channels. Several such guides are available, both in print and on-line. One on-line version simply presents reference tables with different values for different types of channels – for example, a minimum value of 0.025 (normal 0.030 and maximum 0.033) for a clean, straight, main channel at bank full stage with no rifts or deep pools in which the water

---

[12] These three packages were subject to a comparative assessment in a bench-marking study commissioned by Defra in the 2000s and have since functioned as standards for 1D hydraulic modelling in the UK.

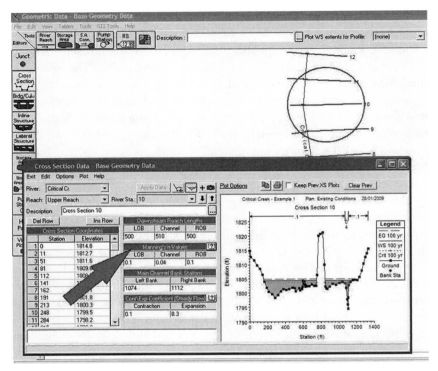

Figure 4.2. Screenshot from HEC-RAS. The inserted arrow indicates the table in which to enter *n*-values for each cross-section LOB (Left Over Bank), Channel and ROB (Right Over Bank) when defining the 'geometry' of the modelled channel. Behind the arrow is a table for cross-section coordinates and to the right a graph representing a cross-section. The window in the background is a representation of the river reach as a line with numbered cross-sections.

flows very fast.[13] At the opposite end of the scale, it gives a minimum value of 0.110 (medium 0.150 and maximum 0.200) for a level floodplain with dense willows in summer, over which water moves very slowly.

We also learned that manipulating the *n*-value is a convenient means of making the simulated flow of the virtual river 'fit' with recorded observations, a procedure known as model calibration. If a run of the model produces water levels in a channel that are much lower than those observed, increasing the *n*-value will slow the movement of water down and raise the level so that the simulated water levels agree with those observed. One of the modellers we interviewed described model calibration in this way:

[13] Values from www.fsl.orst.edu/geowater/FX3/help/8_Hydraulic_Reference/Mannings_n_Tables.htm

... assuming all my rainfall data is right, and I am confident in the hydrology (and that is a big assumption) what that means is that the model is underestimating the levels and it is too late, so what it probably needs is that I need to raise the channel resistance in that reach. You raise the channel resistance which would tend to put the level up a bit, but it might not give you the right timing. Then you reassess the assumption of the hydrology and you say – well, maybe I got the timed peak wrong on the catchment, so I go and reassess the timed peak on the catchment, so what you end up with is something that looks like that. So you think – well, maybe I am overestimating too much so perhaps I will bring my Manning's *n* down a bit. Got the timing right now. So lo and behold you have got your really good match /.../ it is that sort of iterative process. (Interview by C. L., 30th October 2007)

The 'User's manual' accompanying HEC-RAS explains this use of *n*-values as one of the eight essential steps to follow in the calibration of an unsteady flow model (Brunner 2006). The calibration process entails the production of multiple hydrographs at different stages, from which one chooses the ones that correspond best with observed water levels. The manual explains that 'when Manning's *n* is increased the following will occur: (1) stage will increase locally in the area where the Manning's *n*-values were increased; (2) peak discharge will decrease (attenuate) as the flood wave moves downstream; (3) the travel time will increase; (4) the loop effect will be wider (i.e., the difference in stage for the same flow on the rising side of the flood wave as the falling side will be greater)' (Brunner 2006, pp. 8–53). Roughness values work in the same ways in the other two Defra/EA-approved software packages, ISIS and MIKE 11. This means that for a modeller using any of the standard modelling software packages, Manning's *n* works as a tuning device, increasing or decreasing the roughness coefficient such that the water levels in the model can be adjusted to correspond with measurements taken in the physical system. Manning's *n* is here a feature of the computer programme, not a 'real-world' measurement.

It is easy to manipulate model output by changing the *n*-value, but modellers are aware of the problems associated with 'forcing empirical adequacy' this way. For example, the HEC-RAS manual warns users not to

> force a calibration to fit with unrealistic Manning's *n*-values or storage. You may be able to get a single event to calibrate well with parameters that are outside of the range that would be considered normal for that stream, but the model may not work well on a range of events. Stay within a realistic range for model parameters. (Brunner 2006, p. 8–54)

Determining what a 'realistic' *n*-value is requires knowledge not incorporated in the software, but it is an aspect of engineering skill that the hydraulic modeller has to learn. As our HEC-RAS instructor had made us aware, this practical skill has become regularised since the 1960s through the publication of photographic reference works.

4.2 Embodied Skill

The stabilisation of *n*-values to the degree that on-line tables of numbers to employ in modelling software have become useful has been achieved with the development of specialist handbooks for water engineers, which explicate roughness through compilations of photographs of rivers with established *n*-values.[14] We have found works of this type from the early 1960s through the 1990s.[15] The oldest guide that we have come across is a booklet from the United States compiled using previously produced photographs in order 'to illustrate the wide range of the roughness coefficient "*n*" of Manning's formula for channel velocities related to actual channel conditions' (Fasken 1963, p. 3). The author suggests that the '[s]tudy of the pictures and information shown should assist in selecting realistic values of "*n*" for both present and future constructed channels' (ibid) and identifies six key considerations:

1. The material through which the channel will be constructed, such as earth, rock, gravel, and so on.
2. Surface irregularity of the sides and bottom of the channel.
3. Variations of successive cross-sections in size and shape.
4. Obstructions which may remain in the channel and affect the channel flow.
5. Vegetation effects should be carefully assessed.
6. Channel meandering must also be considered.

(Fasken 1963, pp. 3–4)

This guide ties thirty-nine engineered and two natural channels in the United States to specific *n*-values. Each channel is presented in one or more photographs, accompanied by captions that provide information about the location and the photograph, for example – 'Pigeon Creek, Dredged Channel near Cresent, Iowa. Approximate bottom width 15 feet. Picture taken in 1917' (ibid, p. 5). Each black-and-white photograph is followed by a table that lists dates of observation and measurements of average maximum

---

[14] For another case of the importance of visual images in communicating facts about nature, see Merz; and for a contrasting case of how technological facts are packaged for practical use in the field, see Howlett and Velkar (both this volume).

[15] Chow 1959 is referenced by the on-line table and mentioned by many authors discussing Manning's *n*; unfortunately, we have not been able to find it.

depth, average surface width, discharge, average cross-section, mean velocity, mean hydraulic radius, slope of water surface, the roughness coefficient *n* calculated by the author using 'the measured values of slope, hydraulic radius and discharge in the Kutter formula for velocity' (1963, p. 3), plus a detailed description of the hydraulic characteristics of the watercourse. We note that Fasken does not use Manning's formula to calculate the *n*-values he suggests that his readers accept as accurate representations of the characteristics of riverbeds. However, his rationale for producing a guide referring to Manning's *n* and no other parameterisations of roughness is that it is the most widely used because 'it is simpler to apply than other widely recognised formulas and has been shown to be reliable' (1963, p. B.1). His style of presentation suggests that Manning's *n* is a phenomenon that can be observed, which is consistent with his claim that 'Manning's formula is empirical,' an estimation of 'the net effect of all factors causing retardation of flow in a reach of channel under consideration' (op cit). Fasken tries to defend the independence of the *n*-value from the person doing the estimation while recognising that the 'estimation of *n* requires the exercise of critical judgement in the evaluation of the primary factors affecting *n*' (op cit).

The question of exactly what the *n*-value refers to and how to regularise its estimation is handled differently in the second of our 1960s photographic handbooks compiled by Barnes in 1967. Here, the estimation of *n*-values is defined as a skill that has to be honed by practice, in which the 'ability to evaluate roughness coefficients must be developed through experience' (Barnes 1967, abstract). He identifies three ways in which an engineer might improve his estimation of the *n*-value for a particular river reach:

> (1) to understand the factors that affect the value of the roughness coefficient, and thus acquire a basic knowledge of the problem, (2) to consult a table of typical roughness coefficients for channels of various types, and (3) to examine and become acquainted with the appearance of some typical channels whose roughness coefficients are known. (1967, p. 2)

Barnes is more circumspect than Fasken about the applicability of Manning's *n* to natural river systems characterised by non-uniform flow conditions. Nonetheless, his guide presents *n*-values exclusively for natural channels without any particular elaboration. His *n*-values are based on the reverse version of Manning's equation:

$$n = \frac{R^{2/3} S_f^{1/2}}{V}$$

(n = Manning's roughness coefficient, R = hydraulic radius, S_f = energy slope, V = mean flow velocity).

Barnes's examples are also from the United States, and he locates them in relation to permanent gauging stations used by the U.S. Geological Survey to generate stream-flow records. The selected rivers are presented in ascending order by their n-values, starting at 0.024 (Columbia River at Vernita, Washington) and finishing with 0.075 (Rock Creek near Darby, Montana). Each n-value is exemplified by one or more rivers. On the first page of each entry, the author provides information about gauge location, drainage area, date of flood, gauge height, peak discharge and what he calls the 'computed roughness coefficient' (n), together with a description of the channel and a table listing the reach properties. The following page includes sketch plans of the reaches, marking the position of the photographer and cross-sections for each reach. This information is complemented by colour photographs with captions that indicate the direction of flow.

Fasken and Barnes produced their guides for a market of U.S. water engineers in the 1960s, contemporary with the invention of the first computer models in the genealogy of HEC-RAS, but the third reference guide we look at here is much more recent. Produced in New Zealand and published in 1991, this work is presented as a product of sustained research.

> The information presented here is the culmination of a three-year field programme in which roughness and other hydraulic parameters were measured at 78 reaches representing a broad range of New Zealand rivers. The aim of the programme was to provide a reference dataset for use in visually estimating roughness coefficients. This responded to a need for a reference set of reaches representative of New Zealand conditions – our own combination of channel size, gradient, bed material, and vegetation – that would also cater for variations in roughness with discharge. (Hicks and Mason 1991, p. 1)

However, the authors echo the U.S. guides from the 1960s in envisaging that their 'handbook will be used mainly to aid the assignment of roughness coefficients, for example during the application of the slope-area method for estimating flood peak discharge' (1991, p. 11). The format is similar to that of Barnes, with the bulk of the text dedicated to photographs accompanied by descriptions, tables and graphs, but it covers a more extensive range of n-values – from 0.016 to 0.27 (Figures 4.3a and 4.3b).

Despite their similarities, these handbooks imply three different approaches to roughness. For Fasken, it is an empirical fact; for Barnes, a way of seeing a physical phenomenon; for Hicks and Mason, a parameter value that can be generated through research. All three works invoke engineering as an embodied skill, requiring a trained eye to be able to 'see' the

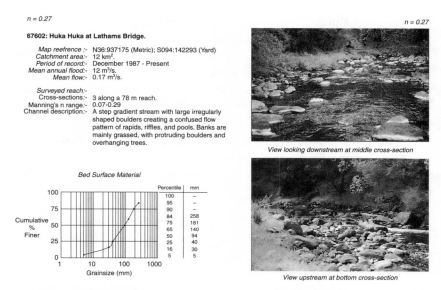

Figure 4.3a. Presentation of *n*-value 0.27 by example of the river Huka Huka. *From*: Hicks and Mason 1991.

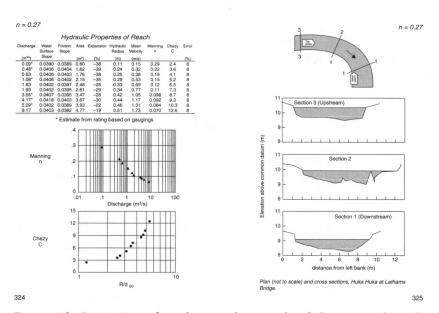

Figure 4.3b. Presentation of *n*-value 0.27 by example of the river Huka Huka. *From*: Hicks and Mason 1991.

n-value of a channel. The existence of these and many similar guides bears witness to the difficulty of estimating the value of *n* for a river reach under varying conditions. *n*-values are supposed to express facts about rivers, but the latter do not seem to behave in a way that allows stability in the former, even though their relationship can be effectively captured in Manning's equation when it is reversed and used for calculating roughness (as shown earlier). Today's on-line reference table, easily accessed by modellers working on their computers, mobilises reference guides like the three discussed earlier in the everyday practices of river engineers in locations all over the world. The small keyboard gesture of the modeller who selects an *n*-value and inserts it into a scroll-down table on the computer screen is enacting decades of engineering knowledge.

The incorporation of Manning's equation and *n*-value in the first hydraulic computer programmes of the 1960s probably owes something to the simplicity of this method of putting roughness to work. Technical constraints on computer power at that time would have favoured simple solutions over more complicated calculations. It might also have resulted from the ways in which *n*-value estimation had already entered into engineering culture as an embodied skill. Whatever the reasons, the hardwiring of *n* into modern technologies of calculation continues to have a hold on hydraulic engineering practice today. 1D modelling software is used everyday in consultant engineering firms undertaking work for a range of UK bodies responsible in law for assessing the impact of their activities on flood risk and the impact of flooding on people and property. Their major challenge is to produce reliable predictions of possible and likely events. In this context, developments in 2D hydrodynamic modelling are enabling better representations of floodplains by calculating the movement of water laterally across topographic maps, with elevations and built structures, derived from digital terrain data.

One software package that is already widely used, and is approved by the Environment Agency, is the Australian TUFLOW. This modelling programme starts with digital terrain data to construct a topographic map across which the flow of water can then be simulated, calculating the energy loss due to friction using Chezy's equation. Despite the sophistication of its treatment of surface topography, the calculation of energy loss still requires the user to input a value called *n*. Unable, as yet, to turn automatically to reference photographs in the estimation of *n*-values for different surfaces, TUFLOW users have begun working out some 'rules of thumb' for common surface coverage, such as grass or multi-level car parks. The desire to develop a set of standard *n*-values for use in TUFLOW floodplain modelling reflects the distance between critical scientists and practising engineers with regard to the utility of Manning's *n*. Where critical scientists regard *n*-values as a dated simplification that obscures important physical processes, practising engineers using TUFLOW continue to find it useful.

5. Conclusions

This chapter has been propelled by a fascination with the apparent hold of a nineteenth-century mathematical formula for estimating hydraulic roughness on the machinery of flood risk science and management today. Our analysis has focused on three related concerns. The first has been to understand hydraulic roughness as a feature of river dynamics constituted differently through changing practices of calculation. The second has been to illuminate the ways in which roughness has been harnessed as a reliable factor in the modelling of river hydraulics in varying contexts. Third and finally, we have sought to highlight the tensile role of *n* in putting roughness to work, at once black-boxing it as an operational standard and representing its effects as a mirror image of 'real' river dynamics.

Robert Manning's work in the nineteenth century was constitutive of 'roughness' as a working estimate of the effects of friction on the velocity of water travelling in a non-uniform channel, a parameter that earlier hydraulic formulations had proved unable to grasp adequately, either mathematically or empirically. Manning's equation and the *n*-value have proved a highly effective packaging of this precarious knowledge claim or, more accurately, method of approximation in terms of its durability and reach in hydraulic science and engineering. What persists above all is a tension in the continued use of Manning's *n* between its practical relevance to the engineer engaged in the urgent business of calculating flood risk and its simplification of a complex physical process to which scientific objections can be raised – a tension that Manning epitomised in his own professional life.

To gain some insight into the durability of this 'fact' of roughness – Manning's *n* – we have examined some of the ways in which it has been more or less successfully packaged and repackaged for use in different techno-scientific regimes of hydraulic engineering. It has travelled from a handy and reliable means of estimating roughness and, thereby, enabling the calculation of velocity for the late-nineteenth-century hydraulic engineer, to a component in automated computer models linked to physical river features via annotated photographs in engineering handbooks. We have argued that in present-day flood-modelling software, *n* is not mainly important as a 'fact,' that is, as an empirical statement about a phenomenon in nature, but as a necessary and reliable means of enabling flood models to work. In contemporary flood-modelling software, *n* is the most amenable lever with which to tune the virtual movement of water through the river channel with that previously observed and, thereby, to validate a model's predictive claims.

Whilst the technologies of calculation and computational power have changed beyond recognition, the purchase of Manning's *n* in the practice of hydraulic science and engineering remains as he originally 'packaged' it – a labour-saving means of estimating roughness. A challenge by twenty-first–century hydraulic scientists claiming that this parameterisation of roughness is an oversimplification both conceptually and empirically has made little dent on its hold on engineering practice. As the apparent lack of success of the Defra/EA programme to replace Manning's *n* suggests, it is not the conceptual or empirical adequacy of this formulation that under-lies its durability. Rather, the more sophisticated method of calculating the effects of friction on conveyance that this programme proposed failed to match its practical appeal, making modelling more time-consuming and complicated by removing the possibility of 'tuning' a model by changing one variable. For the critical scientist, the knowledge claim packaged into *n* may no longer be compelling as a parameter of hydraulic modelling, but for the consultant engineer, it remains the handiest way of putting roughness to work.

Acknowledgements

We are grateful to the librarians and archivists at the Institution of Civil Engineers of Ireland (John Callanan) and the Office of Public Works (Valerie Ingram) in Dublin and at the Institute of Civil Engineers in London for their generous and knowledgeable assistance in locating papers and records related to the work of Robert Manning.

Bibliography

Barnes, H. (1967) *Roughness Characteristics of Natural Channels*. U.S. Geological Survey, Water-Supply Paper 1849, Washington.

Bazin, H. and Darcy, H. (1865) *Recherche shydrauliques*. Paris: Imprimerie Imperiale.

Brunner, G. W. (2006) *HEC-RAS River Analysis System User's Manual*. Version 4.0 Beta, U.S. Army Corps of Engineers, Hydrologic Engineering Center (HEC), Davis.

Chezy, A. de (1775) *Principes d'Hydrauliques*. Paris.

Commissioners of Public Works (1851) *Twentieth Report* (Appendix E 'Extract from the Annual Report of Mr R. Manning, C. E. District Engineer'), pp. 202–3.

Cunningham A. (1883) Recent hydraulic experiments. *Minutes of the Proceedings of the Institution of Civil Engineers*, 71, pp. 1–36.

Dooge J. (1989) *Robert Manning (1816–1897)*. MS Version C, Institution of Civil Engineers of Ireland.

Defra/Environment Agency (2004) *Reducing uncertainty in river flood conveyance. Phase 2, Conveyance Manual.* Flood and Coastal Defence Research and Development Programme, Project Record W5A-057/PR/1.

Du Buat L. G. (1786) *Principes d'hydraulique verifie par un grand nombre d'experiences, faites par ordre de gouvernment.* Monsieur, Paris (2 volumes).

Fasken, G. (1963) *Guide for Selecting Roughness Coefficient "N" Values for Channels.* U.S. Department of Agriculture. Soil Conservation Service. Lincoln.

Fisher, K. and Dawson, H. (2003) *Roughness Review*, Project Report: DEFRA/ Environment Agency Flood and Coastal Defence R&D Programme, Reducing Uncertainty in River Flood Conveyance W5A-057.

Ganguillet, E. and Kutter, W.R. (1889) *A General Formula for the Uniform Flow of Water in Rivers and Canals.* Translated by R. Hering and J. C. Trautwine. London: Macmillan.

Govindaraju, R. and Erickson, L. (1995) Modelling of overland flow contamination due to heavy metals in shallow soil horizons. *Proceedings of the 10th Annual Conference on Hazardous Waste Research*, pp. 287–300.

Hicks, D. and Mason, P. (1991) *Roughness Characteristics of New Zealand Rivers. A handbook for assigning hydraulic roughness coefficients to river reaches by the "visual comparison" approach.* Wellington: Water Resource Survey.

Kutter, W. R. (1876) *The New Formula for Mean Velocity of Discharge of Rivers and Canals.* Translated by L. D'aguilar Jackson. London: Spon.

Lane, S. N. (2005) ESEX Commentary: Roughness – time for a re-evaluation? *Earth Surface Processes and Landforms*, 30, pp. 251–3.

Manning, R. (1866) On the flow of water off the ground. *Proceedings of the Institution of Civil Engineers*, xxv, pp. 458–79.

(1878) Presidential address. *Transactions of the Institution of Civil Engineers of Ireland*, 12, pp. 68–88.

(1882) A method of correcting errors in the observation of the angles of plane triangles, and of calculating the linear and surface dimensions of a trigonometrical survey. *Minutes of the Proceedings of the Institution of Civil Engineers*, 73(3), pp. 289–310.

(1891) On the flow of water in open channels and pipes. *Transactions of the Institution of Civil Engineers of Ireland*, 20, pp. 161–207.

(1895) On the flow of water in open channels and pipes (supplement to the paper read on the 4th December 1889). *Transactions of the Institution of Civil Engineers of Ireland*, 24, pp. 179–207.

McCabe D. (2006) *Major Figures in the History of the OPW (Office of Public Works). Celebrating the first 175 years.* Dublin: The Office of Public Works.

Smith, M., Cox, N. and Bracken, L. (2007) Applying flow resistance equations to overland flows. *Progress in Physical Geography*, 31(4), pp. 363–87.

Zhang, G. and Savenije, H. (2005) Rainfall-runoff modelling in a catchment with a complex groundwater flow system: Application of the Representative Elementary Watershed (REW) approach. *Hydrology and Earth System Sciences*, 9, pp. 243–61.

FIVE

MY FACTS ARE BETTER
THAN YOUR FACTS:
SPREADING GOOD NEWS
ABOUT GLOBAL WARMING

NAOMI ORESKES

1. Introduction

In the late 1980s, planning began for what would become the United
Nations Conference on Environment and Development in Rio de Janeiro,
known informally as the Earth Summit.[1] Scientists had called attention to
the planetary-scale impacts of human activities: acid rain, ozone depletion,
deforestation, biodiversity loss, and now global warming. The latter had
come to public attention as the United Nations Environment Programme
and the World Meteorological Society joined forces in 1988 to form the
Intergovernmental Panel on Climate Change (IPCC) to evaluate the scientific evidence and suggest possible remedies. Scientists had long predicted
that increased greenhouse gases from burning fossil fuels would change
the chemistry of the atmosphere in ways that could affect global climate,
and these predictions were apparently starting to come true.[2] As heat
waves scorched the American Midwest, world-renowned climate modeler
James Hansen declared in testimony to the U.S. Congress that our Earth
had entered a long-term warming trend and that human-made greenhouse
gases almost surely were responsible.[3]

1988 was also the year that George H. W. Bush was elected president
of the United States, and during his election campaign, he had pledged to
combat the greenhouse effect with the "White House effect" – to bring the

[1] UN Conference on Environment and Development 1992.
[2] U.S. President's Science Advisory Committee, 1965; US National Research Council 1966;
Charney et al. 1979; MacDonald et al. 1979; see discussions in Fleming 1998; Weart 2003;
Oreskes 2006.
[3] Hansen 2008.

power of the presidency to bear on the issue of global warming – and to convene a conference on global environmental issues in his first year in office.[4] But like many campaign promises, this one went unfulfilled. One, two, and then three years went by. As June 1992 – the date of the Rio summit – approached and 108 heads of state, 2,400 representatives of nongovernmental organizations (NGOs), and more than 10,000 on-site journalists made plans to converge in Rio (along with 17,000 other individuals who would convene in a parallel NGO forum), it was unclear whether the U.S. president would even attend the meeting.[5]

At the last minute, President Bush went and signed the United Nations Framework Convention on Climate Change, a legally binding Convention that committed the signatories to prevent "dangerous anthropogenic interference" in the climate system.[6] The U.S. president called on world leaders to translate the written document into "concrete action to protect the planet."[7] By March 1994, 192 countries had signed on, and the Convention came into force.

Why had the U.S. president hesitated? Because a significant portion of his constituency was resistant to the idea that human activities might need to be curtailed. Indeed, virtually as soon as a scientific consensus had begun to emerge in the early 1980s that global warming would likely be a problem, a counterargument had emerged that it would not.[8] This counterargument had been fostered by the Reagan administration, which was unsupportive of strict environmental protection and opposed to what it considered excessive government interference in the marketplace.

George Bush was Ronald Reagan's vice-president, and although his political views were more moderate than those of his boss, his political base was similar. Moreover, few in the public knew much about global warming during the Reagan years; by the time of the Bush presidency, many did. Among them were industry leaders and groups concerned about what limits on fossil fuels might mean for them. One such group was the Western Fuels Association, a trade organization of coal producers in the Western United States – primarily Wyoming and Montana – who supplied coal to electrical utilities. In 1991, in anticipation of the Rio meeting, the Western Fuels Association decided to act to protect their interests by challenging the scientific evidence of global warming. To do this work, they created

[4] Thomson 2007.
[5] Brooke 1992.
[6] UNFCCC 2009.
[7] *New York Times* 1992; Natural Resources Defense Council 2002.
[8] Oreskes and Conway 2008; Oreskes, Conway, and Shindell 2008.

two new organizations – *Informed Citizens for the Environment* and *The Greening Earth Society* – to promote an alternative set of facts suggesting that increased atmospheric carbon dioxide (CO_2) would be beneficial, not bad, for the environment and humanity.

The campaign was targeted at ordinary citizens, and it appeared to rest on a scientific theory: that increased CO_2 would enhance photosynthesis. Global warming was good, because it would lead to a greener Earth, so it was something we should welcome, not fear. To ensure that this good news was widely disseminated, the Western Fuels Association produced a half-hour video, sent to public and university libraries around the country.[9] They also enlisted scientists to help prepare press releases and public testimony challenging mainstream climate science: questioning whether warming was occurring, suggesting that it would not be a problem if it were, and promoting the "positive" alternative theory of a greening planet Earth.

Before they embarked on this campaign, however, the Western Fuels Association undertook market research to answer the question of how best to mold public opinion on climate change. In effect, they wanted answers to the questions posed in this volume: What would it take to make their preferred facts travel well? How should these facts be packaged? And who should be their companions? The answers to these questions informed their efforts to make their own facts travel better than the alternatives.

2. The Western Fuels Association and the ICE Campaign

The campaign to promote the Greening Earth theory was launched in 1991 by the Western Fuels Association and largely driven by one man, Fred Palmer, Western Fuels' general manager and chief executive officer. A lawyer by training and businessman by profession, Palmer served on numerous organizations promoting business interests and free-market solutions, including the Global Climate Coalition (a coalition of fossil fuel corporations), the Center for Energy (promoters of coal use), the Alliance for Rail Competition, the National Mining Association, and the Center for the New West (a Denver-based think tank promoting free-market solutions to social

---

[9] This video is briefly discussed in Gelbspan 1997, pp. 36–7. The role of visual media in getting facts to travel is also discussed in this volume by Merz, in the scientific context, and by Schneider, in the artistic context.

problems).[10] A profile in *Range* – a magazine dedicated to the "cowboy spirit on America's outback" – described him as a man who was "determined to defend the coal-fired power plants from an assault launched by professional environmentalists, the United Nations, our own government, and our nation's economic competitors." Above all, Palmer was determined to ensure that the U.S. federal government did not impair the competitiveness of the coal industry.

Acting on behalf of the Western Fuels Association, Palmer retained the services of Bracy, Williams, and Co., a government-relations firm based in Washington, D.C.[11] Their mission was articulated in a series of strategy documents, one of which enumerated ten specific goals and the means to achieve them. Number one was to "reposition global warming as theory (not fact)."[12] They would do this through targeted print and radio media campaigns, which would "start small" and then "build national involvement" as soon as the test market results were in hand in the summer of 1991. The tactics included ensuring that the entire electric utility spoke with a unified voice and had scientists to speak on their behalf.[13]

Western Fuels provided $510,000 for the test market project to run in February–August 1991. Bracy, Williams would run advertising messages in selected radio and print media environments, and would use a polling firm, Cambridge Reports, to test the potential for "attitude change" in listeners and viewers. Four cities were chosen: Chattanooga, Tennessee; Champaign, Illinois; Flagstaff, Arizona; and Fargo, North Dakota. (Later, Bowling Green, Kentucky, was added, while Chattanooga and Champaign were dropped.)

[10] Leonard 2000; Greening Earth Society Press Release, "About Frederick D. Palmer," AMS Archives.

[11] On PR firms challenging scientific evidence, see Rampton and Stauber 2001. My discussion here should not be read to imply that scientists never take their cause "straight to the public." Consider, for example, Bruno Latour's discussion of Louis Pasteur (Latour 1988) or James Hansen's direct communications with public and press: http://www.columbia.edu/~jeh1/. However, often these efforts are viewed with dismay by colleagues who consider it inappropriate to stray outside the institutional mechanisms of vetting and certification, or for those without proper credentials to speak to scientific concerns (Shapin, 1994; see also Mooney and Kirshenbaum, 2009; Schneider, 2009).

[12] ICE, Mission Statement 1991, copy in the archives of the American Meteorological Society (AMS). Most of the materials on the Greening Earth Society were found in the archives of the AMS headquarters in Washington, D.C. (see endnote "A Note on Sources"). The advertising campaign attracted some negative press attention, see: O'Driscoll 1991a and 1991b; Wald 1991; and *The Arizona Daily Sun* 1991. Greenwire, an environmentally oriented news service, suggested that information had been leaked by industry executives who did take the threat of global warming seriously, such as the president of Arizona Public Utilities, who felt that the issue was too complex to be dealt with in "a slick ad campaign" (Greenwire 1991).

[13] ICE, Mission Statement – Strategies: Reposition Global Warming as Theory (Not Fact). AMS Archives.

The cities were chosen using three criteria: (1) that the market derived a majority of its electricity from coal, (2) that it was home to a member of the U.S. House of Representatives Energy and Commerce or Ways and Means Committee, and (3) that it had low media costs.[14] According to a report in the trade magazine, *The Energy Daily*, four of the top fifteen U.S. coal-producing groups contributed funds for the campaign, including ARCO Coal, Peabody Holding Company, Island Creek Coal Company, and Amax Coal Industries. Their requested contributions were based on per-ton coal production: Companies producing more than fifty million tons per year were asked to contribute a minimum of $15,000 each.[15]

The test market proposal summarized the three objectives for the media campaign[16]:

(1) to demonstrate that a "consumer-based awareness program can positively change the opinions of a selected population regarding the validity of global warming";
(2) to "begin to develop a message and strategy for shaping public opinion on a national scale"; and
(3) to "lay the groundwork for a unified national electric industry voice on global warming."

If successful, the results would demonstrate that you could market views about global warming, more or less in the same manner as you could market toothpaste. You could change people's minds about the "validity" of a scientific conclusion in the same way you could get them to switch their laundry detergent.

A crucial element of the campaign would be to create an "independent" organization to promote the alternative message – an organization whose name would suggest concern for the environment and no hint of links to the coal industry. The campaign strategists decided on an acronym before they decided what it stood for – ICE – and an early part of the program used focus groups to test potential names: "Information Council for the Environment," "Informed Citizens for the Environment," "Intelligent Concern for the Environment," and "Informed Choices for the Environment." The focus groups indicated that American citizens trusted scientists more than politicians or political activists – and much less than industry spokesmen – so Western Fuels settled on *Information Council for*

---

[14] ICE, Test Market Proposal, AMS Archives.
[15] O'Driscoll 1991a.
[16] ICE, Test Market Proposal, AMS Archives.

the Environment because it positioned ICE as a "technical" source rather than an industry group. When the advertising campaign was launched, it operated under this name, with a logo of an outstretched hand holding a large plant emerging from the globe.

During the spring and summer of 1991, print and radio advertisements bearing the ICE name and logo saturated the airwaves and local newspapers of the chosen American cities. Print and radio advertisements asked: If the Earth is getting warmer, why is Kentucky getting colder? If the Earth is getting warmer, why is the frost line moving south? If the earth is getting warmer, why is Minneapolis getting colder? Who told you the earth was warming...Chicken Little? And how much are you willing to pay to solve a problem that may not exist?[17] These claims were at odds with mainstream scientific interpretations of the evidence. As American Metrological Society (AMS) officer Anthony Socci noted in a memo to the file, the frost line was *not* moving south, and temperature data showed that the Minneapolis area had warmed 1.0–1.5 degrees in the twentieth century.[18] Other claims were, if not at odds with the scientific evidence, then certainly misleading: Global warming did not imply that every place on earth would get warmer, and there was no reason why Kentucky mightn't have a few cool winters in a row. And the vast majority of climate scientists had concluded that the evidence showed that the problem *did* exist – as the IPCC would soon report.[19]

Several of the print advertisements included a small box, inviting the reader to submit his or her name and address to get more information; radio ads included a toll-free telephone number to call for more information. According to one press report, the campaign produced nearly 2,000 requests to the toll-free number.[20] With this, Bracy, Williams built a mailing list of sympathetic citizens, while creating the impression of a grassroots citizens' organization.[21] They also arranged meetings with local editors and writers, and appearances by sympathetic scientists on local radio and television programs. In Bowling Green, the campaign paid tangible dividends, as the *Bowling Green Daily News* ran a front-page article under the headline: "Hot Debate: Bowling Green now battleground in heated global warming dispute."[22] Western Fuels, under the name of ICE, had persuaded

[17] Press copies found in Greening Earth Society documents, AMS Archives.
[18] Socci 2000.
[19] Houghton et al. 1996.
[20] Wald 1991.
[21] For more on this strategy, see Fritsch 1996.
[22] Baur, "Hot Debate," AMS Archives.

local writers that global warming was not a demonstrated scientific fact, or even a reasonable explanation for certain observed phenomena, but a "heated dispute."

The results of the advertising campaign were assessed through focus groups with citizens in the target communities and showed that you *could* change people's minds, particularly if you presented them with "credible facts" presented by "technical sources." In particular, "older, less educated males" were susceptible to impugning the motives of people talking about global warming, as in the suggestion that "the threat was being exaggerated by members of the media who wanted to increase their audience and influence."[23] Younger, lower-income women were "receptive ...to factual information," and "likely to soften their support for federal legislation [to stop global warming] after hearing new information." And many people were susceptible to the suggestion that the issue was more complex than they had been told. This meant that there was the opportunity to change their minds by presenting them with alternative scientific claims, in effect, with an alternative set of facts.

An important part of the campaign was the construction of the impression that global warming was the subject of active scientific debate, so a crucial component was the use of scientists as spokesmen. Three scientists in particular lent their names to the campaign: Patrick Michaels, a longtime "skeptic" with links to the coal industry; Robert Balling, a professor of geography at Arizona State University; and Balling's Arizona Colleague, Sherwood Idso.[24] On May 15, 1991, a letter was drafted over Michaels's signature. With a side bar listing Michaels, Balling, and Idso as the ICE Scientific Advisory Panel, the letter was sent to the people who had written in to request more information. It thanked them for writing, asserting that climate science was still in its early stages and it was "wrong to predict that higher levels of carbon dioxide will bring a catastrophic global warming." ICE was created, Michaels explained, to "foster better public understanding of global warming and to ensure that any legislation passed by Congress is based on scientific evidence."[25] Michaels, Balling, and Idso were also featured in the radio and television presentations. In each case, the speakers suggested that no one really knew if global warming was a real problem, preparing the ground for the alternative suggestion that global warming was actually *good*.

[23] ICE, Mission Statement – Strategies: Reposition Global Warming as Theory (Not Fact).
[24] For details on Michaels's links to the coal industry, see Gelbspan 1997 pp. 40–4. On Balling, see pp. 44–6, and brief discussion in Gelbspan 2004 pp. 51–2.
[25] Michaels 1991.

3. The Greening Earth Theory and Video

The Western Fuels Association now created a new organization, The Greening Earth Society (although their office address was the same as that of Western Fuels, and the president of the board of directors was Fred Palmer). With funding from Western Fuels, the Greening Earth Society created a half-hour video entitled "The Greening of Planet Earth."[26]

The video was based on the work of Sherwood Idso, a soil scientist at the U.S. Department of Agriculture's Water Conservation Laboratory in Phoenix and long-time skeptic of climate models.[27] In 1980, Idso had published a paper arguing that climate models exaggerated the effects of carbon dioxide on atmospheric heat balance by a factor of ten. At the time, most modelers had come to the conclusion that the likely impact of doubling atmospheric CO_2 would be an increase of global mean temperature of 2–3°C, but Idso argued that the correct value was around 0.3°C – an order of magnitude less than and well within the range of natural variability.[28] If this were true, then the impact of doubling CO_2 would be trivial – certainly nothing to worry about.

Leading climate modelers and geophysicists rejected his argument immediately, and a few years later it was shown that his calculations violated the first law of thermodynamics.[29] But Idso found new ways to make his point. In 1990 he published a paper in *Meteorology and Atmospheric Physics* challenging the link between increased sea surface temperatures and hurricane intensification in the regions of Atlantic cyclogenesis.[30] In the early 1990s, he found another angle: that CO_2 would make the planet greener by enhancing photosynthesis.

The argument rested on the role of CO_2 in plant metabolism. Since plants rely on CO_2 for their basic metabolic processes, increased atmospheric CO_2 could enhance plant productivity – a kind of CO_2 fertilization. If applied to agricultural plants like corn or soybean, this could be highly beneficial to mankind (although perhaps not if it applied to poison ivy).[31] Moreover,

---

[26] Greening Earth Society, *The Greening of Planet Earth*, transcript of video. AMS Archives.

[27] Idso 1980; Schneider et al 1980; Idso 1982; Idso 1988; Idso 1989; Idso et al 1990; Idso et al 1994.

[28] Idso 1980.

[29] Schneider et al 1980; Chamberlain et al., 1980; Cess and Potter 1984.

[30] Idso et al 1990. See also Idso 1982; Idso 1988; Idso 1989; Idso et al 1994. The 1990 work was supported by Kenneth J. Barr, president of Cyprus Minerals, a company that is also cited in public testimony as a funder of Patrick Michaels's work, discussed further below; Hendrey et al., 1999.

[31] For the history of this argument, see Long et al. 2006. In brief, early experiments were "enclosure" experiments – done primarily in greenhouses. When repeated in more realistic

with more CO_2 in the atmosphere, plants might open their stomata for shorter durations, decreasing water loss through stomatal conductance and transpiration, and reducing water demand overall.[32] If plants thrived with less water, they might grow where they could not have grown before, leading to a greener Earth overall.

Like any reasonable argument, the Greening Earth theory contained at least a germ of truth: various experiments had demonstrated this effect.[33] These experiments were reviewed in 1983 by Idso's Department of Agriculture colleague, Bruce Kimball (who had conducted some of these experiments).[34] But Kimball was not merely reviewing the evidence of CO_2 fertilization; he was using it to promote Idso's position that global warming was nothing to worry about.

Kimball's paper began with a discussion of Idso's 1980 argument that the global temperature rise for doubling of CO_2 would be negligible.[35] If this were true, Kimball noted, then there would be no negative effects of global warming, for the simple reason that there would be no warming. Thus, the *only* effect of increased CO_2 would be the positive effect of enhanced photosynthesis. (Kimball thus ignored the refutation of Idso's claim.)

Kimball reviewed the existing literature of relevant evidence, which, he claimed (somewhat anachronistically) went back to de Saussure (famous for his 1804 work demonstrating that plants extracted CO_2 for photosynthesis

conditions – using Free-Air Concentration Enrichment Technology (FACE) – in open fields, the productivity increases were much less. Given that productivity is expected to decrease in a warmer world due to heat, decreased soil moisture, and drought, Long et al. conclude that any small increases from CO_2 fertilization are more than offset by the large decreases due to warming. Moreover, this does not take into account the limits on productivity due to human (social and economic) causes. For discussion of the differential impact of CO_2 on C_3 and C_4 plants, see Lambers, Chapin, and Pons 2008, pp. 518–21. For results of the FACE experiments, see Karnosky et al, 2001; Zak et al., 2003, King et al., 2004, and Hendry et al., 2009.

[32] This is known as stomatal resistance. For what was known on this in the mid-1990s, see Pollard and Thompson, 1997. These authors concluded that although stomatal resistance could decrease water demands, it could also *increase* surface warming of 2°C–5°C by dramatically decreasing transpiration. So much for good news.

[33] Long et al. 2004.

[34] Kimball, 1983.

[35] Kimball 1983; see also Clough et al. 1981. For later work, see also Cure and Acock 1986; Allen et al. 1987; Campbell et al. 1988; Bazzaz 1990; Idso et al. 1994; Mauney et al. 1994; Kimball, et al. 1995. One important result of later work – recognized as early as 1986 (see Cure and Acock 1986) – was the discovery of acclimation: that an initial positive response to increased CO_2 does not persist. This result is generally attributed to a decline in Rubisco activity, due to decreased gene expression or gene products, or to environmental limitations, such as lack of warmth or soil nitrogen. Therefore, genetic capacity and environmental conditions may ultimately limit photosynthetic activity irrespective of ambient CO_2; see Yelle et al. 1989; Sage et al. 1989; Stitt 1991; and Long et al. 2004.

from air, not soil). The available data were diverse and not entirely commensurable – some came from partially ventilated greenhouses in which CO_2 concentrations fluctuated; some experimenters had enriched the ambient CO_2 only for certain stages of plant growth; some did not report what concentrations of CO_2 they had used; and experiments with null results might not have been reported at all – and Kimball admitted that if "only those experiments that had controlled and monitored CO_2 concentrations for their duration were considered, most of the experiments were eliminated."[36] Still, evidence is evidence, even if messy, and the available data suggested an average increase in crop yield around 33 percent. Later studies more or less supported this conclusion: A retrospective review in 2006 concluded that under experimentally doubled CO_2 conditions, crop yields increase by roughly 25 percent for rice, wheat, and soybeans.[37] As one ecologist recently put it, "The evidence for CO_2 fertilization is pretty clear … It does happen."[38]

On the other hand, like all experiments, these faced the question of external validity: whether the observed effects would apply in nature.[39] Kimball asserted that the "response of field-grown crops ought to be as large or larger than greenhouse-grown crops," because the main variable being controlled in greenhouses – temperature – was not a strong control on plant response to CO_2 concentration.[40] But that was only part of the potential story. Nearly all the experiments that Kimball reviewed took place in greenhouses or laboratory-controlled environment chambers, where other requirements (light, warmth, water, and soil nutrients) were met. The argument for a substantial productivity increase in real-life conditions presumed that CO_2 availability was a primary limitation on agricultural productivity, a presumption that was almost certainly incorrect. In natural environments, plant productivity is typically limited by water, warmth, light, and inter- and intraspecific competition; there appears to be no documented example of a

---

[36] Kimball 1983, p. 779. Some of the data came from commercial environments where growers sought to increase yield through enhanced CO_2, so it is not surprising that they might not have kept careful detailed records. Bazzaz 1990 claims that greenhouse growers "have used" CO_2 fertilization to enhance yield, but gives no details or references.

[37] Long et al. 2004, 2006.

[38] Ken Davis, Penn State University, e-mail communication, April 8, 2009.

[39] Cure and Acock 1986 discussed the variables that affect CO_2 response, for example, that the absolute increase in photosynthesis in response to elevated CO_2 was greater in high light than in low light. So these considerations were already known at the time Kimball did his review. The question of whether facts that work in the laboratory will work in the wilds of agriculture is also treated by Howlett and Velkar, this volume.

[40] Kimball 1983, p. 782.

plant dying from lack of CO_2.[41] In agricultural settings, productivity may be limited by social and economic factors, such as the availability and quality of labor; the availability of capital with which to purchase machinery, fertilizers, and pesticides; or the know-how to install irrigation systems.

Moreover, Kimball did not explore the impact of increased CO_2 on natural ecosystems. The relevant studies had been conducted only on a small number of plants, almost entirely C_3 crops. These plants rely on the enzyme ribulose-15 biphosphate-carboxylase-oxygenase to extract CO_2 directly from intercellular air, so it was expected that increased CO_2 would increase photosynthetic CO_2 uptake.[42] But in C_4 plants, which include maize, sorghum, and many wild grasses, the relevant enzymes are not in direct contact with intercellular air space, so one would not expect to see the same effect. Indeed, in nature, C_3 plants might outcompete C_4 plants in enhanced CO_2 conditions, causing undesirable disruptions to ecosystems.[43] So the claim of net benefit was certainly not certain. At best, it was an idea with some experimental support, but as one retrospective study would later put it, "no agrochemical or plant-breeding company would base its business plan solely on greenhouse studies without rigorous field trials."[44]

Still, there was a point. If CO_2 fertilization increased agricultural productivity, then global warming was not necessarily all bad; some positive effects might offset the negative ones. It was at least worth studying. But its advocates did not simply call for more money for research. Rather, they asserted that the future world would, in fact, be better *overall*. This was the message of the video, "The Greening of Planet Earth."

The video opens with bucolic pictures of preindustrial life – men and animals living together in a biblical idyll. Then, the images shift abruptly to those of the Industrial Revolution – machines stamping out metal and cars, factories, and highways springing up – while a narrator describes the inexorable rise of carbon dioxide in the atmosphere. The images begin to move faster – fast-moving people, cars, and machines – and become more and

[41] Ken Davis, e-mail communication, April 8, 2009.
[42] Long et al. 2004.
[43] Johnson et al. 1993 demonstrated that, as expected, at elevated CO_2 levels C_3 plants outcompete C_4 plants (see also Lambers et al. 2008, p. 519). Lambers et al. (2008) conclude that the relevant literature shows evidence of an overall decrease in C_4 plants since the industrial revolution, suggesting that C_3 plants may continue to gain ground as CO_2 increases, although other changes, such as increased temperature and decreased soil moisture, may favor C_4 grasses in some areas.
[44] Long et al. 2006. p. 1919. Recent work suggests that at high elevations, where the atmosphere is thin, CO_2 availability may be a constraint on some plant growth (Shugart 2005). On the other hand, analysis of tree ring data suggests that overall the CO_2 fertilization effect on trees in natural environments is very limited. (Jacoby and D'Arrigo 1997).

more agitating. Then the narrator's voice asks portentously, "What kind of world have we created?" A kindly looking scientist replies: "A better world, a more productive world."[45]

Other scientists follow with similar messages:

"In terms of plant growth, it's nothing but beneficial."

"For citrus, it would be a very, very positive thing."

"With a doubling of CO_2 – why, cotton growers can look forward to yields that are 60 percent and more greater than what they are at present-day levels."

Although the first few speakers refer to specific benefits – cotton, citrus, overall plant growth – one scientist goes even further, declaring without qualification, "Our world will be a much better one." The introduction culminates with Sherwood Idso declaring, "A doubling of the CO_2 content of the atmosphere will produce a tremendous greening of the planet Earth." Triumphant music plays as the title flashes across the screen, superimposed over a map of a visibly greener planet Earth.[46]

The video then presents the Greening Earth theory, strongly emphasizing not only the direct benefit of CO_2 fertilization, as photosynthesis is enhanced, but the additional claim that plants will need much less water – so much so that the entire planet will become more fecund. Sherwood Idso explains that the decreased water demands of plants in an enhanced CO_2 atmosphere will allow plants to flourish in places where at present they cannot – the Arctic, the Sahara, the outback of Australia – as a map of the Earth illustrates the entire planet slowly but surely turning green.[47]

The rest of the half-hour video explores this theme, with extensive discussion by agricultural scientists of the possibilities of beneficial effects. Statistics are presented as to the expected increase in crop yields, greenhouses are shown, and computer models are discussed. As water demands decrease, the video explains, crops will grow more efficiently, and yields will rise. Overall, crop plants will produce "30–40% more than they are currently producing." The motivations of climate scientists are briefly explored and impugned – "garbage in, garbage out," one scientist notes dismissively – but the bulk of the film is focused on the strong positive message that global warming will be a good thing. In essence, it presented an alternative set of facts.

[45] See discussion in Amos 2007.
[46] The gradient in the use of fictions in the transmission of scientific facts in the popular domain is discussed by Adams, this volume.
[47] Greening Earth Society, *The Greening of Planet Earth*, transcript of video. AMS Archives. See also Idso 1995.

But were those facts actually facts?

Climate scientists have long acknowledged that global warming will bring gains and losses, winners and losers, and that agricultural productivity could increase in places like northern Canada and parts of Russia.[48] However, the idea that plants could flourish in the Arctic flies in the face of the limitations presented by the dearth of light or warmth there. At best, one could say that the video exaggerated a real but small potential gain, one that would likely be realized only in areas of the world where the other requirements for high agricultural productivity were already met.

More to the point, the claim that global warming would be a *net* benefit overall was clearly inconsistent with the views of the bulk of the scientific community working on the question at that time.[49] So, if by a scientific fact we mean a conclusion that is broadly accepted by the relevant expert scientific community, then the claims of the video were not, in fact, factual. The video did not explain that the pertinent experiments were performed in highly controlled conditions of uncertain external validity. It did not explain that the views being presented were, at best, a minority position. Rather, it presented the theory as broadly accepted and implied or even asserted the veracity of numerous claims that one might have reasonably said were preliminary results. Dr. Herman Mayeux, for example, a scientist with the Agricultural Research Service of the U.S. Department of Agriculture, says that "the marvellous thing about the increase in water-use efficiency is that *all* plants experience it to some degree or another." This claim was patently unsupported – only a handful of plants had been well studied – and, as explained above, there were good theoretical reasons to suspect that it would not apply to C_4 plants. Moreover, it would be hard to imagine any experiment, or even a set of experiments, that *could* support such an unqualified global assertion.[50]

Other claims were potentially misleading, as when Kimball said that

> The world in which the CO_2 concentration has doubled is one in which plants will enjoy it a lot more. They have, in effect, been eating the CO_2 out of the air for a long time and they are rather starved for CO_2. So the plants are really going to like this high CO_2 world that we're going into now.[51]

[48] IPCC "Impacts, Adaptations and Vulnerability."
[49] Houghton et al 1996.
[50] Greening Earth Society, *The Greening of Planet Earth*, transcript of video, p. 5. Emphasis added. AMS Archives.
[51] Greening Earth Society, *The Greening of Planet Earth*, transcript of video, p. 4. AMS Archives.

This formulation seemed to imply a shortage of atmospheric CO_2 – as if it were a finite resource that plants were depleting. But if CO_2 was rising – and the Greening Earth Society did not dispute it – then CO_2 was *not* being depleted. And who knows what plants enjoy, or if the verb even makes sense in reference to them?

Other scientists made similar claims, as when Dr. Hartwell Allen asserted that "the increase in atmospheric CO_2 is a benefit that will occur around the globe, regardless of where you're located. There'll always be some benefit for somebody, for everybody perhaps."[52] (One only needs to consider the people of the Maldives – whose entire island nation may be swamped by global sea level rise – to realize that whatever benefits might occur from CO_2 fertilization, it cannot possibly be true that *everybody* will benefit.) At best, the claims in the video represented the interpretation of a minority of experts, focused on one particular aspect of the problem to the exclusion of many other aspects, and funded by an organization that had a clear vested interest in that interpretation.[53]

Moreover, the video did not stand alone. By the mid-1990s, The Greening Earth Society had established an office at 4301 Wilson Boulevard in Arlington, Virginia – next door to the U.S. National Science Foundation – from which it distributed numerous reports and pamphlets and press releases. "In Defense of Carbon Dioxide," prepared by New Hope Environmental Services – Patrick Michaels' own consulting company – reiterated the argument that warming would provide a net benefit.[54] The report included a graph showing a steady increase in global crop yields from 1950–87, plotted against the atmospheric concentration of CO_2, seemingly suggesting that the rise in CO_2 had *caused* the rise in food production.[55]

It is well known that food production increased dramatically during the Green Revolution, an increase that is generally credited to increased inputs of energy, fertilizers, and pesticides, as well as improved crop hybrids

---

[52] Greening Earth Society, *The Greening of Planet Earth*, transcript of video, p. 4. AMS Archives.

[53] How many people actually saw the video? It is hard to say, but we do know that it was widely distributed to public and university libraries, to which it was touted as "an enlightening documentary that examines one of the most misunderstood environmental phenomena of the modern age." See for example, the copy held in Oregon State University Library (Greening Earth Society, *The Greening of Planet Earth*, video).

[54] On New Hope Environmental Services as the wholly owned consulting firm of Patrick Michaels, see http://www.prwatch.org/node/8382. PRWatch.org show that New Hope was funded in part by the libertarian CATO Institute, and for many years was listed on their tax returns as an "independent contractor supplying services to CATO." See also Michaels 1998.

[55] New Hope Environmental Services, "In Defense of Carbon Dioxide."

introduced during this period.[56] Yet the suggestion that increased CO_2 *caused* the Green Revolution was made explicit in 1997 in "The Greening of Planet Earth: Its Progression from Hypothesis to Theory," written by Craig D. Idso (son of Sherwood) for the Western Fuels Association.[57] Indeed, the report suggested that plant productivity had *already* increased due to increased CO_2. "[T]he '[G]reen [R]evolution has coincided with the period of recorded rapid increase in concentration of atmospheric carbon dioxide, and it seems likely that some credit for the improved yields should be laid at the door of the CO_2 build-up," Idso wrote. "Yields of soybeans may have been rising since at least 1800 'due to global carbon dioxide increases,'" he quoted, drawing on an article coauthored by Kenneth Boote, one of the experts featured in the "Greening" video. But the crux of the Idso *fils* claim was that the Greening Earth idea was no longer a "mere" hypothesis; it was a theory that had been confirmed by empirical tests. Who did Craig Idso say had reviewed and summarized these tests? His own father, Sherwood Idso.

4. Blocking Others' Facts From Traveling

Most of the work of The Greening Earth Society in the early to mid-1990s focused on promoting knowledge of the CO_2 fertilization effect – in essence, making their preferred facts travel. The documentary records show, however, that the Society's goal was not to advance a research program on CO_2 fertilization, but to change public opinion and decrease political support for legislation to control greenhouse gas emissions. Although this was rarely stated in public, it was implicit in the ICE campaign, with its efforts to change public opinion and "soften support for federal legislation." It was made explicit in a memorandum from Richard L. Lawson, president of the National Coal Association, asking members to contribute to the ICE campaign because of the political stakes: "[M]any policymakers are prepared to act [on global warming]. Public opinion polls reveal that 60% of the American people already believe global warming is a serious environmental problem. Our industry cannot sit on the sidelines in this debate."[58]

It didn't.

By the late 1990s, the coal industry became more open and vocal in trying to counter the facts emerging from mainstream climate science, an effort that was linked to stopping the United States from ratifying the

---

[56] *Encyclopaedia Britannica* 2009.
[57] Idso 1997. The paper is self-described as "Climatological Publications Scientific Paper #25 of the Office of Climatology at Arizona State University," and stamped, "Reproduced at Government Expense." AMS Archives.
[58] Greenwire 1991.

Kyoto Protocol to the U.N. Framework Convention on Climate Change. (The Protocol placed numerical targets on the general principles of the Framework Convention, in effect putting teeth into a declaration of intent.) An additional part of this tactic was to shift attention to suggest that regulating CO_2 would lead to government control of everything.

In a speech to a Coal Industry Conference in Madrid in 1996, entitled "Fossil Fuels or the Rio Treaty – Competing Visions for the Future," Fred Palmer struck a theme that would later become familiar in the community of those resisting governmental action on global warming: that environmental regulation was the first step on the slippery slope to socialism.[59] Regulating fossil fuel use "represents an initial step by government to massively regulate almost all human activity everywhere all of the time." Rio was just the first step down the "path of pervasive government control...."[60]

Palmer linked his position to Christian values, insisting that fossil fuels were God's gift to humanity, indeed, "among God's *greatest* gifts to the human community."[61] To stop using them, or to use them less, would be to look God's gift horse in the mouth. CO_2 fertilization was the clear proof of this. Palmer explained:

> As the ["Greening Earth"] video makes clear, rather than fearing increasing atmospheric CO_2 emissions, we should welcome [them], since, in the words of U.S. Department of Agriculture scientist Dr. Herman Mayeux, a CO_2 enriched atmosphere will bring... "A better world, a more productive world. Plants are the basis for all productivity on Earth. They are the only organisms that can utilize the Sun's energy and create matter, food, and they're going to do that much more effectively, much more efficiently."[62]

In a presentation to the Australian Coal Conference and Trade Exhibition two years later, "We Are One: Kyoto and Our Collective Economic Future," Palmer returned to this theme. CO_2 was an "elixir" whose increased concentrations were producing a "rebirth of the biosphere." "The scientists we work with," he continued, "maintain that it is well-established that some 10% to 15% of the increase in global crop yields that we enjoy around the globe are caused by increasing atmospheric CO_2 concentration."[63]

---

[59] Oreskes and Conway, 2010.

[60] Palmer 1996, p. 2. AMS Archives.

[61] Palmer 1996, (emphasis added), p. 3. AMS Archives. Of course, there is a long history of linking geological science with understanding God's will (see Rupke 1983; Rudwick 2005 and 2008) and even more specifically of seeing evidence of God's beneficence in providing us with natural resources.

[62] Palmer 1996, p. 15.

[63] Palmer 1998, p. 27.

Palmer's speeches consistently told a good-news story – that fossil fuels were a gift from God, that CO_2 is beneficial – and in the late 1990s, it became the official policy of the Greening Earth Society to focus on this "good-news" slant. The Society had been around since 1992, but in April 1998 they relaunched with a press conference in Washington, D.C., announcing themselves as an organization dedicated to promoting "positive environmental thinking."

> What if our planet were getting greener? What if humanity and nature were growing together? It is. We are. The Greening Earth Society is an organization dedicated to promoting this good news.... about our changing climate. Call it the power of environmental positive thinking, says the GES mission statement. Greening Earth Society stands for the proposition that humankind's industrial evolution is good, and not bad, and that humans utilizing fossil fuels to enable our economic activity is as natural as breathing. The truth is out there. ...And it's good, not bad.[64]

The press release also insisted that others were spreading misinformation. In contrast, they, the Greening Earth Society, would stress the facts. "I believe in giving the American people facts, not hype," Palmer insisted.[65] But while he promoted this good-news message in public, in private, Palmer summarized the society's strategy somewhat differently. In an e-mail message to a congressional staffer in 1999, he wrote: "We will bury you in studies."[66]

In the end, it was not so much studies as press releases challenging mainstream scientific studies. In 1999, the Greening Earth Society issued at least thirteen, with titles like "No warming in New Hampshire ground temperature data," "Bristlecone pine data: 20th century warming normal," "Finnish summer saw no warming or cooling over 595 years," "90 years of cooling at Mt. Wilson," "Kitt Peak, Arizona, records reveal short-term warming, longer-term cooling," "Lenin Sea-ice stations show no global warming," "Cooling at China's Three Gorges dam site," and "No warming in Panama."[67] Most of these raised questions about existing scientific data (rather than reporting original data of their own); most quoted men already associated with the society, such as Idso, Palmer, and Michael; and several reported reanalyzes of existing scientific work by the "climate data task force" at Arizona State University – Sherwood Idso's research group.

---

[64] Greening Earth Society Press Release, "New Group Promotes 'Positive Environmental Thinking,'" 1998.
[65] Greening Earth Society Press Release, "New Group Promotes 'Positive Environmental Thinking,'" 1998.
[66] Palmer 1999b. AMS Archives.
[67] Greening Earth Society Press Releases Jan–Dec 1999. AMS Archives.

The contact name on many of the press releases was a man named Ned Leonard, who later became a vice-president at the Center for Energy and Economic Development – another group dedicated to promoting coal use. In 2000, Leonard wrote a piece critiquing climate models that suggested that modelers were deliberately fudging their results.[68] Entitled, "The Greening of Planet Earth," its subtitle ran: "A scientist said, 'Climate modelers have been cheating for so long it almost become respectable.'" The thrust of the piece was that climate modelers had deliberately exaggerated the threat of warming by falsifying their models. Leonard drew on an article by the highly respected science writer, Richard Kerr, discussing constraints in climate modeling. "All the climate disaster scenarios that dominate popular understanding of the threat CO_2 poses to the world's climate are products of computer-based models of atmospheric chemical and physical processes that, in fact, are not well understood," Leonard wrote. "In order to resemble today's climate, the models need to be fudged. 'Climate modellers have been cheating for so long it's almost become respectable,' explains Richard Kerr in a May 1997 *Science* magazine."

Did Richard Kerr say this? Yes, but in an article about a *breakthrough* in climate modeling, entitled "Climate Change: Model Gets It Right – Without Fudge Factors." The full quote was

> Climate modellers have been "cheating" for so long it's almost become respectable. The problem has been that no computer model could reliably simulate the present climate. Even the best simulations of the behaviour of the atmosphere, ocean, sea ice, and land surface drift off into a climate quite unlike today's as they run for centuries. So climate modellers have gotten in the habit of fiddling with fudge factors, so called "flux adjustments," until the model gets it right. No one liked this practice.... But now there's a promising alternative. Researchers at the National Center for Atmospheric Research (NCAR) in Boulder, Colorado, have developed the first complete model that can simulate the present climate as well as other models do, but without flux adjustments. The new NCAR model, says [modeller David] Randall, "is an important step toward removing some of the uneasiness people have about trusting these models to make predictions of future climate."[69]

5. Reinforcing Doubt

The Western Fuels Association promoted the Greening Earth theory through direct approaches to the American people – to make their facts travel – in order to prevent new legislation to curb fossil fuel use. One further tactic

---

[68] Leonard 2000.
[69] Kerr 1997.

for spreading their facts and blocking others was to enlist scientists to testify at public hearings.

In February 1992, the state of Colorado held public hearings on state rules related to resource planning. Fred Palmer submitted testimony on behalf of Western Fuels, which included both his own introduction and formal testimony from three scientists whose names are now familiar: Michaels, Balling, and Idso.

The state was considering whether integrated resource planning (IRP) should include accounting of environmental "externalities": costs to human health, communities, or the environment not included in the market price of that activity or product. (Advocates argue that externalities are real costs that are not expressed by market prices, and are often overlooked in cost-benefit analyses. Mining deaths, for example, are not included in the cost of a ton of coal, but perhaps they should be.) Western Fuels vigorously opposed this idea because coal produces more CO_2 per unit of energy than other fossil fuels, so when viewed in terms of global warming, coal is a more expensive fuel than oil or gas.[70] "Carbon dioxide emissions are often deemed appropriate to be among the environmental externalities considered in IRP hearings," Palmer wrote, but it was "Western Fuels Association's intention…to challenge the basis for the consideration of carbon dioxide externalities…"[71]

The crux of Palmer's argument was that since global warming was speculative, so were any alleged costs associated with it. The "entire foundation [for concern] rests upon the output of general circulation models …The uncertainty is not only embedded in the nature of speculative costs, but extends to the veracity of the analytical tools used to ascertain that there may be any cost at all."[72] Since we didn't even know whether or not global warming was happening, we couldn't calculate its costs. (The benefits of the Greening Earth, however, were not speculative and could be calculated, they insisted.)

Like the "Greening Earth" video, the testimonies were amalgams of empirical claims and large extrapolations from them. Some claims were clearly incorrect, such as when Balling stated that the past century increase

---

[70] Coal also produces very high external costs related to sulphur emissions, mercury, and particulates, see Ming et al. 2005. In an attempt to monetize the external costs of energy production for the year 2005, the U.S. National Research Council put these costs at $120 billion per year, *not* including damages associated with global warming, harm to ecosystems, health effects from mercury, and risks to national security. Of the various energy sources, coal was found to have by far the larger external costs. See United States National Research Council, 2010.

[71] Comments and Prepared Testimony of Western Fuels Association, Palmer testimony, 1992, p. 1. AMS Archives.

[72] Comments and Prepared Testimony of Western Fuels Association, Palmer testimony, 1992, pp. 3–4. AMS Archives.

of carbon dioxide was 40 percent. The correct value was about 25–30 percent, and although this might seem a minor point, it mattered because the more CO_2 had increased in the period when temperature had gone up only half a degree, the more it would look like CO_2 was no big deal. Balling also asserted that variations in stratospheric dust "would appear to account for 0.15°C of the trend," neglecting to say that it was 0.15°C of *cooling*.[73] (Without that dust, warming would have been greater.)

Colorado was just one venue where such hearings were held. Balling and Michaels also testified in Minnesota in 1995, in testimony again sponsored by Western Fuels. Their testimony included a list of their sponsors, which included the British Coal Corporation, the German Coal Mining Association, the Edison Electric Institute, and the Kuwait Foundation for the Advancement of Sciences – as well as a grant from an individual, Kenneth Barr, the president and CEO of Cyprus Minerals.[74]

In 1999, Fred Palmer testified at a United States House of Representatives hearings on National Economic Growth, Natural Resources and Regulatory Affairs. Again he attacked the U.N. Framework Convention on Climate Change, and defended coal. The new information technology (IT) economy required more coal, not less, because IT demanded great amounts of electricity and only coal could supply it. Most people viewed coal as an "old" fuel associated with old industries like railroads, steel-making, and smelting, but this was mistaken. "The Internet begins with coal," he declared, because companies like Dell, Intel, Lucent, Qualcomm, and Yahoo! all "thrive on electricity…. the lifeblood of the information economy."[75] As for Kyoto and the U.N. Framework Convention, they were "the road to nowhere."[76]

6. Conclusion: People Prefer Good News

In focusing on the good news about global warming, Western Fuels recognized a well-known axiom of marketing and public relations – and indeed, of conventional wisdom – that people would rather hear good news than bad.[77] (Of course, bad news can have currency – think gossip – but *ceteris*

---

[73] Comments and Prepared Testimony of Western Fuels Association, Balling testimony, 1992, p. 9. AMS Archives. Maria et al. 2004, and Ramanathan and Carmichael 2008.

[74] Comments and Prepared Testimony of Western Fuels Association, Michaels rebuttal testimony, 1995, p. 15. AMS Archives.

[75] Palmer 1999a. AMS Archives.

[76] Palmer 1999a. AMS Archives.

[77] Greening Earth Society Press Release, "New Group Promotes 'Positive Environmental Thinking'", 1998. AMS Archives. In fact, GES was not new in 1998, but they evidently pushed that idea to promote their Earth Day initiatives that year.

paribus, most of us prefer good news to bad.) By using market research and polling, they were able to determine effective ways to package this good news to reach particular audiences. Omitted from their advertisements, speeches, and testimonies, however, was the acknowledgement that most mainstream scientists rejected this good-news story and did not consider it factual at all. At best, the scientists involved in evaluating global warming and its impacts saw carbon fertilization as a modest positive benefit to be viewed against a much larger landscape of severe detrimental impacts.[78]

But the message of what mainstream scientists believed – the bad news about global warming – did not travel very well at all. Public opinion polls during the 1990s repeatedly showed that a significant portion of the American population did not believe that anthropogenic global warming had been demonstrated, a position at odds with the conclusions of the expert scientific community.[79] Indeed, after President Bush signed the U.N. Framework Convention on Climate Change, the United States government did a *volte face* and refused to participate in the Kyoto Protocol – a position that remains in place as of this writing.

Today, many people expect a change in U.S. policy. Whether this occurs remains to be seen. Historically, however, it will remain the case that, after Rio, the United States stood nearly alone among the world's industrialized nations in rejecting binding limits to carbon dioxide emissions, rejecting government-based incentives to reduce emissions. Many American citizens continue to doubt that global warming is occurring, while others accept its reality but doubt that it is caused by human activities.[80] Of course, these developments might have nothing to do with the story told here; elsewhere, we consider the complex history of resistance to the scientific evidence of global warming and other environmental problems.[81] But at least some evidence suggests that the events described here contributed in some part to the larger history of the resistance by the American people to the conclusions drawn by expert climate scientists about the reality of anthropogenic climate science.

In the summer of 2007, a major opinion poll showed that although a large majority of Americans accepted the reality of anthropogenic climate change, nearly *half* still believed that the scientific jury was out.[82] Nearly thirty years after the National Academy declared that a "plethora of diverse studies ...

---

[78] IPCC 1995.
[79] Nisbet and Myers 2007.
[80] ABC News 2007.
[81] Oreskes and Conway, 2008; Oreskes, Conway, and Shindell 2008; and Oreskes and Conway, 2010. See also McCright and Dunlap, 2010 - the online version of this article can be found at: http://tcs.sagepub.com/cgi/content/abstract/27/2–3/100.
[82] Leiserowitz 2007.

indicates a consensus" that global warming would result from human activities, and more than a decade after the scientific community declared that warming "discernible," a sizable fraction of the American people thought that scientists were still debating the point.[83] This suggests that the various campaigns of resistance – and the one described here was but one of several – were effective in blocking the facts of climate science from travelling to the American lay public.[84] At minimum, the American people came to believe that scientists were still arguing long after they actually were.

One could draw many conclusions from the events described, but one that seems most pertinent here is how Western Fuels studied the question of how facts travel – in fact, paid a good deal of money to ask and answer that question. This contrasts with the behavior of most scientists, who do not generally retain marketing firms or buy advertising space for their views.[85] Indeed, most scientists, fearing the appearance of advocacy and the potential for their objectivity to be compromised, would consider it inappropriate to do so.[86]

Moreover, scientists typically follow a supply-side model, assuming that the information they gather – whether good news or bad – will naturally reach the people who need it. But, as the various studies in this volume show, facts do not travel "naturally"– they have to be *made* to travel. Facts may *not* travel if they encounter active, organized, positive resistance, particularly if that resistance emanates from people who have studied – and understand – how facts travel.[87] The Western Fuels Association studied how facts travel; then they used this information both to create a supply of their own facts – through pamphlets, press releases, advertising campaigns, and the "Greening Earth" video – and to create demand for them through their meetings with editors, their toll-free number, and their creation of memberships in ICE and the Greening Earth Society.

[83] National Academy of Sciences 1979; IPCC 1995.

[84] On other resistance campaigns, see Gelbspan 1997; Mooney 2005; Oreskes and Conway, 2008.

[85] On the other hand, it parallels the work of the tobacco industry, see Brandt, 2007. One tobacco industry document from the mid-1990s, entitled "Communication Principles," states, "Be positive, friendly and firm. ...And don't lie. You don't need to." http://legacy.library.ucsf.edu/tid/gwq45c00/pdf?search=%22communications%20principles%22

[86] On scientists' concern that communicating directly to the public may compromise – or be seen as compromising – their objectivity, see Oreskes 2007; see also Mooney and Kirshenbaum, 2009; Schneider 2009.

[87] For more on this, see Oreskes and Conway, 2010. There, we trace the various campaigns to challenge scientific evidence on global warming, acid rain, ozone depletion, secondhand smoke, and the harms of DDT to the tobacco industry in the 1950s, where the idea appears to have first been implemented. See also Brandt 2007 and Proctor 1995.

As the twentieth century turned into the twenty-first, Americans received the message that global warming was real, its consequences not so good.[88] How exactly they received this message is another question. Perhaps it was because climate records continued to be broken around the world; perhaps because hurricanes Katrina and Rita brought home the vulnerability of even wealthy, highly industrialized nations to the kinds of dislocations that climate change can bring; or perhaps because the widely viewed film, *An Inconvenient Truth*, won an Academy Award and was shown in cinemas and discussed in homes across the nation. Or perhaps because the American mass media mostly stopped presenting the issue as one with two opposite and more or less equal scientific sides.[89] Yet the public view may prove unstable, for such experiences and events (e.g., a few long, cold winters), may alter lay beliefs regardless of the science base of knowledge.

But, however well the scientific facts travelled or not, one thing is clear: The scientific message of global warming was separated from its scientific roots. For polls showed that even among those who accepted the scientific message that global warming was happening, many did not know that scientific experts had a consensus on the matter.[90] In the summer of 2007, a Yale University-Gallup poll found that 72 percent of Americans were completely or mostly convinced that global warming was happening, but 40 percent thought that there remained "a lot of disagreement" among *scientists* on the issue.[91] This, of course, was precisely the message that the Western Fuels campaigns had tried to convey. The expert consensus on global warming did reach the American people, but without those people understanding that this *was* the expert consensus. Put another way, the facts – as defined by the scientific community – traveled, but were separated from their origins in scientific research.

By definition, traveling involves separation from roots to some degree, so we might not expect the public to fully understand the scientific roots of scientific facts. Still, generally, when people travel, they retain their national and cultural identity; when books travel, they retain their copyright and publisher; when objects travel, information about their origins is embedded in them. So when scientific facts travel, we might expect their recipients to understand – at least in an approximate way – their origins in scientific research. This has turned out not to be the case for global warming, perhaps in part because of the events described here.

---

[88] Leiserowitz, http://www.climate.yale.edu/People/Anthony-Leiserowitz/; see esp. http://environment.yale.edu/news/5305/american-opinions-on-global-warming/

[89] On the practice of "equal and opposite," see Boykoff and Boykoff 2004.

[90] On this consensus, see Oreskes, 2004, and Doran and Kendall Zimmerman, 2009.

[91] Leiserowitz 2007.

Moreover, studies suggest that the American people remained divided over how serious the problem of global warming is, even as the scientific community increasingly argued that it was already having serious negative consequences around the globe.[92] The scientific "bad news" did not triumph over the alternative possibilities that global warming was either good or at least not entirely bad, as Western Fuels had argued. This suggests that resistance campaigns were effective in creating a lasting impression of scientific debate and discord, and sowing doubt among the American people about the consequences of anthropogenic warming. Most American citizens believe – for whatever reasons – that global warming is real, but many doubt its human causes and still believe that scientists have not settled the matter. In essence, the facts traveled, but their lineage did not. If we think of scientific facts as the progeny of the scientists who produce them, we can say that when the facts finally traveled, they left their parents behind.

The story told here is consistent with the arguments of barrister Sir Neil MacCormick, who has argued that good stories push out bad ones – even if the bad ones are true.[93] This is an observation, Sir Neil notes, that lawyers and prosecutors ignore at their peril. Our story suggests that scientists ignore it at their peril, too.[94]

Acknowledgments

I am indebted to all the members of the "Facts" project, who commented on earlier versions of this work. I wish to thank particularly Mary Morgan for her consistent enthusiasm for this project and Erik Conway, Harro Maas, Richard Somerville, Ken Davis, and an anonymous reader for helpful comments on the manuscript. Thanks also to my remarkable assistants, Charlotte Goor and Afsoon Foorohar, to Anthony Socci for bringing the Greening Earth Society documents to my attention, and to Christopher Patti for sage legal advice.

A Note on Sources

Many of the materials on the Greening Earth Society were found in the archives of the American Meteorological Society (AMS) headquarters in Washington, D.C. These archives consist simply of a set of filing cabinets, thus explaining the absence of the normally expected box-and-folder

---

[92] IPCC "Impacts, Adaptation, and Vulnerability".

[93] The role of good news stories in helping false facts to travel is paralleled in Haycock (this volume), and contrasted by a bad-news story that gets well attested facts to travel in Ramsden (also this volume).

[94] MacCormick 2007.

numbers for citations to these materials. Scholars wishing to consult these materials should contact the AMS.

Abbreviations

The following abbreviations are used

ICE   Informed Citizens for the Environment
IPCC   Intergovernmental Panel on Climate Change

Bibliography

ABC News. "Exclusive: Cheney on Global Warming – Vice President's Views at Odds with Majority of Climate Scientists." *ABC News - Science and Technology*. Feb. 23, 2007. Jan. 20, 2009 http://abcnews.go.com/Technology/Story?id=2898539&page=1.

Allen, L.H., et al. "Response of vegetation to rising carbon dioxide: photosynthesis, biomass, and seed yield of soybean." *Global Biogeochemical Cycles*. 1:1 (Mar. 1987):1–14.

Amos, Deborah. *Transcript: Hot Politics*. Apr. 24, 2007. *Frontline*. Accessed Jan. 20, 2009 http://www.pbs.org/wgbh/pages/frontline/hotpolitics/etc/script.html.

Arizona Daily Sun "Global Warming ad campaign termed irresponsible." *Arizona Daily Sun* 45:281 (May 24, 1991).

Baur, David C. L. "Hot Debate: Bowling Green now battleground in heated global warming dispute". *Bowling Green Daily News* (Archive, American Meteorological Society).

Bazzaz, F. A. "The response of natural ecosystems to the rising global CO_2 levels." *Annual Review of Ecology and Systematics* 21 (1990):167–96.

Boykoff, Maxwell T and Jules M.Boykoff. "Balance as Bias: Global Warming and the U.S. Prestige Press." *Global Environmental Change* 14 (2004):125–36.

Brandt, Allan M. *The Cigarette Century: The Rise, Fall, and Deadly Persistence of the Product that Defined America*. New York: Basic Books, 2007.

Brooke, James. "The Earth Summit; President, in Rio, Defends His Stand on Environment." *The New York Times*. Jun. 13, 1992. Accessed Jan. 20, 2009 http://www.nytimes.com/1992/06/13/world/the-earth-summit-president-in-rio-defends-his-stand-on-environment.html?scp=4&sq=Bush+Earth+Summit&st=ny.

Campbell, W. J., et al. "Effects of CO_2 concentration of Rubisco activity, amount, and photosynthesis in soybean leaves." *Plant Physiology* 88 (1988):1310–6

Cess, Robert D. and Gerald L. Potter. "A Commentary on the Recent CO_2–Climate Controversy." *Climatic Change* 6:4 (Dec. 1984):365–76.

Chamberlain, J., et al. "Comments on S. B. Idso's Paper: 'Climatological Significance of a Doubling of Earth's Atmospheric Carbon Dioxide.'" JASON Technical Report JSN-80–01, SRI International, August 1980.

Charney, J., et al. "Carbon Dioxide and Climate: A Scientific Assessment." *Report of an Ad Hoc Study Group on Carbon Dioxide and Climate*. Woods Hole, Massachusetts, July 23–27, 1979. To the Climate Research Board, Assembly of Mathematical and Physical Sciences, National Research Council. Washington, D.C.: National Academy of Sciences, 1979.

Clough J. M., et al. "Effects of high atmospheric CO_2 and sink size on rates of photosynthesis of a soybean cultivar." *Plant Physiology* 67 (1981):1007–10.

Comments and Prepared Testimony of Western Fuels Association, Inc. Doctor Robert Balling in the Matter of: Rules Concerning Resource Planning. Docket No. 91R-642EG. Colorado Public Utilities Commission. Feb. 14, 1992. Copy in American Meteorological Society Archives, Washington, D.C.

Comments and Prepared Testimony of Western Fuels Association, Inc. Doctor Patrick Michaels in the Matter of: Rules Concerning Resource Planning. Docket No. 91R-642EG. Colorado Public Utilities Commission. Feb. 14, 1992. Copy in American Meteorological Society Archives, Washington, D.C.

Comments and Prepared Testimony of Western Fuels Association, Inc. General Manager and Chief Executive Fredrick D. Palmer in the Matter of: Rules Concerning Resource Planning. Docket No. 91R-642EG. Colorado Public Utilities Commission. Feb. 14, 1992. Copy in American Meteorological Society Archives, Washington, D.C.

Comments and Prepared Testimony of Western Fuels Association, Inc. Rebuttal Testimony of Dr. Patrick J. Michaels before the Minnesota Public Utilities Commission, In the matter of the Quantification of Environmental Costs Pursuant to Laws of Minn. 1993, Chapter 356, Section 3. Docket No. E-999/CI-93-583. Sponsored by Western Fuels Association, Inc., Lignite Energy Council, Center for energy and Economic Development, State of North Dakota. Mar. 15, 1995. Copy in American Meteorological Society Archives, Washington, D.C.

Cure J. D. and Basil Acock. "Crop responses to carbon dioxide doubling: A literature survey." *Agricultural and Forest Meteorology* 38 (1986):127–45.

Doran, Peter D. and Maggie Kendall Zimmerman. *EOS* 90:3 (Jan. 20, 2009):21–2.

Encyclopædia Britannica. "Green Revolution." 2009. Encyclopædia Britannica Online. Accessed Feb. 3, 2009 http://www.britannica.com/EBchecked/topic/245058/Green-Revolution.

Fleming, James R. *Historical Perspectives on Climate Change.* New York: Oxford University Press, 1998.

Fritsch, Jane. "Sometimes, Lobbyists Strive to Keep Public in the Dark." *The New York Times on the Web.* Mar. 19, 1996. Accessed Jan. 20, 2009 http://www.nytimes.com/1996/03/19/us/sometimes-lobbyists-strive-to-keep-public-in-the-dark.html?scp=1&sq=Sometimes+lobbyists+strive+to+keep&st=nyt.

Gelbspan, Ross. *The Heat Is On: The High Stakes Battle over Earth's Threatened Climate.* Reading, Mass: Addison-Wesley Publishing Company, 1997.

   Boiling Point: How Politicians, Big Oil, and Coal, Journalists and Activists Are Fueling the Climate Crisis, and What We Can Do to Avert Disaster. New York: Basic Books, 2004.

Greening Earth Society. Membership materials and press releases. Copy in American Meteorological Society Archives, Washington, D.C.

Greening Earth Society. *The Greening of Planet Earth: The Effects of Carbon Dioxide on the Biosphere.* Transcript of Video recording. The Institute for Biospheric Research, Inc., Tempe Arizona. Copy in American Meteorological Society Archives, Washington, D.C.

Greening Earth Society. *The Greening of Planet Earth: The Effects of Carbon Dioxide in the Biosphere.* VHS video. Oregon State University Library: Multimedia Collection. http://osulibrary.oregonstate.edu/video/met4.html.

Greening Earth Society Press Release. "About Frederick D. Palmer." Contact: Marianne
    Brewster. Copy in AMS archives.
"New Group Promotes 'Positive Environmental Thinking.' " Contact: Marianne
    Brewster, Executive Director. Apr. 21, 1998. Copy in American Meteorological
    Society Archives, Washington, D.C.
"ASU Climate Data Task Force Report Finds no Warming in New Hampshire Ground
    Temperature Data." Contact: Ned Leonard. Jan. 20, 1999. Copy in American
    Meteorological Society Archives, Washington, D.C.
"Lenin's Sea-Ice Stations Show No Global Warming." Contact: Ned Leonard. Feb. 17,
    1999. Copy in American Meteorological Society Archives, Washington, D.C.
"1437 Years of Bristlecone Pine Data: 20th Century Warming Normal, CO_2 Enhanced
    Growth." Contact: Ned Leonard. Mar. 25, 1999. Copy in American Meteorological
    Society Archives, Washington, D.C.
"ASU Climate Data Task Force Finds 90 Years of 'Cooling' At MT. Wilson
    Observatory Outside LA." Contact: Ned Leonard. May 18, 1999. Copy in American
    Meteorological Society Archives, Washington, D.C.
"Kitt Peak, Arizona, Records Reveal Short-Term Warming, Longer-Term Cooling."
    Contact: Ned Leonard. June 28, 1999. Copy in American Meteorological Society
    Archives, Washington, D.C.
"Finnish Summer Saw No Warming or Cooling Over 595 Years." Contact: Ned Leonard.
    Oct. 5, 1999. Copy in American Meteorological Society Archives, Washington, D.C.
"Balling Finds Cooling at China's Three Gorges Damsite." Contact: Ned Leonard. Oct.
    25, 1999. Copy in American Meteorological Society Archives, Washington, D.C.
"No Warming in Panama." Contact: Ned Leonard. Dec. 13, 1999. Copy in American
    Meteorological Society Archives, Washington, D.C.
Greenwire "Inside Track: Sowing the seeds of doubt in the greenhouse." *Greenwire*, June
    19, 1991.
Hansen, James. *Global Warming 20 Years Later: Tipping Points Near*. 2008. Accessed Jan.
    20, 2009 http://www.columbia.edu/~jeh1/2008/TwentyYearsLater_20080623.pdf.
Hendrey, G. R., et al. "A free-air enrichment system for exposing tall forest vegetation to
    elevated atmospheric CO_2." *Global Change Biology* 5 (1999):293–309.
Houghton, J.T., et al. Eds. *Climate Change 1995: IPCC Second Assessment Report:
    Contribution of Working Group I to the Second Assessment of the Intergovernmental
    Panel on Climate Change, Summary for Policy Makers*. Cambridge University
    Press: UK, 1996.
Idso, Craig D. "The Greening of Planet Earth: Its Progression from Hypothesis to
    Theory." *Climatological Publications Scientific Paper #25* Office of Climatology.
    Tempe Arizona: Arizona State University, Jan. 1997.
Idso, Sherwood B. "The Climatological Significance of a Doubling of Earth's Atmos-
    pheric Carbon Dioxide Concentration." *Science*. 207:4438 (May 28, 1980): 1462–3.
    Carbon Dioxide: Friend or Foe? Tempe, AZ: IBR Press, 1982.
"CO_2 and Climatic Change." *BioScience* 38:7, Conservation Biology. (Jul.–Aug.
    1988):442.
Carbon Dioxide and Global Climate: Earth in Transition. Tempe, AZ: IBR Press,
    1989.
"CO_2 and the Biosphere: The Incredible Legacy of the Industrial Revolution."
    Department of Soil, Water and Climate, University of Minnesota, St Paul, MN.
    1995.

Idso, Sherwood B., et al. "Effects of free-air CO_2 enrichment on the light response curve of net photosynthesis on cotton leaves." *Agricultural and Forest Meteorology* 70 (Aug. 1994):183–8.

Idso, Sherwood B., R. C. Balling, and R. S.Cerveny. "Carbon Dioxide and Hurricanes: Implications of Northern Hemispheric Warming for Atlantic/Caribbean Storms." *Meteorology and Atmospheric Physics* 42:3–4 (Dec. 1990):259–63.

Informed Citizens for the Environment (ICE): Mission Statement. 1991. Copy in American Meteorological Society Archives, Washington, D.C.

    Mission Statement – Strategies: Reposition Global Warming as Theory (Not Fact). Copy in American Meteorological Society Archives, Washington, D.C.

    Test Market Proposal. Copy in American Meteorological Society Archives, Washington, D.C.

Intergovernmental Panel on Climate Change (IPCC). "Second Assessment Synthesis of Scientific-Technical Information relevant to interpreting Article 2 of the UN Framework Convention on Climate Change." 1995. http://www.ipcc.ch/pdf/climate-changes-1995/2nd-assessment-synthesis.pdf.

    Fourth Assessment Report. Working Group II Report: "Impacts, Adaptation and Vulnerability." http://www.ipcc.ch/ipccreports/ar4-wg2.htm.

Jacoby, Gordon C. and Rosanne D. D'Arrigo. "Tree rings, carbon dioxide, and climatic change." *Proceedings of the National Academy of Sciences* 94 (1997): 8350–53.

Johnson, H. B., et al. "Increasing CO_2 and plant-plant interactions: Effects on natural vegetation." *Vegetatio.* 104/105 (1993):157–70.

Karnosky, D. F., et al. "FACE systems for studying the impacts of greenhouse gases on forest ecosystems." In: D. F. Karnosky et al. (eds.), *The Impact of Carbon Dioxide and Other Greenhouse Gases on Forest Ecosystems*, CABI, Wallingford, U.K., 2001, pp.297–324.

Kerr, Richard A. "Climate change: Model gets it right – Without fudge factors." *Science* 276:5315 (May 16, 1997):1040–2.

Kimball, B. A. "Carbon dioxide and agricultural yield: An assemblage and analysis of 430 Prior Observations." *Agronomy Journal* 75 (1983):779–88.

Kimball B. A., et al. "Productivity and water use of wheat under free air CO_2 enrichment." *Global Change Biology* 1 (1995):429–42.

King, John S., et al. "A multi-year synthesis of soil respiration response to elevated CO_2, from four forest FACE experiments." *Global Change* 10.6 (2004):1027–42.

Lambers, Hans, F., Stuart Chapin III and Thejs L. Pons. *Plant Physiological Ecology.* New York: Springer, 2008.

Latour, Bruno. *The Pasteurization of France.* Trans. Alan Sheridan and John Law. Cambridge, Massachusetts: Harvard University Press, 1988.

Leonard, Ned. "The Greening of Planet Earth." *Range Magazine.* Fall 2000. Accessed Jan. 20, 2009 http://www.rangemagazine.com/archives/stories/fall00/greening_planet_earth.htm.

Leiserowitz, Anthony, Principal Investigator. "American Opinion on Global Warming." *Yale University/Gallop/ClearVision Institute Poll.* July 2007. Accessed Jan. 20, 2009 http://environment.research.yale.edu/documents/downloads/a-g/Americans GlobalWarmingReport.pdf.

Long S. P., et al. "Food for thought: Lower-than-expected crop yield stimulation with rising CO_2 concentrations". *Science* 312:5782 (June 30, 2006):1918–21.

    "Rising atmospheric carbon dioxide: Plants face the future." *Annual Reviews of Plant Biology* 55 (2004):591–628.

MacCormick, Sir Neil. "On narrative coherence." *Enquiry, Evidence, and Facts: An Interdisciplinary Conference of the British Academy*, 2007.

MacDonald G.F., et al. *The long term impact of atmospheric carbon dioxide on climate. JASON Technical Report JSR-78–07*. Prepared for U.S. Dept of Energy. Arlington, Virginia: SRI International, 1979.

Maria, Steven F., et al. "Organic aerosol growth mechanisms and their climate-forcing implications." *Science* 306:5703 (Dec. 10, 2004):1921–4.

Mauney, Jack R., et al. "Growth and yield of cotton in response to a free-air carbon dioxide enrichment (FACE) environment." *Agricultural and Forest Meteorology* 70 (Dec. 1994):49–67.

McCright, Aaron M. and Riley E. Dunlap. "Anti-reflexivity: The American Conservative Movement's Success in Undermining Climate Science and Policy." *Theory Culture Society* 27:2–3 (2010):100–133.

Michaels, Patrick J. Draft Letter, May 5, 1991. Greening Earth Society, ICE Campaign. Copy in American Meteorological Society Archives, Washington, D.C.

   State of the Climate Report: Essays on Global Climate Change. Published by New Hope Environmental Services, Inc. Funded by Western Fuels Association. 1998. Copy in American Meteorological Society Archives, Washington, D.C.

Ming, Yi., et al. "Health and climate policy impacts on sulphur emission control." *Reviews of Geophysics* 43, RG4001. 2005.

Mooney, Chris. *The Republican War on Science*. New York: Basic Books, 2005.

Mooney, Chris and Sheril Kirshenbaum. *Unscientific America: How Scientific Illiteracy Threatens Our Future*. New York: Basic Books, 2009.

National Academy of Sciences Archives. "An Evaluation of the Evidence for CO_2-Induced Climate Change," Assembly of Mathematical and Physical Sciences, Climate Research Board, Study Group on Carbon Dioxide, 1979, Film Label: CO_2 and Climate Change: Ad Hoc: General.

Natural Resources Defense Council Press Release. "NRDC Calls on President Bush to Fulfill His Father's Promises at Rio." Aug. 30, 2002. Accessed February 9, 2009. http://www.nrdc.org/media/pressreleases/020830.asp.

New Hope Environmental Services. "In Defense of Carbon Dioxide: A Comprehensive Review of Carbon Dioxide's Effects on Human Health, Welfare, and the Environment." Prepared by for the Greening Earth Society. Copy in American Meteorological Society Archives, Washington, D.C.

New York Times."The Earth Summit; Excerpts From Speech by Bush on 'Action Plan.'" *New York Times on the Web*. June 13, 1992, accessed Jan 20, 2008. http://query.nytimes.com/gst/fullpage.html?res=9E0CE6DD1F3CF930A25755C0A964958260&n=Top/News/Science/Topics/Environment.

Nisbet, Matthew C. and Teresa Myers. "The Polls—Trends: Twenty Years of Public Opinion about Global Warming." *Public Opinion Quarterly* 71:3 (Fall 2007): 444–70. http://poq.oxfordjournals.org/cgi/content/abstract/71/3/444.

O'Driscoll, Mary. "Greenhouse ads target 'low income' women, 'less educated' men." *The Energy Daily* 19:120 (June 24, 1991a):1–2.

   "ICE gets cool reception at meeting on global warming." *The Energy Daily* 19 (July 2, 1991b):126.

Oreskes, Naomi. "Beyond the ivory tower: The scientific consensus on climate change." *Science* 306:5702 (Dec. 3, 2004):1686.

Testimony before the Committee on Environmental and Public Works. United States Senate. Dec. 6, 2006. http://www.stanford.edu/dept/cisst/ORESKES.Senate%20 EPW.FINAL.pdf.

"The scientific consensus on climate change: How do we know we're not wrong?" In: Joseph F. C. DiMento and Pamela Doughman (eds.), *Climate Change: What It Means for Us, Our Children, and Our Grandchildren*, MIT Press, 2007, pp. 65–99.

Oreskes, Naomi, and Erik Conway. "Challenging knowledge: How climate science became a victim of the Cold War." In: Robert N. Proctor and Londa Schiebinger (eds.), *Agnotology: The Making and Unmaking of Ignorance*, Stanford: Stanford University Press, 2008. pp. 55–89.

Merchants of Doubt: How a Handful of Scientists Obscured the Truth on Issues from Tobacco Smoke to Global Warming. London: Bloomsbury Press, 2010.

Oreskes, Naomi, Erik M. Conway, and Matthew Shindell. "From Chicken Little to Dr. Pangloss: William Nierenberg, global warming, and the social deconstruction of scientific knowledge." *Historical Studies in the Natural Sciences* 38:1 (2008):109–52.

Palmer, Fredrick D. "Fossil Fuels or the Rio Treaty – Competing Visions for the Future." Speech by Frederick D. Palmer, General Manager and Chief Executive Office, Western Fuels Association, Inc. Coaltrans 96. Madrid Spain. 1996.

"We Are One: Kyoto and Our Collective Economic Future." Speech by Frederick D. Palmer, General Manager and Chief Executive Office, Western Fuels Association, Inc. President - Board of Directors, Greening Earth Society. 1998 Australian Coal Conference and Trade Exhibition. Brisbane, Queensland. May 18, 1998.

General Manager and Chief Executive Officer Western Fuels Association, Inc. Testimony before the Subcommittee on National Economic Growth. Natural Resource and Regulatory Affairs. U.S. House of Representatives. July 15, 1999. July 20, 1999a. Copy in American Meteorological Society Archives, Washington, D.C.

"Re: CO₂ as 'pollution'" e-mail to Bryan Hannegan. Nov. 22, 1999b. Copy in American Meteorological Society Archives, Washington, D.C.

Pollard, David and Starley L. Thompson. "The climatic effect of doubled stomatal resistance," Chapter 13 in Wendy Howe and Ann Henderson-Sellers (eds.), 1997, *Assessing Climate Change: Results from the Model Evaluation Consortium for Climate Assessment*, http://www.springerlink.com/content/k658r3n533600228/.

Proctor, Robert. *Cancer Wars: How Politics Shapes What We Know and Don't Know About Cancer*. New York: Basic Books, 1995.

Ramanathan V. and G. Carmichael. "Global and regional climate changes due to black carbon." *Nature Geoscience* 1:4(Apr. 2008): 221–7. Accessed Jan. 20, 2009. http:// www.nature.com/ngeo/journal/v1/n4/pdf/ngeo156.pdf.

Rampton, Sheldon and John Stauber. *Trust Us, We're Experts! How Industry Manipulates Science and Gambles with Your Future*. New York: Tarcher/Putnam, 2001.

Rudwick, Martin, J. S. *Bursting the Limits of Time: The Reconstruction of Geohistory in the Age of Revolution*. Chicago: University of Chicago Press, 2005.

Worlds before Adam: The Reconstruction of Geohistory in the Age of Reform. Chicago: University of Chicago Press, 2008.

Rupke, Nicolaas. *The Great Chain of History: William Buckland and the English School of Geology, 1814–1849*. Oxford: Clarendon Press, 1983.

Sage R.F., et al. "Acclimation of photosynthesis to elevated CO₂ in five C3 species." *Plant Physiology* 89(1989):590–6.

Schneider, Stephen H. *Science as a Contact Sport: Inside the Battle to Save Earth's Climate.* National Geographic, 2009.

Schneider, Stephen H., et al. "Carbon Dioxide and Climate." *Science* 210:4465 (Oct. 3, 1980):6–8.

Shapin, Steven. *A Social History of Truth: Civility and Science in Seventeenth-Century England.* Chicago: University of Chicago Press, 1994.

Shugart, Herman H. "Remote Sensing Detection of High Elevation Vegetation Change." *Advances in Global Change Research* 23 (2005): 457–65.

Socci, Tony, Associate Director, US Global Climate Research Program. Re: The Heat is Online – The Industry's "ICE" Campaign. E-mail to self. July 7, 2000. Copy in American Meteorological Society Archives, Washington, D.C.

Stitt M. "Rising CO_2 levels and their potential significance for carbon flow in photosynthetic cells." *Plant, Cell and Environment* 14 (1991):741–62.

Thomson, A. C. "Timeline: The Science and Politics of Global Warming." Apr. 24, 2007. *Frontline.* Accessed Jan. 20, 2009 http://www.pbs.org/wgbh/pages/frontline/hotpolitics/etc/cron.html.

United Nation Conference on Environment and Development (1992). May. 23, 1997. United Nations. Dept. of Public Information. Accessed Jan. 20, 2009 http://www.un.org/geninfo/bp/enviro.html.

United Nations Framework Convention on Climate Change (UNFCCC). *Facing and Surveying the Problem.* Accessed Jan. 20, 2009 http://unfccc.int/essential_background/feeling_the_heat/items/2914.php.

United States National Research Council. Panel on Weather Modification. "Weather and Climate Modification Problems and Prospects; Final Report of the Panel on Weather and Climate Modification." Washington: National Academy of Sciences-National Research Council, 1966.

United States National Research Council. Committee on Health, Environmental, and Other External Costs and Benefits of Energy Production and Consumption, 2010. *Hidden Costs of Energy: Unpriced Costs of Energy Production and Use* (Washington DC, National Research Council). http://www.nap.edu/catalog.php?record_id=12794.

United States President's Science Advisory Committee. Environmental Pollution Panel "Restoring the quality of our environment: Report of the Environmental Pollution Panel." Washington: The White House, 1965.

Weart, Spencer R. *The Discovery Of Global Warming.* Cambridge: Harvard University Press, 2003.

Wald, Matthew. "Pro-Coal Ad Campaign Disputes Warming Idea." *The New York Times* on July 8, 1991.

Yelle S., et al. "Acclimation of two tomato species to high atmospheric CO_2." *Plant Physiology* 90 (1989):1473–7.

Zak, Donald R., et al. "Soil nitrogen cycling under elevated CO_2: A synthesis of forest FACE experiments." *Ecological Applications* 13:6 (2003):1508–14.

SIX

REAL PROBLEMS WITH FICTIONAL CASES

JON ADAMS

To convey science to a wider non-specialist audience, it is usually necessary to "translate" the content of specialist scientific publications into so-called "popularisations" (e.g., see Royal Society 1986). Popularisations aim to make scientific facts and theories available to audiences who do not have scientific training (see Shinn and Whitley 1985; Burham 1987). Although they are not exclusively written for nor only read by non-specialists,[1] the popularisation is characterised by a more broad-based accessibility than would typically be found within a specialist technical publication. The nominal target audience is not assumed to possess specialist knowledge.

Therefore, popularisations are vehicles by which scientific facts travel to a wider audience. These audiences are not an homogenous mass, but comprise a variety of "publics" with different needs and different levels of scientific training. Scholarship within science studies eschews the notion of a unitary "science" and a unitary "public" in favour of more nuanced, multiple conceptions of both "science" and "public." As Silverstone states: "There is no such thing as *the* communication of science; ... There is no such thing as *the* public" (Silverstone 1991, p. 106). This view has become conventional among scholars working within science studies (e.g., Hilgartner 1990; Locke 1999; Yearly 2000; Mellor 2003; Einsiedel 2007).

One consequence of this pluralistic conception of multiple "publics" is that because there is no single "public," there is no single level of "popularisation." Stephen Hilgartner has pointed out that although the "culturally-dominant view" of science popularisation seeks to contrast "pure, genuine scientific knowledge" with popularisation as a "'distortion' or 'degradation' of the original truths" (Hilgartner 1990, p. 519), it is not

---

[1] For example, in conversation with the author, Leonard Smith had said of his contribution to the Oxford's *Very Short Introduction* series that he fully expected the book to be read by fellow mathematical scientists, and hoped that it would clarify aspects of the topic for them (see Smith 2007).

practically possible to demarcate a clear and unambiguous point between "professional" and "popular" discourse. In its place, Hilgartner suggests a "spectrum ranging from laboratory 'shop talk,' to technical seminars, to scientific papers in journals, to literature reviews, grant proposals, text-books, policy documents, and mass media accounts" (Hilgartner 1990, p. 524). Different "levels" or "strata" of this explanatory spectrum will typically employ different approaches to more directly address the audience at which it is aimed (for analyses of this process of dissemination, see Fahnestock 1986; Lewenstein 1995). Consequently, there are multiple ways in which popularisations of science aim to make scientific facts and theories interesting and comprehensible to these multiple audiences. Techniques used by authors of popularisations might include simplification (e.g., Krukonis and Barr 2008), illustrations (e.g., Hawking 1996), analogies and metaphors (e.g., Dawkins 1988, Ridley 1994; for analogies, see Greene 1999; and see Leane 2007), even the use of humour (e.g., Pratchett 1999, Bryson 2004). The particular focus of this paper is on the use of *fictional* devices and techniques as a means of more successfully communicating facts through popularisations.

Following Hilgartner's identification of the "spectrum" or gradient between professional and popular discourse in science writing, this paper will argue that an analogous gradient exists with respect to the boundary between popularisation and fiction. In order to demonstrate this gradient, this paper will examine how authors of science popularisations use fictional techniques and devices in order to communicate to a non-specialist audience facts that originate and usually circulate only within specialist scientific discourses.

The structure of the paper aims to draw attention to the incremental character of this gradient by presenting a series of brief cases illustrating the use of fictional techniques in science popularisation. At the start of the paper, the cases used are factual popularisations of science. In successive examples, the use of fictional techniques and devices becomes incrementally greater or less easily extricable, with the ratio of fact to fiction gradually "tipping" in favour of fiction. By the end of the paper, the examples given are nominally and explicitly fiction, yet retain much of the educational, expository and persuasive intent that characterises a work of science popularisation. Through presenting these cases as incrementally more reliant on fictional techniques and devices, this paper shows how this gradient makes it practically impossible to demarcate a clear point or figurative "line" between popularisation and fiction. Fiction proves to be an effective but problematic means of making facts travel.

This paper will conclude that a categorical confusion may occur, with readers unsure of the genre of writing they are reading: factual or fictional. This confusion may result in an inability on the reader's behalf to successfully demarcate factual information from fiction, and therefore, an inability to separate real facts from fictional claims, leading to the introduction of "false facts" (see Haycock, this volume). The problems presented by such generic ambiguity are exacerbated by the special vulnerability of the popularisations' "target audience": Non-specialists are less qualified to distinguish real facts from false facts.[2]

The central claim is that texts that employ fictional devices enable the efficient conveyance of information, but at the expense of epistemic authority. Using fictional devices can result in confusion on the part of the reader, who may consequently direct excessive credulity toward claims with a weak epistemic status ("false facts" are believed to be true), or excessive scepticism toward claims possessing strong epistemic status (true facts are quarantined as false).

1. What is a Fictional Device?

By "fictional device," I refer to rhetorical techniques more prevalent within fiction written primarily for entertainment than within factual expository writing. It is important to note that the distinction between fictional and factual styles of writing is conventional and normative. Although some techniques are more commonly associated with fictional writing and some with factual writing, no techniques are the exclusive preserve of either genre. Recorded speech, for example, is more common in fiction, but it also occurs in many factual cases: for example, ethnographic field reports and police statements. Meanwhile (with the tacit consent of their readers), fiction writers sometimes aim to present their work as if it were factual, and may choose to avail themselves of a broad spectrum of rhetorical styles and devices in order to present the semblance of reality – newspaper reports, legal documents, academic papers (e.g., Dos Passos 1996; McEwan 1999). A given piece of writing is factual or fictional with respect to an ontological relation that may or may not hold with the world, not with respect to a rhetorical style: "Fictionality" is an ontological relation, not a rhetorical category, and thus "fictional device" is a normative term.

---

[2] This is because non-specialists have a more limited knowledge base against which to cross-reference claims, and because non-specialists inevitably possess a restricted understanding of the normative standards of plausibility that familiarity with a particular area of knowledge brings.

2. Using Narrative as an Organising Structure

Narrative is not a fictional technique per se. It is an organising device: the ordering of events into a structure that readers can "follow" by presenting a series of events and imputing a motive (in the case of agency) or mechanism by which two or more chronologically sequential events stand in causal relation (causal ligatures need not be explicit; narrative is the *sequential ordering* of events, and thus causality may remain implicit). Thus, narrative links events causally in a chronological structure.[3] Narrative is used widely in factual writing. For example, historians tell "stories" in factual writing to establish the coherence of a particular causal sequence of events. In so doing, they are seeking to persuade others of the plausibility of their account over rival accounts.[4]

There are further advantages. Narrative is a mnemonic tool of extraordinary power, enabling the comprehension and retention of more items than when the same items are presented in a random string. A well-known case is George A. Miller's "magical number seven" (Miller 1956). Miller's claim is that (on average) humans can memorise only seven (plus or minus two) "chunks" of information at one time. As Miller formally recognised, narrative is one method of extending this capacity. Presented as an abstract number string, we have a (mean) limit of seven; but framed within a narrative, the number of chunks we can recall and sequence is significantly larger. The apparent ubiquity of this cognitive phenomenon makes narrative attractive to writers aiming to convey more than seven chunks of information.

Narrative structures vary in the control they exert upon the material. The first example this paper will introduce is Richard Dawkins's *The Ancestor's Tale* (2004). Richard Dawkins is a zoologist and, since the publication of *The Selfish Gene* in 1976, a prominent science populariser. *The Ancestor's Tale*, published in 2004, uses narrative as an organising device to convey thematic unity on its material. The title of *The Ancestor's Tale* is a reference to Chaucer's *Canterbury Tales*. As Chaucer's pilgrims exchanged stories on their way to Canterbury, so Dawkins's "pilgrims" are increasingly distant

---

[3] "Conventional narrative" is a normative category acknowledging the pattern followed by the majority of cases, whereby the structure of the text reflects this antecedent-descendent structure by placing (within the text) events that are chronologically prior *before* events that are chronologically sequent. An author's decision to absent such causal links, or the reversal of the chronological sequence, is definitive of an "unconventional" narrative (e.g., Faulkner 1991; Amis 1991).

[4] Narrative transcends both "story" and "plot." "Story" is the chronological sequence; "plot" is the textual ordering and presentation of that story chosen by the author. Historians compete to have their story accepted as being the case.

ancestors of modern humans, each taking their turn to tell their "tale" as the book (and the ancestors' pilgrimage) wends its way back through time towards its conclusion in the very origins of organic life. Despite the elaborate-seeming structure, the "tales" are really just linked articles on natural history and do not require the narrative frame in order to be understood. Instead, the narrative simply acts to establish by figurative means the literal continuity of life (as Chaucer's stories are passed along a journey through time and space, so genes are the information passed along through evolutionary time – the stepwise structure emphasising the incremental changes). Thus, although the framework and title of the book make explicit reference to fiction, the (factual, expository) content is kept quite apart.

Of course, given this frame, the opportunity presents itself to tell each of the ancestor's "stories" in first person – a conceit Dawkins reports deciding against (2004, p. 11). Yet the device of inhabiting the subject matter's perspective (even though the subject matter is insensible) is something Dawkins had already advocated with his "gene's eye" view of evolution, and imagining what they might say if only animals could speak is something other zoologists have less successfully resisted.

3. Inhabiting the Perspective of the Subject Matter

One case – well known among entomologists and myrmecologists (insect and ant specialists, respectively) – is William Morton Wheeler's *The Termitodoxa, or Biology and Society*, published in early 1920 in the generalist but respectable *Scientific Monthly* (which was absorbed into *Science* in 1957)[5]. Here, Wheeler purports to be in the possession of a letter he has received from "Wee-Wee," the king of a West African termite colony. After briefly explaining its provenance, Wheeler "quotes" the letter (which describes the organisation of termite society) in its entirety. In common with fiction writers who have employed the same device (a classic example is Mary Shelley's *Frankenstein*, which uses an epistolary frame narrative to embed the story and distance the author's identity from that of the narrator), Wheeler finds that the narrative frame works in his favour in a number of ways.[6] The capricious setup means

---

[5] This section has been stimulated by Abigail Lustig's research on writings about insect communities, reported in part at our Workshop on "Facts at the Frontier: Crossing Boundaries between Natural and Social, Animal and Human," 16th–17th April 2007.

[6] The frame narrative allows Wheeler to endorse eugenics from the termites' point of view: "by promptly and deftly eliminating all abnormalities, we have been able to secure a mentally and physically perfect race" (1920, p. 119). Though the analogies to human society are intended (the point is made in direct comparison to "the great problems of reproduction"

that although he is open to criticism for ventriloquising a termite in the first instance, it would be an especially joyless audience that would seek to pick out factual errors in his setup.[7] Caprice apart, he nonetheless succeeds in communicating a good deal of factual information about the organisation and development of termite colonies – dietary habits, caste system, evolutionary history:

> According to tradition our ancestors were descended in early Cretaceous times from certain kind-hearted old cockroaches that lived in logs and fed on rotten wood and mud. Their progeny, the aboriginal termites, ... chanced to pick up a miscellaneous assortment of Protozoa and Bacteria and adopted them as an intestinal flora and fauna, because they were able to render the rotten wood and mud more easily digestible. (1920, pp. 114–5)

In personifying his subject matter, Wheeler has generated a rhetorical space in which he can have his facts heard while reducing the chances of serious challenges: If anyone does choose to contest a factual claim, Wheeler retains the defence that this was simply an entertainment, the work of Wee-Wee the termite king. Levity here is a valve that seems to allow the delivery of facts whilst acting to effectively limit their contestation. Of course, the facts are perhaps not taken so seriously as were they to appear in a more sober scientific publication, but Wheeler has nonetheless succeeded in exposing an audience to zoological data about termites. Because the scenario (a letter written by termites) is absurd, it is easy to distinguish the real facts from the fictional frame, and thus the facts about termite life can travel well. Using light-hearted packaging to conceal a serious message is a technique that popularisers have often found effective (see, for example, the way in which Steven Pinker often introduces his serious arguments with jokes, such as the "Mallifert Twins" case in Pinker 2002). David Kirby argues that Hollywood motion pictures often do similarly serious work (e.g., Kirby 2008a, 2008b).

E. O. Wilson, who inherited Wheeler's position as the most prominent ant specialist of his time, would later write his own first-person version of termite society:

> Since our ancestors ... learned to write with pheromone script, termitistic scholarship has refined ethical philosophy. It is now possible to express the

---

with human societies), Wheeler retains the defence that these are not *his* wishes but those of the termites. For the stigma of eugenics, see Ramsden 2006.

[7] It is worth noting that the piece was originally read at a meeting of the American Society of Naturalists during Christmas week – so we might also assume some added goodwill toward Wheeler in respect of this. Wheeler even inserts a disclaimer at the end of the letter where he claims to have noticed "a great deal of inaccuracies and exaggerations" (1920, p. 124).

deontological imperatives of moral behavior with precision. These imperatives are mostly self-evident and universal. They are the very essence of termity. They include ... the centrality of colony life amidst a richness of war and trade among the colonies ...; the evil of personal reproduction by worker castes ...; rejection of the evil of personal rights; ...; the joy of cannibalism and surrender of the body for consumption when sick or injured... (Wilson as discussed by McEwan 2005, pp. 11–2)

Wilson cleaves to a slightly different goal than Wheeler: Rather than simply describe termite society, he wants to dramatically demonstrate the great differences in social organisation between termites and humans. The purpose of his vignette is to force in the reader a recognition that the organisation of human society is not inevitable but contingent. Wilson is not seriously suggesting that termites have mental experiences even loosely comparable to those in his fancy; nor is his termite monologue an entirely whimsical enterprise. Social and moral norms, Wilson is saying, are *species-specific*; and had our evolutionary heritage favoured a slightly different route, our "self-evident" rights and duties may have been drastically otherwise. The right and the good are not transcendent qualities, but emerge as compromises between the organism and its environment. And thus the central premise of sociobiology (and latterly evolutionary psychology): culture and psychology, no less than physiology, are evolved. By playing along with Wilson's capricious scenario, we have (unwittingly) absorbed one of the fundamental maxims of his thought system.

Adopting the perspective of the subject matter doesn't always require such bold imaginative leaps – especially when the subject matter is itself already conscious and sentient. Thus, the *scientific biography* presents the populariser with an opportunity to employ characterisation (and thus supply the reader with characters) in a far less contentious form – though even this is not without its problems.

4. Fictional Devices in Scientific Biography

Biography does not immediately suggest itself as a fictional genre, but categorical lines between (auto-) biography, memoir and the novel are difficult to establish, and tracing their history sees convergence upon a common origin. Ian Watt's 1987 account of *The Rise of the Novel*, well known in literature departments, conceives of the novel as an outgrowth of biography. The novel, as Watt has it, reflects (and perhaps catalysed) a political shift in the importance we attach to individuals (against deities, collectives, ideals).

Early novels were almost all fictionalised memoirs,[8] many still are, and their authors made (and continue to make) efforts to convince us of the veracity of their fiction. Stylistically, the differences between novels and biographies are slim. Biographers of all types avail themselves of fictional devices, and the scientific biography is no exception. This section looks first at examples from the work of science writer and journalist James Gleick (Gleick 1987 & 1992), whose style is heavily dependent on characterisation. By foregrounding the personalities of his subjects, Gleick sets the stage for the exposition of his factual material by developing characters in much the same way as novelists develop characters.

The aim of Gleick's 1987 book, *Chaos: Making a New Science*, is to introduce readers to the "new science" of chaos theory. Gleick might well have simply begun his book: "Chaos is the name for a branch of mathematics. Much of this mathematics was originally formulated by a man called Mitchell Feigenbaum...." Instead, our introduction to Feigenbaum is oblique, novelistic: a description of a man living twenty-six-hour days, seen walking in his New Mexico campus at night. In *Chaos*, Feigenbaum becomes a character whose idiosyncrasies are sufficient to hold the reader's attention long enough for Gleick to explain who this man is, why he is walking at night, what mystery he is trying to solve. By setting a problem in the reader's mind ("Why is this man living twenty-six-hour days?"), Gleick employs his narrative to raise questions, the answers to which will supply the book's substantive factual content. A narrow focus on the biographical facts of a personality is used as preparatory material for a broader focus on facts about the history of science.

Gleick employs a similar structure with his subsequent book, *Genius: Richard Feynman and Modern Physics* (1992). As the word order of the subtitle suggests, the book follows the life story of Richard Feynman and in so doing, contextualises Feynman's contributions to science. On account of Feynman's working relations with many prominent twentieth-century physicists, and his own central role in the discipline's development, Feynman's life story becomes a spindle from which to tell the story of modern physics. We read for the biography, and learn (at least the history of) the science almost incidentally. Scientific facts become embedded within the personal development of a named individual, and obstacles in the path of physics are not obstacles for the discipline, but for the character.

---

[8] *Robinson Crusoe's* full title assures us that the account is "written by himself," and Defoe would go on to fabricate his *Journal of the Plague Year*. *Crusoe* also contains much expository factual information on, for example, bread making, boat building, bookkeeping.

The connections with fiction run further. As Elizabeth Leane (2007, esp. pp. 142–50) has persuasively argued, Gleick borrows much from the hard-boiled detective genre (especially Raymond Chandler): not simply in the trivial similarities present between detectives and scientists as individuals on a quest to discover hidden knowledge, but even down to the manner in which the actors are described – Feigenbaum the lone night-owl, Feynman the radical maverick – socially exiled by their own brilliance, performing their intellectual work on the margins of society. By centralising and elaborating upon the personalities of his subjects, Gleick aims to make the experience of reading history of science an approximation of the experience of reading fiction.

5. Biography and Novelistic Form

As Gleick's work suggests, the step from a novelistic but essentially factual biography to a straight novel is a short one.[9] One need only consider the gradient on which sit Dava Sobel's *Longitude* (1995), Clare Dudman's *Wegener's Jigsaw* (2003) and Daniel Kehlmann's *Measuring the World* (2007). Categorically, *Longitude* is a narrative history of science, focussing on clockmaker John Harrison's development of a sea-going timepiece; but the book's success owes much less to the subject matter than to Sobel's skilful presentation of it in what is an essentially novelistic form.[10] One move from that is Dudman's book: a novel, replete with fictionalised dialogue and interior monologues, it nonetheless presents an historically accurate and meticulously researched account of Alfred Wegener's conception and development of the tectonic plate theory (the "jigsaw" of her title). Dudman offers the work as a fiction, but stresses that she retains a pedagogical intent.[11] Kehlmann's book goes a step further: employing real characters from the history of science, but more in order to comment on Germanic intellectual culture than to educate his readership. The account Kehlmann offers of Alexander von Humboldt and Carl Friedrich Gauss not only embellishes their life and work, but falsifies and amends known historical facts in order to create a more successful fiction – as one reviewer

---

[9] The distinctions between straight history and the historical novel are similarly fuzzy. A good discussion of the topic is Strout (1992).

[10] That the generic ambiguity of Sobel's (1995) book was viewed by many academic historians of science as transgressive is apparent from responses such as Miller (2002); discussed in Govoni (2005).

[11] Dudman confirmed this at a workshop on the "Fact/Fiction Ratio" held at the London School of Economics, 12–13 April 2006.

has it, "sacrificing 'factual correctness' for the truth of his story" (Anderson 2007).[12] Ultimately, Kehlmann's work is perhaps closer to Pynchon (1997) or Banville (2001) than it is to a scholarly scientific biography, but the continuities between *Longitude* and *Measuring the World* are such that it would be difficult to point to exactly where the tipping point into fiction occurs – how many omissions, conflations, fabrications?

The sense of co-dependence these treatments nurture between the scientist's biography and their work plays into the constructivist's emphasis on the importance of the social conditions of discovery and of the contingency of progress. Yet if there is scope to tease out from these accounts the substance of the science without the story of its discovery (and the existence of biographically sparse textbooks confirms there is), there are nonetheless areas of scientific work where the subject matter is entwined with biographical characterisations. An example is the psychiatric or neurological case study.

6. Scientist as Character

Oliver Sacks is a neurologist whose case studies (Sacks 1987, 1995) use the bizarre effects of malfunctioning, abnormal (neuro-)psychology to illustrate and explain the smooth running of the normal (neuro-)psychology we usually take for granted. Although not ideologically aligned, Sacks's work follows a tradition begun with Freud: the psychological case study as short story. As with Gleick's work, Sacks's stories are *fictionalised* in the manner of their exposition: Sacks maintains narrative tension through the withholding of information, much as a mystery or thriller writer would. Before he sits down to write (and reconstruct) one of his cases, he already knows the conclusion, but lets the details unfold – that is, he plots his narratives – in such a way that the reader is invited to share the puzzlement Sacks himself felt upon first encountering the patients he describes.

Although they stylistically resemble fiction – scene-setting, dialogue, suspense – Sacks certainly does not intend that we read his case studies *as* fiction. (Indeed, if the stories were not advertised as and believed to be true, they would lose much of their appeal: the Borgesian, Kafkaesque qualities are present in too low a concentration to merit attention. As a fiction, even the title story from Sacks's 1987 collection, *The Man Who Mistook His Wife*

---

[12] Note how "truth" here is used in an honorific rather than epistemic sense. It ought also to be noted that Kehlmann never denies that he has altered the historical facts – there is no accusation here of deliberately misleading his readers.

for a Hat, would simply be a piece of surrealism. But as a true case history, it is a testament to the way our brains enable us to make sense of sense data.) Rather, he wants to exploit the capacity of fiction to deliver information to a far wider readership than articles in professional journals would reach. A reader unlikely to peruse a neurological article on visual agnosia or deformations of the basal ganglia may yet learn much of the same factual information from Sacks's stories. As with Gleick's work, the reading experience aims to approximate the experience of fiction whilst supplying facts about science: here, facts about the neurological processes that are guarantors of normal psychology and how precarious that normality is.

There is yet a further muddling of genre. On account of his actually being the neurologist whose patients we are learning about, Sacks is also a character in his own stories. This tendency runs to full autobiography in *Uncle Tungsten*; while, more problematically in terms of generic allocation, in *Awakenings* he is represented by a fictionalised version of himself. Again, as with Sobel and Dudman's work, the degree of entwinement between the expository and the merely illustrative makes the task of extracting "pure" factual information from the fictional ore increasingly difficult.

7. Parables and Illustrations: An Heuristic Role for Fiction

Minor fictional embellishments within a biographical novel (especially one such as Dudman [2003]) may be difficult to extricate from the factual material. But (ontologically, as opposed to stylistically) fictional sections are not always so enmeshed. Fiction may also be employed illustratively as an heuristic device, a parable, to explain an abstract principle through a fictive scenario. The factual truths picked out here are not discreet empirical claims but general principles. Douglas Hofstader's *Gödel, Escher, Bach* (1979) is exemplary. The book's chapters alternate between explanatory, informative sections ranging over contrapuntal music, recursive systems, DNA and computer programming, and in between these, humorous dialogues between Achilles and the Tortoise. Hofstader's explicit model here is Lewis Carroll, less well known as a Cambridge mathematician, whose *Alice* books present problems of philosophical and mathematical logic "disguised" as fanciful encounters.

Unlike Carroll's storybooks, however, Hofstader's fictional sections are interspersed with much lengthier substantive chapters. The dialogues demonstrate and dramatically perform the logico-mathematical themes and paradoxes these substantive chapters discuss. They do so in conjunction with the text but remain "detachable": They merely illustrate and do not replace the expository sections. Instead, they function as heuristic devices to close the gap (and demonstrate isomorphisms) between mathematical logic and what might be called our "folk logic" – the more familiar logic of common sense.[13]

To some extent, because the facts here are (in the main) facts about abstract logical relations rather than specific claims about what objects exist in the world, the fictional dialogues are simply an algebraic substrate on which the principle is illustrated. Achilles and the Tortoise are merely representative placeholders. In this regard, and significantly, the "fiction" of these dialogues has much in common with the "fiction" used to educate children (that is, scenarios where a particular concrete and countable noun, for example, "apple," represents and realises an abstract property – "five apples"), and is of a part with the familiar mathematical "fiction" of algebra, which lets an arbitrary symbol represent an unspecified value. The characters of Achilles and the Tortoise are fictional insomuch as they do not (and are not supposed to) exist, but their exchanges enact logico-mathematical facts that have the same ontological status as mathematical facts elsewhere.[14] Had Hofstader used symbolic notation, his examples might have been less elaborately fictional, but it is not clear that they would have been any more *true* – such is the character of mathematical facts. Hofstader makes his dialogues amusing to at once camouflage and embody the mathematical logic he is explaining, in much the same way as "if I have five apples…" both camouflages and embodies the arithmetic operations such examples demonstrate for children.

8. Prescribing Fiction to Deliver Facts

In the personifications used by Wheeler and Wilson, in the various forms of novelistic scientific biography looked at previously, in the curious case studies of Sacks and in Hofstader's humorous dialogues, the authors aim to educate covertly: to supply a reader with factual information even as that

---

[13] The mechanism by which this contextualising works is the subject of Peter Wason's well-known experiments with the same logical problem couched first in abstract terms and then again as a narrative. Respondents baffled by the logical test were nonetheless able to solve a logically identical problem when contextualised as a story with characters. This can be read as an endorsement of the power of stories to help us organise causal and logical connections. What the so-called Wason Selection Task shows is not that our brains are illogical, simply that (most of us) are better able to employ logical reasoning when we have a narrative framework on which to work that operation. (Wason 1971.)

[14] The ontological status of logico-mathematical relations is a wider issue, independent of this paper. See, for example, Tymoczko (1998).

reader is being entertained. Characterisation relocates the subject matter to a less abstract framework (Wason 1971), while the arrangement of this information into a narrative structure increases the reader's retention capacity (Miller 1956). That this form of learning seems possible is clearly of considerable value to the populariser, but it also poses a considerable threat. The very mechanisms that permit the uptake of factual material through the reading experience of narratives allow the fiction writer to dispense information with no less efficiency than the populariser. Just as expository writing borrows from fiction, so fiction could play an expository role: If fictions are able to deliver factual information, they could do useful educative work.[15] Might (suitable) fictions be embraced as a means of delivering non-fictional content – could fictions deliver facts?

Recently, scholars in development studies have called for just this: asking that (suitable) literary fictions be considered "valid" and even "authoritative" sources of knowledge.[16] Noting that several recent fictions provide complementary accounts of the conditions of poverty and cultural difference that development studies aims to describe and ultimately ameliorate, David Lewis, Dennis Rodgers and Michael Woolcock make the case that fictions are not only more widely read, but often provide more detailed accounts of the same material described in the academic field of development studies. In a 2008 paper discussing "literary representation as a source of authoritative knowledge," they argue that (some) fictions could be a useful tool: "[I]t is arguable that [Monica Ali's novel] *Brick Lane* has contributed to wider public understanding of development in ways that no academic writing ever has" (Lewis et al. 2008, p. 208). However, novels such as Ali (2004), Hosseini (2004) or Mistry (2006) – all of which they hold up as good examples of the types of books they have in mind – are only useful insofar as the arguments they make and the situations they describe accord with the existing academic research. The fictions possess evidential value only to the extent that they have been ratified by the type of quantitative, empirical research to which they are putatively equivalent. The claim is less radical than it first seems, and they leave the issue much as they found it: The role they allot fiction is ultimately illustrative.

---

[15] Jane Gregory has analysed how Fred Hoyle used fiction to disseminate his ideas (Gregory 2003).

[16] As discussed later, their title is a little misleading: There is no claim that fiction offers *authoritative* knowledge (which really would be a radical argument), but only the more modest claim that fictional material can play a useful illustrative role in support of empirical, quantitative work.

If Lewis et al. do wish to employ fiction directly in the service of development studies, two possibilities present themselves. Rather than scour existing literature for novels that capture some aspect of the development studies
message, the most efficient route that they could take would be to produce
their own fictions.[17] Fiction written to highlight a cause or draw attention
to human suffering under oppressive regimes is nothing new – it's clear that
the fictions of, for example, Koestler (1994), Levi (1991) and Solzhenitsyn
(1970) are not meant to be read in isolation from the real-world events
they depict. The second option, which is what Lewis et al. effectively settle
upon, is to offer a reading list or canon of acceptable texts. The important
point to take away from this is that not just *any* fiction will play the role of
a vehicle for delivering facts, but *only* those fictions whose contents accord
with accepted notions within development studies. Nonetheless, Lewis et
al prescribe fiction as a complementary but legitimate means of acquiring
facts, and that is significant.

9. Fictions as Vehicles for Travelling Facts

The introduction by Lewis et al. of fiction as a means by which an audience including non-specialist readers might acquire facts (here, facts about
life in impoverished countries) marks a turning point in this paper. The
examples looked at so far (with the exception of Kehlmann, and possibly
Dudman) are all cases where, in spite of the sometimes considerable use of
fictional devices, the material is presented and intended to be read as factual. However, the previous examples have demonstrated that the stylistic
boundaries between fictional and factual writing are unclear. The following
examples demonstrate that this permeability has not gone unnoticed, and
that fiction can and is being used both as a means of supplying an audience with factual information and even as a particular form of evidence in
itself.

In the *Voyages* (Sauer 1971) anthology, published by a group called the ZPG,
a claim about (at least the future of) the real world is made based upon the
content of fiction. The ZPG – which stands for "Zero Population Growth" –
were a population control advocacy group, formed in direct response to
Paul R. Ehrlich's 1968 book *The Population Bomb* – in which Ehrlich predicted dire consequences for a crowded earth. Limited to the United States,
most of the ZPG's work involved standard campaigning: distribution of

---

[17] For an example of an academic discipline using fiction to provide research evidence, see
      Schell, this volume, on Gorry.

contraceptives, pro-choice petitioning and the production of educational films and public lectures. But in 1971, they published an anthology of short stories called *Voyages: Scenarios for a Ship Called Earth* (Sauer 1971). The collection includes contributions from J. G. Ballard, Doris Lessing and Ray Bradbury. None of the stories were written specifically for the ZPG, but in one way or another, all imagine a world where over-population has resulted in great discomfort. The function of these stories is to provide an imaginative scenario, akin to a thought-experiment or counterfactual, against which the reader is invited to measure their own (intuitive) sense of what would be a desirable life. As the stories all imagine dystopic futures, it is expected that readers will agree with the ZPG that such a future would not be desirable. In this respect, the stories do not propose to supply facts about the actual future or even the actual present, but rather, to describe (preventable) conditions under which the world would be a less tolerable habitat.

But there is also a stronger argument at work – a claim to the effect that the scenarios these stories describe deserve to be taken *as seriously as* the scenarios described by the predictions of economists and demographers. Ehrlich himself introduces the *Voyages* anthology by predicating the stories on an argument from induction:

> It is fairly well known that recent visits to the moon by astronauts and all the earlier preparatory space experimentation were described with remarkable accuracy a good deal more than a decade in advance by science fiction writers. It is less generally known, except perhaps by aficionados, that many science fiction writers during the last thirty years or more have been writing about the problems associated with overpopulation. (Ehrlich, in Sauer 1971, p. ix)

Thus, we are invited to read these stories as possessing a comparable predictive accuracy to those stories of imaginary rocketships and satellites that were the precursor to real rockets and real satellites: That is, science fiction got it right before, and will do again. So in this case, the fiction really is offered as evidence, albeit of a quite loose and speculative type. Ehrlich says of the stories: "Hopefully, they will provide the reader with many fresh insights into today's problems and tomorrow's possible solutions" (Ehrlich, in Sauer 1971, p. ix). We are being asked to alter our behaviour based on the contents of J. G. Ballard's imagination, even though it is just a short story and never really happened. In the case of the ZPG, the *content* of the fiction is offered as a plausible outcome extrapolated from present concerns. Fiction is being used as a means of supplying a non-specialist audience with information on which to decide what is, in effect, a matter for scientific debate: Will population growth outstrip the capacity of the planet to sustain tolerable human life? Again, these are not facts, but they are presented as rivals to the claims of those who predicted that continued population growth would *not* yield unbearably unpleasant conditions (e.g., Simon 1996, p. 22).

For obvious reasons, contesting the probability of events that have not yet occurred – that is, challenging predictions – is much easier than challenging documented *past* events or *present* situations (including processes, e.g., disease transmission – see Mansnerus and Oreskes, both this volume). This is not to say that predictions made by fiction writers actually enjoy comparable epistemic status to predictions made by scientists, simply that, in the absence of the specialist knowledge necessary to check the workings, their differential probability of transpiring is less easily discerned. In the next case, we see how this situation is capitalised upon by a fiction writer who seeks both to challenge predictions from the sciences (at least partially by means of his own fictional writings), and also to exploit (again in his own fictional writings) the fact/fiction ambiguity discussed previously in order to increase the plausibility of information held by scientific consensus to be of dubious epistemic status.

10. Subversive Exposition: Using Fiction to Convey Facts

Michael Crichton's last two novels offer an unusual "hybrid" case, where the generic distinction between factual and fictional material is sufficiently elusive that the readership may be left unsure of whether they are consuming factual or fictional material. Crichton began writing novels in the 1960s while training to be a doctor. His novels became so successful that he did not pursue a medical career. He died in 2008. The focus here is on Crichton's last two novels: *State of Fear*, from 2004; and *Next*, from 2006. Like his previous fictions, such as *Jurassic Park*, the books dramatized the consequences of modern techno-science going awry. In order to achieve this, Crichton typically introduced his readers to some intriguing element of contemporary science through lengthy sections of expository writing, usually spoken by characters whose dramatic function within the novel is sometimes limited to the delivery of data. An exemplary scene from *State of Fear*, between "John Kenner," an heroic climber, prodigious MIT scientist and climate-change sceptic, and "Peter Evans," a young lawyer, shows how Crichton used dialogue to embed facts:

> "Well, let's consider that. When Hansen announced in the summer of 1988 that global warming was here, he predicted temperatures would increase .35 degrees Celsius over the next ten years. Do you know what the actual increase was?"
> "I'm sure you'll tell me it was less than that."

"*Much* less, Peter. Dr. Hansen overestimated by three hundred percent. The actual increase was .11 degrees."
"Okay. But it *did* increase."
"And ten years after his testimony, he said that the forces that govern climate change are so poorly understood that long-term prediction is impossible."
"He did not say that."
[…] "Proceedings of the National Academy of Sciences, October 1998"*
("*" indicates a full bibliographic reference included at the foot of the page, Crichton 2004, p. 293)

Note that (at least during these sections of text) "Evans" and "Kenner" are functionaries in the same way that "Wee-Wee the Termite King" or "Achilles and the Tortoise" are functionaries. Elsewhere in the novels, chase scenes and romances drive the plot forward, as per the thriller genre.

Crichton was so effective at this blend of thriller-fiction and expository writing that he seemed to be operating at the boundary between popular fiction and popular science. Indeed, so distinctive was this generic blurring that Søren Brier called for the inauguration of a new genre, which he suggests we call "Ficta" (Brier 2006). Brier is generally positive about the conflation of fictional and factual writing, praising Crichton's combination of "dramatic effect and scientific sobriety" (2006, p. 163). Of *Jurassic Park*'s treatment of chaos theory, he says: "This is excellent, brief, and to-the-point popular science … [an] amazingly effective, correct and dramatic introduction to some of the most significant insights of modern science" (2006, p. 163). Brier's endorsement embeds an important condition: The method is effective as a form of expository writing, providing Crichton is *correct*. Brier acknowledges this, but sees it as a problem not with his so-called Ficta, but with popularisations as a whole: "The problem of the mixed genres," he writes (rehearsing an argument of Stephen Turner's [2001, p. 129]), "is that lay people, as with all popular science, do not have a real chance to evaluate the quality of the scientific knowledge the book is based on" (2006, p. 171). Thus, the audience for popularisations are more susceptible to being misled by biased or false claims. Needless to say, the target audience for popular fiction (such as Crichton's novels) are if anything even less qualified and therefore even more vulnerable than the target readership of popular science books.

With Crichton's work this is a significant problem, compounded because not all of the "facts" are facts. Crichton's novels are not marketed to a scientifically literate readership, but they engage with live scientific issues: *State of Fear* makes the case against anthropogenic climate change, and *Next* against patented genetic engineering. It is unlikely that *Jurassic Park* was written to seriously dissuade billionaires from investing in dinosaur theme parks, but both *State of Fear* and *Next* clearly have intentions to persuade. Both novels achieve this through a combination of factual and fictional material. Real news reports are included alongside faked news reports; real graphs are cited and reprinted in the novels, and given interpretative glosses spoken by the fictional characters (e.g., Crichton 2004, pp. 100–6). Vitally, the distinction between the factual and fictional material is, for the main part, concealed – the epigraph to *Next* elliptically claims: "Everything here is fiction, except for the parts which aren't."

How this vacillation between the use of factual and fictional material works can be illustrated by looking in a little more detail at Crichton's case against the environmentalist movement. The plot of *State of Fear* requires that the dangers posed by climate change have been greatly exaggerated. One of the novel's supplementary arguments in support of this is that, far from being beneficent stewards of the planet's ecological health, the Green Movement is a cynical operation that exhausted its initial remit in the late 1970s. Bereft of purpose, but now a profitable venture, they needed to identify new problems on which they might be legitimately grounded. Hence, they "invented" global warming. Our current panic over climate change is an induced "state of fear," manufactured as a means of sustaining the charity machine. This may seem unlikely, but our willingness to take such a scenario seriously might be contingent on the *character* of those involved in such a cynical ploy. As a novelist, Crichton is now in a position to manipulate exactly this aspect: Hence, we see grasping and dishonest charity workers. This is not data in itself, but it certainly plays a part in creating the impression of conspiracy – it supplies the important missing element: motive. For despite the plentiful graphs and charts (e.g., Crichton 2004, pp. 439–52) contain seven footnotes and fully twenty graphs displaying data on climate change), Crichton's argument does not stand on this data alone. Rather, it is only when the data are arrayed in conjunction with the scurrilous character of the charity workers that his case begins to look plausible. Even if the expository dialogues and graphs can show us gaps in the data, the crime still needs a culprit. Thus Crichton (as novelist) can supply exactly the type of people who would use a charity to fleece our generosity. The facts about climate change are cast into doubt by their association with fictional villains. Having selected the data that suits him, character sketches are the "missing part" that (as a novelist) he is able to fill in.

The novelist's ability to supply missing motives and invent missing data points also exposes one of the principle problems with the quality of the "evidence" that fiction might provide: What happens in fiction is entirely under the control of the author. If even the best-laid plans go awry, it's because the author wants them to do so. As Noël Carroll puts it: "Fictions that are intended to advance theses ... are typically designed in such a way that the story content supports its putative truth claims. ... the evidence, if that is what the story is, is, so to speak, cooked from the get-go" (Carroll 2002, p. 5). In an account about events in the real world, for example, a legal case, coherence is a good indicator that the account is reliable and accurate.[18] In a fiction that aims to propound a theory about the real world, that same criterion is not a useful indicator because coherence is one of the features of a text that an author of fiction can control.

Muddying the categories further, Crichton doesn't just include fiction in his books; he also employs supplementary material that, although bound within the same covers, is no part of the fiction. *State of Fear* (2004) and *Next* (2006) both possess scholarly apparatus and appendices, and both are appended by a short non-fiction essay from Crichton where, stepping out of his role as an entertainer to speak as an activist, his serious intent to impart information is made explicit. These factual appendices also work to confer respectability on the fact-like information that the preceding fiction contained. They show, for example, that Crichton has done research, increasing the trustworthiness of claims made by characters in the story (e.g., "The planet warmed about .3 degrees Celsius in the last thirty years" [Crichton 2004, p. 439]) – that is, increasing the chances that the claims made in the fiction about the world are true of the actual world. Crichton also sources legitimacy from his real-world familiarity with scientific practices – his medical training. In 2005, Crichton testified as an expert before a Senate Committee on the environment.[19] He manoeuvred himself into a position where he found he could wield much of the epistemic authority of a scientist, whilst operating with the imaginative latitude allowed to a novelist.[20]

---

[18] Rather as the individual facts of new medical cases have to hang together before they can be taken as defining a new disease, see Ankeny, this volume.

[19] Committee on Environment and Public Works, 28 September 2005.

[20] The prefatory note to *State of Fear* (2004) is a good example of how Crichton uses the scholarly apparatus to sustain rather than resolve ambiguity about the factual status of claims: "This is a work of fiction. Characters, corporations, institutions, and organizations in this novel are the product of the author's imagination, or, if real, are used fictitiously without any intent to describe their actual conduct. However, references to real people, institutions, and organizations that are documented in footnotes are accurate. Footnotes are real." Crichton can supply a footnote when a claim has a source, but needn't admit when no source exists.

Yet with less latitude than most: For those seeking to steer an unwilling public towards action on climate change, Crichton is seen as a real threat.[21] Naomi Oreskes, whose research underpins much of the statistical argumentation in Al Gore's *An Inconvenient Truth*, complains: "Crichton is a novelist, and he knows how to write fiction. But he should leave the scientific facts to scientists, the historical facts to historians." (Oreskes 2005). Were it not for the dual role as educator and entertainer, this type of reaction might seem strange: After all, Crichton *is* writing fiction (in spite of Brier's reclassification), and fiction seems by definition to be that kind of writing that deals with what is not the case. Crichton's case is interesting insofar as by explicitly predicating his fictions on real science and admixing the fictional narrative with real data, he invites us to treat his books more like the novelised histories of Sobel (1995), Dudman (2003) and Kehlmann (2007). He invites and encourages his readers (through, for example, the appendices) to accept facts they encounter in his fictional world as facts about the actual world. And it doesn't matter if the "facts" so-disseminated and absorbed are true – only that the fiction acts a means both of delivering them and (through supplying a framework into which they achieve a logical coherence) making them seem plausible.

Similarly, Crichton is also able to erode the security of specialist knowledge claims that would stand in his way. For example, he is able to make the business of computer modelling seem more like guesswork:

> "Then how do they make computer models of climate?" Evans said.
> Kenner smiled. "As far as cloud cover is concerned, they guess."
> "They *guess?*"
> "Well, they don't call it a guess. They call it an estimate, or parameterization, or an approximation. But if you don't understand something, you can't approximate it. You're really just guessing."
> Evans felt the beginning of a headache. He said, "I think it's time for me to get some sleep." (Crichton 2004, pp. 222–3)

Although the climate scientists (unlike the soporific Evans) have good grounds on which to respond to such a charge, that debate won't be something that Crichton's readership will necessarily involve themselves in. This is a problem crucial for the popular debate. It may be the case that no *academic* will take Crichton's arguments seriously, but academics aren't the

[21] Al Gore is widely assumed to have Crichton in mind when he remarked: "The planet has a fever. If your baby has a fever, you go to the doctor. If the doctor says you need to intervene here, you don't say, 'Well, I read a science fiction novel that told me it's not a problem.'" Gore's comments were made to the House Energy Committee on 21 March 2007.

audience that this contest is being fought over. In terms of the *public under-standing* of these issues, the contestants do not need to secure an academic victory,[22] but simply to inject sufficient doubt into the public arena. This is largely because, as Brier noted, the limited evidence popularisations sup-ply underdetermines their conclusions: A popularisation doesn't persuade on the same grounds as a professional paper (Brier 2006, p. 171). Stephen Turner formulates it neatly: "[T]he basis on which experts believe in the facts or validity of knowledge claims of other experts … is different from the basis on which non-experts believe in the experts" (Turner 2001, p. 129). To an audience unable to check the workings, fake science and real science will look very much alike. The situation is exacerbated by the widespread use of fictional techniques by popularisers. Fictions may not rank high as credible sources of factual information, but an audience exposed to popularisations that look like fictions and fictions that look like popularisations may find it increasingly difficult to tell the difference. If Crichton's mixture of fact and fiction succeeds, it is at least in part because there is already a background of generic blurring in which science writers, through seeking to make their facts travel, are complicit.

11. Conclusion: Making Facts Travel in the Marketplace of Ideas

Of the examples briefly looked at here, it is clear that science popularisa-tion benefits greatly from using fictional devices. Narrative is a means of organising chunks of information into a causal sequence and allowing more chunks to be retained in sequence. Characterisation likewise supplies a rea-son to keep turning pages by capitalising on our innate interest in other lives – with education piggy-backing on what Raymond Tallis once called "this ordinary gossipy interest" (Tallis 1995, p. 8) that fiction feeds into and upon. But these effects are neutral with respect to the epistemic status of the information being conveyed: True and false facts travel just as well. Along with characterisation, narrative coherence is also a means of establishing plausibility, which has an intuitive authority, but again, the neatness of the structure does not reflect the truth status of the information so arranged – as Clifford Geertz put it, "[T]here is nothing so coherent as a paranoid's delusion or a swindler's story" (Geertz 1973, p. 18).

For an audience who cannot be assumed to possess the requisite special-ist knowledge to corroborate "fact-like" claims (see Turner 2001) or for the

---

[22] Oreskes's paper (this volume) discussed an explicit attempt to inject doubt into the public's view of the climate science facts.

same reasons recognise which contributor is the "expert" (see Lynch and Cole 2005), making the difference between fictional and factual literature stylistically indistinct is perilous. A reader unable to tell the difference is apt to adopt a stance of either exaggerated credulity or exaggerated scepticism. This ambiguity can be valuable for those seeking to contest the authority of scientific facts. Against a background of scholarship within science studies that problematises the notion of expertise, it is no surprise that Søren Brier celebrates Crichton's categorical gerrymandering as a viable adaptation to the modern information market, and declares support for a laissez faire epistemology: "It is no longer publish or perish, it is *Agora or agony.* Popularize and tell your story in the Agora, or lose the game of knowledge and power" (Brier 2006, p. 171). But outside of the insular concerns of science studies, Brier's confidence in the ability of the marketplace to competently arbitrate epistemic value (or rather, the scepticism about truth and expertise that this position implies) is troubling, for the popularisations can significantly shape both the public image and self-identity of science (a point acknowledged even within science studies: see, for example, Lewenstein 1995, 2002; Shermer 2002).

For all the available gains, perhaps the populariser intent on having their work taken seriously would do well to minimise the extent to which they avail themselves of fiction's tricks. Although the use of fictional techniques and devices is a successful means of making scientific facts and theories more accessible, it can eventually contribute to a diminution of the educative value of the popularisation. Although fictional devices can be a successful way of communicating scientific facts, the use of fiction techniques and devices can be deleterious to the reliability of the material. It may ultimately undermine the original agenda of the science populariser: the communication of useful and reliable scientific facts.

Acknowledgment

This work was supported by the ESRC/Leverhulme Trust project, "The Nature of Evidence: How Well Do 'Facts' Travel?" (grant F/07004/Z) at the Department of Economic History, LSE.

Bibliography

Ali, Monica. 2004. *Brick Lane.* London: Black Swan.
Amis, Martin. 1991. *Time's Arrow: or, The Nature of the Offence.* Harmondsworth: Penguin.
Anderson, Mark M. 2007. "Humboldt's Gift," *The Nation* (April 12).

Banville, John. 2001. *The Revolutions Trilogy: Doctor Copernicus, Kepler, Newton Letter – An Interlude*. London: Picador.

Brier, S. 2006. "Ficta: Remixing generalized symbolic media in the new scientific novel," *Public Understanding of Science* 15: 153–74.

Bryson, Bill. 2004. *A Short History of Nearly Everything*. London: Black Swan.

Burnham, John C. 1987. *How Superstition Won and Science Lost: Popularizing Science and Health in the United States*. New Brunswick, NJ: Rutgers University Press.

Carroll, Noël. 2002. "The Wheel of Virtue: Art, Literature, and Moral Knowledge," *The Journal of Aesthetics and Art Criticism* 60,1: 3–26.

Crichton, Michael. 1990. *Jurassic Park*. New York: Ballantine-Random House.

   2004. *State of Fear*. New York: HarperCollins.

   2006. *Next*. New York: HarperCollins.

Dawkins, Richard. 1976. *The Selfish Gene*. Oxford: Oxford University Press.

   1988. *The Blind Watchmaker*. Harmondsworth: Penguin.

   2004. *The Ancestor's Tale: A Pilgrimage to the Dawn of Evolution*. New York: Houghton Mifflin.

Dos Passos, John. 1996. *U.S.A.: The 42nd Parallel/1919/The Big Money*. Townsend Ludington, ed. New York: Library of America.

Dudman, Clare. 2003. *Wegener's Jigsaw*. London: Sceptre.

Ehrlich, Paul R. 1968. *The Population Bomb*. New York: Ballantine.

Einsiedel, Edna. 2007. Editorial, *Public Understanding of Science* 16: 5–6.

Fahnestock, J. 1986. "Accommodating Science: The Rhetorical Life of Scientific Facts," *Written Communication* 3: 275–96.

Faulkner, William. 1991. *The Sound and the Fury*. New York: Vintage.

Geertz, Clifford. 1973. *The Interpretation of Cultures: Selected Essays*. New York: Basic Books.

Gleick, James. 1987. *Chaos: Making a New Science*. Harmondsworth: Viking-Penguin.

   1992. *Genius: Richard Feynman and Modern Physics*. London: Abacus-Little, Brown.

Govoni, Paola. 2005. "Historians of Science and the 'Sobel Effect,'" *Journal of Science Communication* 4,1: 1–17.

Greene, Brian. 1999. *The Elegant Universe: Superstrings, Hidden Dimensions, and the Quest for the Ultimate Theory*. New York: Norton.

Gregory, Jane. 2003. "The Popularization and Excommunication of Fred Hoyle's 'Life-from-Space' Theory," *Public Understanding of Science* 12: 25–46.

Hawking, Stephen. 1996. *The Illustrated Brief History of Time*. London: Bantam Press.

Hilgartner, Stephen. 1990. "The Dominant View of Popularization: Conceptual Problems, Political Uses," *Social Studies of Science* 20: 519–39.

Hofstader, Douglas. 1979. *Godël, Escher, Bach: An Eternal Golden Braid*. New York: Basic Books.

Hosseini, Khalid. 2004. *The Kite Runner*. London: Bloomsbury.

Kehlmann, Daniel. 2007. *Measuring the World*. Translated by Carol Brown Janeway. New York: Vintage-Random House.

Kirby, D.A. 2008a. "Hollywood Knowledge: Communication Between Scientific and Entertainment Cultures." in D. Cheng, et al. (eds.), *Communicating Science in Social Contexts*. New York: Springer, pp. 165–81.

2008b. "Cinematic Science: The Public Communication of Science and Technology in Popular Film," in B. Trench & M. Bucchi (eds.), *Handbook of Public Communication of Science and Technology*. New York: Routledge, pp. 67–94.

Koestler, Arthur. 1994. *Darkness at Noon*. London: Vintage.

Krukonis, Greg and Tracy Barr. 2008. *Evolution for Dummies*. Hoboken, NJ: Wiley Publishing, Inc.

Leane, Elizabeth. 2007. *Reading Popular Physics: Disciplinary Skirmishes and Textual Strategies*. Aldershot: Ashgate.

Levi, Primo. 1991. *If This Is A Man/The Truce*. London: Abacus.

Lewenstein, Bruce V. 1995. "From Fax to Facts: Communication in the Cold Fusion Saga," *Social Studies of Science* 25.3: 403–36.

2002. "How Science Books Drive Public Discussion," in Gail Porter (ed.), *Communicating the Future: Best Practices for Communication of Science and Technology to the Public*. Gaithersburg, MD: National Institute of Standards and Technology, pp. 69–76.

Lewis, David, Dennis Rodgers, and Michael Woolcock. 2008. "The Fiction of Development: Literary Representation as a Source of Authoritative Knowledge," *Journal of Development Studies* 44,2: 198–216.

Locke, Simon. 1999. "Golem science and the public understanding of science: from deficit to dilemma," *Public Understanding Science* 8: 75–92.

Lynch, Michael, and Simon Cole. 2005. "Science and Technology Studies on Trial: Dilemmas of Expertise," *Social Studies of Science* 35,2: 269–311.

McEwan, Ian. 1999. *Enduring Love: A Novel*. New York: Anchor-Random House.

2005. "Literature, Science, and Human Nature," in Jonathan Gottschall and David S. Wilson (eds.), *The Literary Animal: Evolution and the Nature of Narrative*. Evanston, IL: Northwestern University Press.

Mellor, Felicity. 2003. "Between fact and fiction: Demarcating science from non-science in popular physics books," *Social Studies of Science* 33: 509–38.

Miller, David P. 2002. "The 'Sobel Effect': The Amazing Tale of How Multitudes of Popular Writers Pinched All the Best Stories in the History of Science and Became Rich and Famous While Historians Languished in Accustomed Poverty and Obscurity, and How This Transformed the World. A Reflection on a Publishing Phenomenon," *Metascience* 2 (July): 185–200.

Miller George A. 1956. "The Magical Number Seven, Plus or Minus Two: Some Limits on Our Capacity for Processing Information," *The Psychological Review* 63: 81–97.

Mistry, Rohinton. 2006. *A Fine Balance*. London: Faber and Faber.

Oreskes, Naomi. 2005. "'Fear'-Mongering Michael Crichton Is Wrong," *The San Francisco Chronicle*, February 16: B11.

Pinker, Steven. 2002. *The Blank Slate: The Modern Denial of Human Nature*. Harmondsworth: Penguin.

Powers, Richard. 1995. *Galatea 2.2*. New York: Farrar, Straus and Giroux.

Pratchett, Terry, Ian Stewart, and Jack S. Cohen. 1999. *The Science of Discworld*. London: Ebury Press.

Pynchon, Thomas. 1997. *Mason & Dixon*. New York: Henry Holt.

Ramsden, Edmund. 2006. "Confronting the Stigma of Perfection: Genetic Demography, Diversity and the Quest for a Democratic Eugenics in the Post-War United States," *Working Papers on the Nature of Evidence: How Well Do "Facts" Travel?* 12/06. Department of Economic History, LSE.

Ridley, Matt. 1994. *The Red Queen: Sex and the Evolution of Human Nature.* Harmondsworth: Penguin.

Royal Society. 1986. "Public Understanding of Science: The Royal Society Reports," *Science, Technology, and Human Values* 11,3: 53–60.

Sacks, Oliver. 1987. *The Man Who Mistook His Wife for a Hat.* London: Picador.

1991. *Awakenings.* London: Picador.

1995. *An Anthropologist on Mars.* London: Picador.

2002. *Uncle Tungsten: Memories of Chemical Boyhood.* London: Picador.

Sauer, Rob, ed. 1971. *Voyages: Scenarios for a Ship Called Earth.* New York: ZPG/ Ballantine.

Shermer, Michael B. 2002. "This View of Science: Stephen Jay Gould as Historian of Science and Science Historian, Popular Scientist and Science Populariser," *Social Studies of Science* 32,4: 489–525.

Shinn, Terry, and Richard Whitley. 1985. *Expository Science: Forms and Functions of Popularisation.* Dordrecht: Reidel.

Simon, Julian. 1996. *The State of Humanity.* Wiley-Blackwell.

Silverstone, Roger. 1991. "Communicating Science to the Public," *Science, Technology, & Human Values* 16,1: 106–10

Smith, Leonard. 2007. *Chaos: A Very Short Introduction.* Oxford: Oxford University Press.

Sobel, Dava. 1995. *Longitude: The True Story of a Scientific Genius Who Solved the Greatest Scientific Mystery of His Time.* Harmondsworth, Penguin.

Solzenhitsyn, Aleksandr. 1970. *One Day in the Life of Ivan Denisovich.* Harmondsworth: Penguin.

Strout, Cushing. 1992. "Border Crossings: History, Fiction, and *Dead Certainties*," *History and Theory* 31: 153–62.

Tallis, Raymond. 1995. *Newton's Sleep: Two Cultures and Two Kingdoms.* London: MacMillan.

Turner, Stephen. 2001. "What Is the Problem with Experts?" *SSS* 31,1: 123–49.

Tymoczko, Thomas, ed. 1998. *New Directions in the Philosophy of Mathematics: An Anthology.* Revised and expanded. Princeton, NJ: Princeton University Press.

Wason, P. C. 1971. "Natural and Contrived Experience in a Reasoning Problem," *Quarterly Journal of Experimental Psychology* 23: 63–71.

Watt, Ian. 1987. *The Rise of the Novel: Studies in Defoe, Richardson and Fielding.* London: Hogarth.

Wheeler, William Morton. 1920. "The Termitodoxa, or Biology and Society," *The Scientific Monthly* 10,2: 113–24.

Yearly, Steven. 2000. "Making Systematic Sense of Public Discontents with Expert Knowledge: Two Analytical Approaches and a Case Study," *Public Understanding of Science* 9: 105–22.

PART THREE

INTEGRITY AND FRUITFULNESS

SEVEN

ETHOLOGY'S TRAVELING FACTS

RICHARD W. BURKHARDT, JR.

1. Introduction

In 1949, the fledgling ethologist Robert Hinde observed a happy interchange between Konrad Lorenz and Niko Tinbergen, the founders of ethology, in their first days together after World War II. The location was Cambridge, England. The occasion for the ethologists being in Cambridge was a special symposium on "Physiological Mechanisms in Animal Behaviour," hosted by the Society for Experimental Biology. The interchange in question happened outside of the official proceedings. As Hinde recalled:

> We were walking down Jesus Lane in Cambridge, and Tinbergen and Lorenz were discussing how often you had to see an animal do something before you could say that the species did it. Konrad said he had never made such a claim unless he had seen the behaviour at least five times. Niko laughed and clapped him on the back and said 'Don't be silly, Konrad, you know you have often said it when you have only seen it once!' Konrad laughed even louder, acknowledging the point and enjoying the joke at his own expense. (Hinde 1990, p. 553)

This story is instructive for what it tells about Lorenz and Tinbergen and their relationship to one another. It is also helpful in drawing attention to the kinds of facts in which ethologists were interested and how these facts were identified or constructed. In particular, it highlights the ethologists' concern with behavioral differences among species, the metafact, so to speak, that species differ from one another in behavioral characters in much the same way that they differ in physical characters, and that behaviors may thus serve to distinguish one species from another just as effectively as structures do. This metafact means that behavioral facts observed in any one species may or may not be good candidates for travel when it comes to thinking about behavior in other species. The appropriateness or inappropriateness of the travel of specific behavioral facts from one context to another has thus been one of the recurring issues in the history of animal behavior studies.

2. Lorenz and Tinbergen and Ethology's Identity

Before examining specific cases of behavioral facts traveling either within ethology or beyond the discipline's boundaries, it will be helpful to survey the kinds of facts, issues, and approaches by which the science of ethology identified itself. The science of ethology started taking shape in the 1930s under the leadership of the Austrian zoologist Konrad Lorenz, ably complemented by his Dutch counterpart Nikolaas Tinbergen. Their subject was animal behavior, focused in a way that distinguished their work from that of other investigators who were also interested in behavior studies. They were especially keen to differentiate their work from two different psychological traditions: the European animal psychologists on the one hand and the American behaviorist or comparative psychologists on the other.

What interested the ethologists above all were instinctive behavior patterns –patterns that appeared to be innate rather than learned and that were just as characteristic of a species as its physical structures. The problem with the Continental animal psychologists, as the ethologists saw it, was that they defined instinct too broadly and were furthermore primarily interested in its subjective dimensions, that is, in the topic of the animal mind. The ethologists, in contrast, wanted to promote an objectivistic study of instinct, where instincts were defined as specific motor patterns, and where (at least as Tinbergen saw it) speculations on the animal's subjective experience had no place. The problem the ethologists saw with the American behaviorist psychologists was that they concentrated almost exclusively on the study of learning in white laboratory rats. The ethologists cared little for facts generated in a behaviorist's experimental laboratory. They were interested in facts observed out in the field (Tinbergen's preference) or in an aviary or other such animal-rearing facility (Lorenz's preference), sites where animals could display the full range of their natural behavior patterns unhindered by fear or the pernicious effects of domestication.

The ethologists were additionally keen to point out that the questions they asked about instinctive behavior were *biological* questions. Tinbergen offered what is now regarded as the classic expression of this in 1963 in a paper identifying what thereafter came to be called "the four questions of ethology." For any given behavior pattern, he maintained, one could (and indeed *should*) ask about its physiological causation, its function or survival value, its evolutionary history, and how it developed in the individual. It is worth noting, however, that this formulation emerged over time. In the 1930s and 1940s, the ethologists typically stressed one or two of the later "four questions" – sometimes even three – but never the four of them all at once. In 1942, in contrasting the objectivistic nature of ethology with the subjectivist approach of his country-man Bierens de Haan, Tinbergen indicated that ethology's primary interest was behavioral causation, that is, understanding innate behavior in physio-logical terms (Tinbergen 1942). Meanwhile, Lorenz, even while providing the conceptual foundations for studying behavioral causation, typically insisted that it was his field's comparative, evolutionary perspective that made it dis-tinctive. The founding insight of the field – its "Archimedean point," as he liked to call it (e.g., Lorenz 1978, p. 3) – was the notion that innate behavior pat-terns, just like claws, teeth, or other body parts, needed to be examined from the comparative, evolutionary viewpoint. In other words, instinctive behavior could be used like physical structures, not only in identifying species but also in reconstructing phylogenies and assessing genetic affinities. Indeed, when referring to his field, Lorenz seems to have preferred the phrase "comparative behavior study" (*vergleichende Verhaltensforschung*) to the word "ethology."

The diversity of questions addressed by the ethologists necessarily had a bearing on how ethological facts traveled both within ethology and beyond it. A fact bearing on the physiological mechanisms of instinct, for exam-ple, might or might not be seen as particularly relevant to the enterprise of reconstructing evolutionary histories. And when it came to reconstructing phylogenies, one still needed to evaluate whether similarities between spe-cies should be interpreted as matters of homology (the result of a common evolutionary history) or analogy (the convergence of characters resulting from adaptation to similar circumstances).

Also relevant to the traveling of animal behavior facts was the question of who was a reliable witness of them. Lorenz once allowed that he would only give credence to facts reported by other people whom he regarded as being among "the limited number of genuine animal observers" – ideally, people with whom he had "a close personal acquaintanceship." Otherwise, he said, he found it difficult to read an account of an animal's behavior and feel he was getting a clear account of what the animal actually did. Thus, he was willing to consider facts from his ornithologist friends Horst Siewert and Oskar Heinroth, but not those reported, say, by the British psycholo-gist Conwy Lloyd Morgan (Lorenz [1935] 1970b, p. 113). His own status as ethology's founder would make Lorenz on balance a very good companion for the facts he reported – except in certain postwar situations where allega-tions about his wartime Nazi sympathies made him a bad companion.[1]

---

[1] The potentially problematic role of the companion can also be seen in the examples of Michael Crichton and Dr. William Harvey discussed, respectively, by Adams and Haycock, both this volume.

3. Putting Ethology's Facts in Motion

With respect to behavioral facts moving beyond ethology's own boundaries, it is understandable that an ethological fact bearing on the understanding of the physiology of a motor pattern or the genetic affinities between different species might not travel to comparative psychology, where laboratory psychologists were pursuing problems in animal learning. Nevertheless, the ethologists were certainly *interested* in demonstrating to the practitioners of other disciplines that ethology was a science whose findings merited broader attention. At the 1949 Cambridge conference, the ethologists were keen to show physiologists that ethologists could talk meaningfully about physiological mechanisms in behavior. In the 1950s, the ethologists felt the need to make their discoveries and ideas known above all to American comparative psychologists (though ethologists looked to other audiences as well, the general public included). The ethologists presented their work to nonethologists through interdisciplinary seminars and conferences, public lectures, articles, books, films, television appearances, interviews, and so forth. Some of their cherished facts were boosted in their travels by the images, theories, or stories associated with them. Others were aided by the activity of individuals friendly to the ethologists' cause. Still others failed to make permanent inroads when the baggage that was accompanying them proved unwelcome.

The ethologists were interested in transmitting more than just facts. At the Cambridge conference of 1949, Lorenz and Tinbergen each presented models of instinctive action. Simultaneously, though, they stressed the factual foundations of their models. Thus, Lorenz acknowledged the "extreme crudeness and simplicity" of his own, "psycho-hydraulic" model, but he insisted that the model symbolized "a surprising wealth of facts really encountered in the reactions of animals" (Lorenz 1950, p. 255). Beyond that he emphasized the strong, empirical inclinations of ethology's forefathers. Identifying the American biologist Charles Otis Whitman and the German ornithologist Oskar Heinroth as the two great pioneers of comparative ethology, Lorenz allowed that they achieved what they did because they were first and foremost animal lovers and empiricists: "Happily ignorant of the great battle waged by vitalists and mechanists on the field of animal behaviour, happily free from even a working hypothesis, two 'simple zoologists' were just observing the pigeons and ducks they loved, and thus kept to the only way which leads to the accumulation of a sound, unbiased basis of induction, without which no natural science can arise" (Lorenz 1950, p. 222).

Almost everyone in Lorenz's audience must have recognized this as hyperbole. Whitman was indeed a lover of pigeons, but he was also thoroughly engaged with the broadest questions of biology. Issues of evolution, heredity, and development constituted the *raison d'être* of his pigeon studies. The portrait of Whitman as a happy empiricist simply does not fit. However, it suits Heinroth somewhat better. In their classic study on the birds of central Europe, Heinroth and his wife Magdalena operated on the assumption that what was innate and what was learned in different bird species could only be determined by means of experiments conducted on a species-by-species basis. Their painstaking multiyear enterprise involved rearing individuals of every different central European bird species by hand, from the egg, and watching how each bird behaved from the time it hatched all the way to its adulthood (Heinroth and Heinroth 1924–1934). Even Heinroth, though, was capable of looking up from his facts to see a broader vision. In 1910 he expressed what might be called the "sooner or later" motif of animal behavior studies, that is to say, the belief that such studies would ultimately have something of value to offer for understanding human behavior. At the international ornithological congress of 1910 he closed his paper on the ethology of ducks and geese with the following prediction: "The study of the ethology of the higher animals – unfortunately a still very untilled field – will bring us ever closer to the realization that in our conduct with family and strangers, in courtship and the like, it is more a matter of purely inborn, more primitive processes than we commonly believe" (Heinroth 1910, p. 702).[2]

A generation later, Lorenz embraced the goal of applying insights from animal behavior studies to the understanding of humans. In 1931, not long after becoming acquainted with Heinroth and Heinroth's work, Lorenz wrote ecstatically to the older man, saying: "Who knows what will become of today's human psychology if one can only know what is instinctive behaviour and what is rational behaviour in humans? Who knows how human morals with their drives and inhibitions would look if one could analyze them like the social drives and inhibitions of a jackdaw" (Heinroth and Lorenz 1988, p. 42). From the 1930s onward, Lorenz was keen to proclaim that the study of animal instincts would shed light on human instincts. One of the reasons he welcomed the German takeover of Austria in the spring of 1938 was that he had come to believe that his career as a scientist was unlikely to flourish under Austria's Catholic educational establishment, which wanted

---

[2] For further discussion of human–animal comparisons based on experimental studies of behavior, see Ramsden, this volume.

no part of his ideas about the animal roots of human behavior. He imagined that the Third Reich would provide a more receptive *Weltanschauung* for his ideas. In lectures and publications, he sought to demonstrate the relevance of his work to the Reich's concerns with race hygiene.[3] After the war, these efforts on his part were neither entirely forgotten nor forgiven. The American comparative psychologist Daniel Lehrman noted them in 1953 in his widely debated critique of Lorenzian ethology.

4. How Ethological Facts Traveled

We turn now to a number of specific ethological facts to consider how they traveled. We will begin with facts related to the mechanisms of instinctive action. Then we will look at the metafact of interspecific behavioral differences.[4] Finally, we will consider a number of instances where facts about animal behavior were used specifically to illuminate human behavior, and we will consider some of the resistance that these efforts met.

We begin with a kind of fact that Lorenz felt deserved special status, a fact that manifested itself unexpectedly, without any special reason to be looking for it, and was thus untainted (or so he liked to claim) by any preconceptions. Such facts fit well with his preferred method of research, which was to raise wild birds in a state of semicaptivity and observe them over the course of months and even years. This allowed him, he said, not only to come to know the whole of a bird's normal behavior patterns, but also to witness rare but instructive behavioral events that some people watching birds in nature would most likely never see. He observed a pet starling, for example, perform "the entire fly-catching behavioural sequence" – even though the bird had never caught a fly in its life and there was no fly present at the time. He recounted the bird's behavior in detail:

> It would fly up to an elevated look-out position (usually the head of a bronze statue in our living-room), perch there and gaze upwards continuously as if searching the sky for flying insects. Suddenly, the bird's entire behaviour would indicate that it had spotted an insect. The starling would extend its body, flatten its feathers, aim upwards, take off, snap at something, return to its perch and finally perform swallowing motions. (Lorenz [1932] 1970a, p. 93)

However, as Lorenz insisted, "there really were no insects to be seen."

---

[3] For discussions of Lorenz's career as a scientist under the Third Reich, see Föger and Taschwer (2001); Taschwer and Föger (2003); and Burkhardt (2005).

[4] Schell, this volume, also considers the issue of the innateness or instinctual character of behavior in the context of the alpha-male hero in romance novels.

The importance of this behavioral fact for Lorenz's later theorizing was that he identified it as a case of "threshold lowering" carried to such an extreme that the instinctive behavior pattern was performed "in vacuo," that is, without there being any stimulus to elicit it. This, in turn, became part of his idea that every instinct has its own "action-specific energy" that builds up within the organism until it is ultimately released. When he first cited the fact in question, however, his point was that the behavior pattern he had witnessed was entirely innate and not guided by any external stimuli. He also mused on what the bird might be experiencing subjectively: "When observing such behaviour, one is immediately conscious of the question as to what subjective phenomena are experienced by the animal, since this behaviour is so reminiscent of that of certain human psychopaths who experience hallucinations" (Lorenz [1932] 1970a, p. 93). Five years later in a major paper on "the instinct concept," he again cited the starling's behavior and again drew the parallel between the bird's behavior and that of mentally ill patients suffering from hallucinations (Lorenz [1937] 1970c, p. 277).

The fact involving Lorenz's pet starling traveled beyond Lorenz's own writings. It reappeared in Tinbergen's 1951 book, *The Study of Instinct* (the first overview of ethology as a field of study). There Tinbergen cited Lorenz's account of "vacuum activities" in the pet starling in support of the idea that "a drive may even become so strong that its motor responses break through in the absence of a releasing stimulus" (Tinbergen 1951, pp. 61–2). Significantly, however, Tinbergen omitted any reference to Lorenz's reflections on what the starling might have been experiencing subjectively. In effect, the traveling fact was shorn of its subjective companion, at least in Tinbergen's presentation of it. Tinbergen, as we have seen, wanted to exclude from ethology's purview the question of animal subjective experience (Burkhardt 1997).

Let us consider now some facts that appeared not as fortuitous observations but instead as the result of targeted experiments. (What impressed Lorenz most about Tinbergen when they first met was that Tinbergen, unlike Lorenz, seemed gifted at conducting experiments.) Tinbergen's early ethological experiments used various sorts of "dummies" to identify which stimuli were most effective in eliciting particular instinctive reactions on an animal's part. For example, he and his students at Leiden University found in their studies of the three-spined stickleback that the males of this fish, which have red undersides during the breeding season, would attack dummies with red undersides, even if the dummies in other respects barely resembled fish at all. Tinbergen later reported the charming story of how the students in his lab observed that when the red Royal Mail van drove past the lab, the male sticklebacks in their row of tanks beside the laboratory's large windows "dashed toward the window sides of their tanks and followed the van from one corner of the tank to the other," responding just as they would have done to a rival male (Tinbergen 1953b, p. 66). Tinbergen told this story in his book, *Social Behaviour in Animals*, in describing how he tested the stimuli eliciting fighting behavior in sticklebacks. A few years later in his book, *Curious Naturalists*, he mentioned it as an example of how a "mistaken" response to an external stimulus could help the investigator untangle "the significant sign stimuli which release certain responses" (Tinbergen 1958, p. 270). It is not surprising that a fact as appealing as this one should have continued to travel. It is also not surprising to find the immediate context for citing the fact varying from one instance to another. When the author Robert Ardrey retold the story in his book, *The Territorial Imperative*, his emphasis was not on releasing mechanisms or mistakes in responding to stimuli, but instead on territoriality – here exemplified by the stickleback male's furious defense of his territory against anything red (Ardrey 1966, p. 96).

The appeal of the starling and stickleback cases was not simply that they offered amusing stories about the "mistakes" animals can make. The facts traveled well because they straddled the line between fact and theory. The facts instantiated ideas about how instincts work – how they build up internally and how they involve releasers and innate releasing mechanisms – and also, in the case of the sticklebacks' territorial defense, the functions instincts serve.

5. The Hawk–Goose Experiments

Let us consider now a set of experiments conducted jointly by Tinbergen and Lorenz over a span of nearly three months in spring 1937 at Lorenz's home in Altenberg, Austria. Once again the facts related to the ethologists' theories about innate releasing mechanisms and the stimuli that trigger them, but this story is more complicated than the previous stories were. The presentation of the facts was somewhat uneven, and the interpretation of the facts was contested in different quarters, but the facts continued to travel, and the further one got from the original accounts, the greater became the inaccuracies in the accounts or interpretations of what it was that the ethologists had originally reported.[5]

Tinbergen and Lorenz never wrote up in detail the experiments in question, in contrast to their study of the egg-rolling behavior of the graylag goose (Lorenz and Tinbergen 1938). The results of the avian predator

---

[5] A similar point is made in the context of architecture by Schneider, this volume.

Figure 7.1. Tinbergen's figure of some of the different shapes he and Lorenz used for dummies when testing the innate fear responses of juvenile birds. "Only the models with a short neck (marked +) released escape reactions."
From: Tinbergen 1948, p. 7, © The Wilson Ornithological Society, reprinted with permission.

experiments, when all was said and done, may have seemed too fragmentary, too incomplete, and too lacking in controls. Lorenz, however, summarized the experiments briefly in a paper in 1939, and Tinbergen described them even more briefly – but with eye-catching (and mind-catching) illustrations – in a popular article in 1939, a scientific paper in 1948, and in two books: *The Study of Instinct* (1951) and *The Herring Gull's World* (1953a).[6] The experiments followed up on some observations by Heinroth and an experimental study by the ornithologist Friedrich Goethe. Goethe found that young grouse, when presented with models moved by wires, showed more fear toward a model of a predator than toward other models of similar complexity (Goethe 1937). Tinbergen and Lorenz proceeded to test the reactions of hand-reared fowl of various species to different simulated flying predators, the latter being cardboard dummies of a variety of shapes. The experimenters strung up a rope between two tall trees some fifty to one hundred yards apart (Tinbergen said fifty yards in one account, one hundred in another) and pulled the dummies along the rope to mimic the motion of birds in flight (see Figure 7.1).

The experiments were conducted on a variety of birds, including young graylag geese, turkeys, and numerous species of ducks – virtually all the young fowl that Lorenz had at Altenberg in the spring of 1937. The cardboard dummies were pulled along the rope above the birds at different speeds and in both directions. The results were not identical from one species to the next. The fear reactions of young geese, for example, were elicited by slowly moving models *no matter what the models' shape*. For the young turkeys, however, the shape of the moving dummy apparently made

---

[6] Tinbergen's field notes from the experiments, dated from March 16 to June 11, 1937, are among the Tinbergen papers preserved at Oxford University at the Bodleian Library, Department of Special Collections and Western Manuscripts.

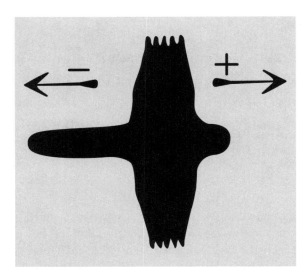

Figure 7.2. Tinbergen's figure of "a card board dummy that elicits escape reactions [in young turkeys] when sailed to the right ('hawk') but is ineffective when sailed to the left ('goose')." *From*: Tinbergen 1948, p. 34, © The Wilson Ornithological Society, reprinted with permission.

a difference. Dummies with "short necks" elicited the turkeys' alarm calls much more readily than did dummies with "long necks." Most remarkably, the investigators found they could evoke these results with a single, relatively crude dummy constructed with the "wings" located toward one end of the body in such a way as to make one end of the body short and the other end long. Which end appeared as the neck and which end appeared as the tail depended on the direction in which the dummy was pulled. The young turkeys displayed more alarm when the dummy was moved slowly above them with the short end (short neck) forward, and less alarm when the dummy was moved above them with the long end (long neck) forward. The naturalists concluded that the young turkeys were frightened when the model resembled a hawk (short neck first), but not frightened when it resembled a goose (long neck first). (See Figure 7.2.)

Tinbergen and Lorenz understood the young turkeys' response to the gestalt of the slowly moving, short-end-forward shape as an innate response, forged by natural selection, to an environmental cue signaling "predator." To use the terms that would come to be prominent in ethological writings, an environmental "sign-stimulus" triggered a special central nervous system mechanism (called an "innate schema" or "innate releasing mechanism" [IRM]), thereby releasing the performance of an instinctive motor action. The conceptual appeal of this case to Lorenz and Tinbergen was that it showed how well the IRM was tuned to the stimulus situation that triggered it. As Tinbergen took care to insist, the young birds were not responding to the shape of the model in and of itself; they were responding to the particular configurational stimulus of the model when it was *moving* short end first (as would a hawk) (Tinbergen 1948, p. 35).

Here, certainly, was an experiment involving facts[7] – and with an explanation to go with them. The facts traveled far enough beyond the bounds of ethology to elicit a challenge in 1955 from three American psychologists: Jerry Hirsch, R. H. Lindley, and E. C. Tolman. The psychologists undertook to replicate the Tinbergen-Lorenz experiment using white leghorn chickens. Failing to get similar results, the psychologists summarized their findings as follows: "The Tinbergen hypothesis that certain specifically shaped sign stimuli innately arouse a fear response was tested on the white Leghorn chicken and found to be untenable under controlled laboratory conditions." (Hirsch et al. 1955, p. 280) This conclusion drew critical, indeed scornful, responses from both Tinbergen and Lorenz. As Tinbergen put it: "Whatever the shortcomings of 'ethological' studies may be, one thing they have demonstrated convincingly: the fact that different species usually behave differently in the same situation." The obvious implications of this were that "*Facts found in one species, or hypotheses formed about one species, simply cannot be disproved by testing another species*, under however well 'controlled laboratory conditions.'" He additionally observed that the white leghorn chicken was a poor choice as a test animal, stating: "[I]t is known that the behaviour of domesticated forms often differs considerably from that of the wild ancestral forms." (Tinbergen 1957, pp. 412–3). Lorenz echoed Tinbergen's criticism of the American psychologists' paper. It was as if, Lorenz said, one scientist reported finding melanins in the fur of wild hamsters and another scientist claimed to refute this by saying that his own studies on white laboratory rats showed the hamster results to be untenable. (Lorenz 1965, p. 100).

The ethologists' response was valid. It does not negate it to note that some of Tinbergen's accounts of the 1937 avian predator experiments were vague

[7] Lorenz was much more explicit than Tinbergen in noting that the different species involved in the avian predator experiments responded differently to the stimuli presented to them. In several of his publications (1939, 1948, 1957) Tinbergen noted or implied such interspecific differences, but in the most accessible of his publications, *The Study of Instinct* (1951) and *The Herring Gull's World* (1953a), he did not. Not until his response to the critique by Hirsch et al. did Tinbergen underscore the importance of interspecific differences. A new discussion of the hawk/goose experiments by Schleidt et al. (accepted for publication) stresses the differences between Lorenz's and Tinbergen's published reports. For antecedents to the ethologists' experiments and for additional discussion, see Gray 1966.

about interspecific differences or did not mention them at all,[8] or that the paper by Hirsch et al. was more nuanced in its claims than its concluding quotation would suggest (for Hirsch's response to Tinbergen, see Hirsch 1957), or that Lorenz in some of his writings was inclined to leap from one species to another in a way that often left other scientists uncomfortable. Nor does it negate it to note that in 1961, Wolfgang Schleidt, one of Lorenz's students, reported new experiments on the reactions of young turkeys to simulated predators and offered a different conclusion than the one that both Lorenz and Tinbergen had offered. Schleidt explained the behavior of the Lorenz-Tinbergen turkeys as a result of habituation rather than as proof of an innate releasing mechanism adapted to a moving, short-necked shape. By his account, the young turkeys had seen flying ducks and geese more often than they had seen flying predators prior to the experiments with the dummies, and they were thus more frightened by the less familiar, short-neck models than by the more familiar, long-neck models. (Schleidt 1961). Tinbergen decided that Schleidt's habituation explanation was correct, and he employed it in his 1965 Time-Life book, *Animal Behaviour.*

We are not yet done, however, with the hawk–goose story, which has still more to offer with respect to the theme of facts traveling. Despite the doubts that Schleidt's turkey experiments cast on the interpretation of the original experiments of 1937, images from Tinbergen's 1948 paper and from his 1951 book, *The Study of Instinct,* continued to be reproduced, and prominently so. They appeared on the covers of two important animal behavior textbooks of the mid-1960s: Peter Marler's and William J. Hamilton III's *Mechanisms of Animal Behaviour* (1966) and Aubrey Manning's *An Introduction to Animal Behaviour* (1967). The cover of the Marler and Hamilton book featured the various shapes of the dummies that Tinbergen had reported using in the experiments. Manning's cover featured one of Tinbergen's versions of the "hawk-goose" model (see Figure 7.3). (The author thanks Wolfgang Schleidt for calling his attention to these covers.) Manning was familiar with Schleidt's study, and he cited it. Manning's overall conclusion, nonetheless, was that "there is evidence that wild birds do possess an IRM which enables them to respond to birds of prey on the first occasion that they see them. This IRM probably has different properties in different species but short neck and relative speed of movement are among them." (Manning 1967, p. 53).

---

[8] In *The Study of Instinct,* referring to the results of other experimenters as well as to his own experiments with Lorenz, Tinbergen stated: "The reactions of young gallinaceous birds, ducks, and geese to a flying bird of prey are released by the sign-stimulus 'short neck' amongst others" (Tinbergen 1951 p. 77).

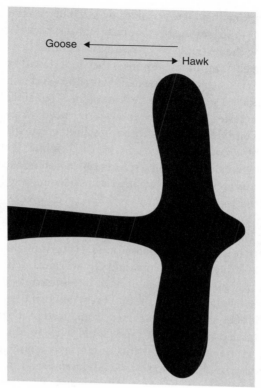

Figure 7.3. The book cover of Manning [1967].
From: © Manning, reprinted with permission.

Not only was the (now disputed) fact from the Tinbergen/Lorenz experiments still traveling, with images accompanying it, the fact moved into the public arena. As reported by Schleidt (personal communication with the author), S. Dillon Ripley, the distinguished ornithologist who was secretary of the Smithsonian Institution and an admirer of the work of Lorenz and Tinbergen, became a proponent of the idea that the silhouette of a raptor could deter songbirds from flying into windows. Under his direction the Smithsonian Museum shop began marketing raptor stickers to put on windowpanes. Today on the Internet, where facts seem to travel freely even when disclaimers of them are traveling as well, one can find web sites that continue to advertise raptor silhouette stickers as a means of preventing birds from flying into windows, even though more authoritative web sites (of such organizations as the Audubon Society, the Cornell Laboratory of Ornithology, or the United States Fish and Wildlife Service) indicate that ornithologists have concluded that raptor silhouettes do not serve as effective deterrents against birds flying into windows.

It bears remarking that there was nothing in Tinbergen's and Lorenz's original experiments to suggest that a *stationary* raptor silhouette would have any
effect in eliciting the instinctive fear responses of other birds. Motion was a
critical factor in eliciting all the responses they reported. If raptor silhouettes
showed up for sale in the Smithsonian's gift shop (and at numerous other
sites), this was not the result of close scrutiny of the original Tinbergen/Lorenz
experiments but instead a response to a serious need, that of reducing the
high number of bird deaths caused by birds colliding with glass windows.

What about the continuing fortunes of the original Tinbergen/Lorenz
hypothesis that young birds of some species possess an innate recognition of
the shapes of birds of prey? That hypothesis has been revisited since Schleidt
offered his alternative explanation in 1961. Several different studies of young
ducks (including Green et al. 1968; Mueller and Parker 1980; and Canty and
Gould 1995) all found that ducklings respond more strongly to a moving
"hawk" model than to a moving "goose" model (measured in terms either of
escape behavior or heart rate). Canty and Gould concluded on the basis of
their own experiments and their analysis of the "mixed results" of previous
studies that "the original suggestion by Lorenz and Tinbergen, that young
Anseriformes [ducks and geese] recognise some feature diagnostic of birds
of prey, seems to be correct." (Canty and Gould 1995, p. 1095.) The abstract
of the paper states more specifically that Lorenz and Tinbergen found "that
goslings respond more to moving hawk silhouettes than to moving goose
shapes." As we have seen, however, Lorenz and Tinbergen found that
although the young *turkeys* responded differently to the hawk model versus
the goose model, the young *geese* did *not*. What had traveled best over time,
it seems, were not the facts of the original experiments, but a more general
claim along with a striking image – the hawk/goose model – enhanced by
the association with ethology's founders, Lorenz and Tinbergen.[9]

6. Traveling Beyond Ethology's Borders

Let us take a brief but broader look now at the metafact of interspecific behavioral differences to see how this metafact traveled beyond ethology's borders.
This fact overarched the ethologists' project of developing detailed descriptions – "ethograms" – of the behavioral repertoires of the different species
they studied. It was also central to the ethologists' complaint that American
comparative psychology was not really comparative at all, insofar as it focused
almost exclusively on studies of the white rat. That this posed a problem for

---

[9] Merz, this volume, also considers what and how facts travel with scientific images.

psychology was signaled not only by the Continental ethologists but also by a few American comparative psychologists as well. Prominent among them was the Yale psychologist Frank Beach. In his 1950 article, "The Snark was a Boojum," Beach indicted American comparative psychology for not having been genuinely comparative for a good many years. He proved his point by analyzing the papers published in the field's primary journals over the previous four decades. In addition to displaying his findings graphically, he characterized the predicament with a cartoon, inspired by the story of the Pied Piper of Hamelin. In the cartoon, the familiar roles of humans and rodents were reversed. A rat (a white one) played the tune, while the people, a crowd of scientists, followed eagerly behind the rat, unaware they were being led to their doom. When Beach went on to discuss the potential benefits of a genuinely comparative approach, the first two authors he cited (even though he was talking about learning rather than instinct) were Tinbergen and Lorenz. He cited Tinbergen for his studies on learning in the hunting wasp. He cited Lorenz for his observations on imprinting in precocial birds.

Beach's case deserves a more extended examination than can be offered here. He played an important role in bringing the Continental ethologists and American comparative psychologists together for their mutual enlightenment. He was enlisted by Tinbergen as the first American on the editorial board of the field's new journal, *Behaviour*, and he became a regular participant in the international ethological congresses and a member of the organizing board for these conferences. In a 1955 letter to the Cambridge ethologist W. H. Thorpe, Lorenz bragged how he had made a convert out of Beach by showing him films. In Lorenz's words: "The best means to convince people that there is such a thing as instinctive movements is the film. I played duck films to Frank Beach until he nearly fainted, he got seriouser and seriouser and in the end he said in a small voice: 'You know I did not believe a word of it and now I believe everything'" (Burkhardt 2005, pp. 401–2). Judging from a paper Beach published the same year, however, it is hard to countenance the idea that he had come to believe *everything* that Lorenz wanted him to believe about instinct. In his 1955 paper, entitled "The Descent of Instinct," Beach suggested that when the development of behavior in the individual came to be properly analyzed, the concept of "instinct" would not be needed. Clearly if Beach felt there were things that American comparative psychologists could learn from Continental ethology, he likewise thought there were insights that ethologists could gain from American comparative psychology.

Between 1950 and 1970, the notion of interspecific differences in behavior (together with other biological notions) made major inroads into psychology. This was no small matter, because the ethologists' attention to interspecific differences in behavior had the potential of destabilizing the comparative psychologists' whole premise of general laws of learning, as well as destabilizing a whole research tradition that focused too exclusively on learning in the white rat. Here we can simply note a few benchmarks in the story. Among these was when Keller Breland and Marian Breland, two professional animal trainers, reported (1961) that the cherished principles of operant conditioning that they had been taught simply failed to work when an animal subject's instincts got in the way. The Brelands recounted various examples of what they called "instinctive drift." Instead of completing the tasks they had been conditioned to do, raccoons preferred "washing" the objects with which they were dealing, while pigs preferred rooting with objects. In the Brelands' words, "After 14 years of continuous conditioning and observation of thousands of animals, it is our reluctant conclusion that the behaviour of any species cannot be adequately understood, predicted, or controlled without knowledge of its instinctive patterns, evolutionary history, and ecological niche" (1961, p. 684).

A decade later, the *American Psychologist* published a long essay by Robert B. Lockard entitled, "Reflections on the decline of comparative psychology." Lockard maintained that comparative psychology's premises about animal learning had become wholly outmoded by the findings of ethology and behavior genetics and by the general incursion of biological thinking into behavior studies. He concluded with the admonition that if psychologists wanted to have an animal-based understanding of human behavior, they would need to study behavioral homologies (as exhibited in related groups of animals) and behavioral analogies (as exhibited by animals that had adapted to comparable ecological settings). "For this reason," he said, "animals of all sorts suddenly become relevant to psychology, relevant for the sharpening of scientific tools, not for casual and direct extrapolations to human behaviour" (Lockard 1971, p. 177).

7. From Animals to Humans

What then of the ethologists' own extrapolations from animal behavior to human behavior?[10] Here, characteristically, Lorenz rushed forward, and Tinbergen had to offer cautionary notes in his colleague's wake. In the 1950s and 1960s, Lorenz offered a number of pronouncements about the

---

[10] Historically, the extrapolation process has gone both ways, but the present paper does not attempt to address the rich and complex issue of how assumptions about human behavior have historically influenced the formulation of "facts" of animal behavior.

biological bases of human behavior and about the light this shed on the human condition. Two of these were especially striking. The first was that the human species is unique among higher animals in that it lacks innate inhibitions against killing its own kind. The second was that aggression is an instinct, and that, as such, it builds up internally, like a fluid in a reservoir, eventually requiring release. Lorenz presented both of these claims essentially as facts, though neither is credited with that status today, nor were they universally regarded as such when Lorenz first elaborated them. We will look at the first of these claims in some detail but simply note the other in passing.

Lorenz highlighted the first claim with a striking contrast between wolves and doves, offered at the end of his popular book, *King Solomon's Ring* (1952). Wolves, he allowed, have been equipped by evolution not only with fearsome weapons – their strong jaws and their sharp teeth – but also powerful, instinctive inhibitions against using these weapons against other wolves. When two wolves fight and one gets the better of the other, Lorenz explained, if the loser submissively exposes its neck to its adversary, the victor cannot finish the loser off. Instinctive inhibitions prevent it from doing so.

Doves, in comparison, have no powerful natural weapons. Because of this, they have not had to develop inhibitions against hurting their own kind. In nature, by Lorenz's account, when two doves fight, the bird that loses can simply fly away. If the birds are confined to a cage, however, fleeing is impossible, and the weaker bird is in danger of being killed because the winner has no innate inhibitions against continuing the fight to the end. Lorenz described how he placed a male turtledove and a female African blond ringdove together in the same cage, hoping they would mate. When he returned, he found that the ringdove had nearly pecked the turtledove to death.

Were there implications here for the human species? Lorenz believed there were. Indeed, he had offered the same biological morality tale as early as 1935, when he introduced it in an article in the *Neues Wiener Tagblatt*. (There, interestingly enough, his comparison had been between humans and hares, not doves. He evidently decided subsequently that doves, as "the very symbol of peace," made a better story.) The human species, he explained in *King Solomon's Ring*, is more like the dove than the wolf when it comes to dealing with its own kind. Humans do not have powerful natural weapons, like wolves do, and thus until relatively recently, evolutionarily speaking, humans have had no need to develop strong instinctive inhibitions against killing one another. Unfortunately, in the latest stages of our history, our science and technology have far outpaced our biological evolution. We have developed artificial weapons of tremendous destructive power without developing instinctive inhibitions against using them.

There are several problems with the story Lorenz constructed. In citing the example of a ringdove nearly pecking to death a bird of another species, a turtledove, he was begging the question of whether animals of the same species kill each other. Beyond this, he was completely ignoring the testimony of one of the twentieth century's leading expert on dove behavior, Wallace Craig, whose work Lorenz knew, and who had explicitly insisted that doves of the same species do *not* go on fighting each other to the death.[11] Most damaging to Lorenz's story, however, was the arrival of new facts showing that Lorenz was wrong.[12] Field studies of animal behavior in the years after Lorenz made his claim have established that higher animals other than humans do indeed kill members of their own species, and not simply as an occasional accident, as Lorenz maintained, but more systematically, as in the case of male lions killing off the cubs of other sires, or male chimpanzees killing the infant chimps or other members of other chimp tribes. These last findings would have made little sense to Lorenz, given his predilection for "good of the species" type arguments. They become more understandable – and hence more mobile themselves – in the context of the kind of "selfish gene" thinking that has developed since the 1970s.

As for Lorenz's theory of aggression, it was controversial from the time he enunciated it in his best-selling book in 1966, *On Aggression*. Without rejecting the fact of man's aggressiveness, the critics were inclined to reject Lorenz's claim that aggression is an instinct that builds up internally and requires release. Among the critics was Niko Tinbergen. In his inaugural lecture as professor of animal behavior in the Department of Zoology at Oxford, he addressed the theme: "On War and Peace in Animals and Man" (Tinbergen 1968). There he expressed reservations, albeit gently and diplomatically, regarding Lorenz's *On Aggression* and another recent bestseller, *The Naked Ape*, written by Tinbergen's own former student, Desmond Morris (1967). Tinbergen maintained that it was unwise to extrapolate from a few selected animal species to humans. What ethology could offer of benefit to the study of the human species, he said, was its general biological

---

[11] The difference between Craig's account and Lorenz's account is discussed in more detail in Burkhardt (2005, pp. 451–3).

[12] Before Lorenz's claim of man's uniqueness as a killer of members of his own species was shown to be wrong, it was repeated by other authors (e.g., Storr 1968). Tinbergen in 1968 wrote, "Man is the only species that is a mass murderer, the only misfit in his own society" (Tinbergen 1968, p. 1412).

approach (its integrated attack on the "four questions" of behavioral survival value, causation, development, and evolution), not specific results from studies of other animal species. (As Tinbergen saw it, psychologists continued to be inclined to neglect questions of the survival value and evolution of behavior, and their analyses suffered accordingly.)

8. MACOS

Tinbergen's hope that one might use the *approaches* of animal behavior in understanding human behavior (rather than extrapolating directly from animal behavior to human behavior) was shared by certain educators in the United States in the mid-1960s. These were the educators who developed the federally funded social science curriculum for elementary school students entitled "Man: A Course of Study" – MACOS. The prime mover of the curriculum was Jerome Bruner, the cognitive psychologist who was cofounder and, as of 1960, director of the Center for Cognitive Studies at Harvard. Bruner's desire was to "form the intellectual powers" of the students whom the MACOS curriculum was intended to serve, namely elementary school students in the fifth and sixth grades. He wanted students to become self-conscious about their strategies of thought. The content of MACOS was identified in 1965 as "man: his nature as a species [and] the forces that shaped and continue to shape his humanity." The three recurring questions of the course were to be: (1) "What is human about human beings?" (2) "How did they get that way?" (3) "How can they be made more so?"[13]

Bruner regarded the exploration of contrasts as a particularly effective tool for learning. Early on, the developers of MACOS thought of using a single animal species, the savannah baboon, as a contrast with humans. The trouble with this approach, as it turned out, was that the elementary school pupils saw baboons as being so similar to humans they had trouble identifying strong differences between the two (Education Development Center 1976, p. 26). To underscore certain differences more carefully, the educators introduced two more species: the Pacific coast salmon and the herring gull. Young salmon must do without parental protection in their struggle

[13] J. S. Bruner, *Man: A Course of Study*. Occasional Paper No. 3, The Social Studies Curriculum Program, Educational Services, Inc. (Cambridge, MA, 1965), p. 4. Cited in Dow, Peter B., "Man: A Course of Study: A continuing exploration of man's humanness," in Education Development Center, Inc., *Man: A Course of Study. Talks to Teachers. 1983 Edition* (Curriculum Development Associates: Washington, D.C., [1970a] 1983), p. 4.

to survive. Their story was used to highlight the significance in humans of the length and the quality of the *human* infant's dependence on its parents. Herring gull chicks, unlike young salmon, are taken care of by their parents. The gull story, based on the work of Tinbergen, was used to examine more closely the *causes* of animal behavior. Observations of how the gull chicks must peck at the red spot on their parent's beak if they are to be fed provided an entry to the discussion of innate versus learned behavior. The herring gull section also helped introduce the idea that behavior patterns, like physical structures, should be understood in terms of their adaptiveness or survival value (see Figure 7.4). Beyond this, the herring gull study was intended to give children the opportunity to study territoriality, fighting, and communication. The authors of MACOS suggested that children are intrigued by the idea of an aggressive instinct and that the gull study would allow them to "consider the ways a human handles his aggressive feelings without really fighting." They recommended that children be given a chance to act out scenes of adult male fighting in herring gulls, where the use of particular bodily gestures enables the antagonists to escape serious harm. They also suggested that the teachers go to Tinbergen's *The Herring Gull's World* and Lorenz's *On Aggression* for helpful background reading. As an "optional reading assignment," the curriculum developers suggested to the teachers: "You or one of the better readers in your class might read to the children parts of the last chapter of Lorenz's *King Solomon's Ring*. This chapter describes many instances of animals fighting each other, and Lorenz discusses the gestures they use to keep from inflicting serious harm upon each other" (Education Development Center 1970b, pp. 21–4.)

It bears emphasizing that the curriculum designers were not interested primarily in the transmission of facts per se, though facts were inevitably part of the curriculum (e.g., when the chick pecks at the red spot on the bill of its herring gull parent, the parent regurgitates food for the chick, or "from every six or seven eggs laid, only one chick survives to reproduce"). Rather, they were interested in developing the students' thinking strategies. Teachers were thus instructed how to set the scene before showing a filmstrip on gull behavior:

> [T]he students are scientists, students of animal behaviour, beginning study of herring gull behaviour. They are on an island off the New England coast in the early spring. In order not to be seen by the birds, they have erected a "hide" from which to observe the birds. At this point, you can show the last frame of the filmstrip, which pictures the hide, and students can agree upon rules for observing the birds, such as the need for silence. (Education Development Center 1976, p. 33)

The instructions continue: "Now, with the room darkened, the filmstrip is shown frame by frame. Students should note what they think they know about the birds and any questions they have as they observe." After the first viewing, the filmstrip would be viewed again, "this time with discussion and questions."

The teachers (and students) had abundant access to facts about gull behavior via the filmstrip just mentioned, a film, a workbook, two books of gull photos, two books by Tinbergen (*The Herring Gull's World* and the Time-Life book *Animal Behaviour*), and additional materials. Discussions of innate versus learned behavior in conjunction with the gull unit helped the students learn to avoid anthropomorphic explanations of the gull's behavior (again, see Figure 7.4). All in all, the facts, concepts, and interpretive strategies of ethology appear to have traveled very well to this new curriculum.

The MACOS teachers were introduced to the concept of natural selection by a short piece written by the evolutionary theorist Robert Trivers. Trivers's concluding observation was that one could not legitimately talk about higher versus lower animals, or more evolved versus less evolved animals. As he put it, "[In] different environments, different characteristics are adaptive." Expressing a theme that would recur at different levels through the course, Trivers wrote: "There are no traits in this scheme that have an absolute value, an absolute value irrespective of the environment."[14]

The notion of no traits having an absolute value irrespective of the environment was what ultimately caused trouble for the MACOS curriculum. Perhaps no one would have objected if the story had stopped with herring gulls or even baboons, but when it was applied to human behavior, as exemplified by the lives of Netsilik Eskimos (who practiced senilicide and infanticide), this was too much for those who believed that human values are God-given.

In her book, *Science Textbook Controversies and the Politics of Equal Time*, Dorothy Nelkin describes what transpired. United States Congressman John Conlan of Arizona characterized MACOS in 1974 as "a Godawful course," "almost always at variance with the beliefs and values of parents and local communities." He urged that National Science Foundation appropriations for MACOS be terminated because of its "abhorrent, repugnant, vulgar and

---

[14] Robert Trivers, "Natural Selection," in Education Development Center, Inc., *Man: A Course of Study. Talks to Teachers* (Curriculum Development Associates: Washington D.C., [1970a] (1983 edition), pp. 35–41, quotation on p. 41. The booklet *Talks to Teachers* also included a section by Tinbergen entitled "The Study of Animals," extracted from his 1965 Time-Life book *Animal Behaviour* and a section by Irven Devore, with the assistance of R. Trivers and I. Rothman, entitled "Innate and Learned Behaviour."

The urge to mate is aroused by the sight of a familiar gull tossing its head.

The urge to incubate is aroused by the sight of eggs.

The urge to feed chicks is aroused by the pecking of the chicks.

Figure 7.4. A page from the MACOS student workbook on herring gulls, woodcut illustrations with accompanying explanations (p. 13).
From: © EDC, reprinted with permission.

morally sick content" (Nelkin 1977, p. 112). Federal funds were withdrawn, and textbook sales dropped sharply between 1974 and 1975. One can only assume that information about gulls was an incidental casualty of these events and thereafter did not appear so often (i.e., no longer "traveled well") within the social science curriculum. Pupils not taught the MACOS curriculum would have little likelihood of learning about chicks pecking at the red spot on their parents' bills, the different habits of the ground-dwelling herring gull versus the cliff-dwelling kittiwake, or the threat postures with which gulls defend their territories.

9. Meaningful Comparisons

While MACOS stirred up one angry group, Edward O. Wilson's book, *Sociobiology: The New Synthesis*, published in 1975, stirred up another. The most vocal protesters in the second case were not conservative, fundamentalist Christians but instead the radical scientists who constituted the Sociobiology Study Group of Science for the People. In their attack on Wilson, they lumped him together with Lorenz as a biological determinist, noting, too, Lorenz's past association with the Nazis.

For his part, Wilson had praised Lorenz and other recent writers of popular ethology books (Lorenz 1966; Morris 1967; Ardrey 1970; Tiger and Fox 1971) for "calling attention to man's status as a biological species adapted to particular environments," but he had also sought to put some distance between his science and theirs. Their problem, he indicated, was that "They selected one plausible hypothesis or another based on a review of a small sample of animal species, then advocated the explanation to the limit." The correct way to use comparative ethology, Wilson insisted, "is to base a rigorous phylogeny of closely related species on many biological traits." Even this, however, by Wilson's assessment, did not allow one to deduce with confidence how human social behavior had evolved. One could identify character traits that persisted throughout the primates, he said, and then expect these characters to be conserved relatively unaltered in *Homo sapiens*, but he also specifically noted, "the comparative ethological approach does not in any way predict man's unique traits" (Wilson 1975, p. 551).

Such statements would seem to suggest that Wilson would be very careful in making comparisons between species. At least some of his critics, however, did not think so. One of his critics (appropriately enough from the standpoint of this paper) was the American psychologist Frank Beach. The man who had chastised his fellow comparative psychologists a quarter of a century earlier for failing to pay sufficient attention to interspecific differences found occasion to use this charge again in commenting upon Wilson and his followers. Prominent among Beach's complaints was "[the sociobiologists'] apparent omission or disregard of facts concerning inter-specific similarities and especially interspecific differences." Explaining the flaws in Wilson's discussions of what Wilson blithely termed "homosexual-ity" in a wide range of animals, Beach wrote:

> There is a fundamental rule that applies to all such cases whether the compar-ison is between animals and humans or between different species of animals. *The validity of interspecific comparison is limited by the reliability of intraspecific analysis.* Meaningful comparisons between Species A and Species B simply are not possible until the behaviour in question has been analyzed with equal care, objectivity, and precision *in both species.* (Beach 1978, p. 131)

10. Conclusion

Meaningful comparisons – these were the basic idea behind Lorenz's identification of his field as "comparative behaviour study." Comparisons constituted the "next step" after one did one's detailed gathering of facts about the complete behavioral repertoires of diverse, individual species. Demonstrating the metafact of interspecific *differences* in behavior was an important achievement in itself, and, as we have seen, it took some time for this metafact to make its way from ethology to comparative psychology. Beyond this, students of behavior remained faced with the task of deciding when, how, or if the facts of behavior identified in one species could help understand the facts of behavior of another species. In this paper we have observed a variety of instances of facts traveling, justifiably or not, from one setting to another. Among other things, we have seen that association with an authority, a story, a theoretical claim, an image, a moral lesson, or a desired practical application could have a bearing on a fact's traveling power.[15] We have also had glimpses of the opposite, where the association of facts with an unwelcome theory or a tainted authority stood to constrain the facts' further travels.

Inevitably, the comparisons that humans find most fraught with sig-nificance are those that involve humans themselves. The question thus posed to students of animal behavior is this: What can the study of animal behavior contribute to an understanding of human behavior? Earlier in

---

[15] For other examples of such associations that allow facts to travel, see in this volume the chapters by Haycock (traveling with an authority), Adams (a story), Merz (an image), and Whatmore and Landström (a desired practical application).

this paper, when discussing Oskar Heinroth's comments on animal and human social instincts, we spoke of this as the "sooner or later" motif in animal behavior studies. But paired with this is the other motif that we saw Frank Beach expressing – namely, that in seeking implications for humans in the study of animal behavior, close scrutiny is critical in constituting the facts in the first place, for all the species concerned, before judging which kinds of comparisons are meaningful – that is, how far the facts in question might appropriately travel. These motifs are mirror images of each other. The ongoing tension between them has much to do with the perennial fascination of animal behavior studies, where one must continually address the question of traveling facts. Those facts that travel far are not always those whose travels might be considered legitimate in terms of offering or enabling meaningful comparisons. So the question for such studies is not which facts are traveling well, but which facts are good candidates for doing so – and which are not.

Acknowledgments

Elements of this paper are based on the author's book (Burkhardt 2005), the research for which was supported by funding from the National Science Foundation (SOC78–05922 and SBE9122970), the John Simon Guggenheim Foundation (1992–1993), and the Research Board of the University of Illinois at Urbana-Champaign. The author is also pleased to acknowledge the helpful suggestions of all the participants in LSE's "Facts" project, and he thanks especially Edmund Ramsden, Mary Morgan, and Peter Howlett for involving him in the April 2007 workshop "Facts at the Frontier."

Bibliography

Ardrey, Robert. 1966. *The Territorial Imperative*. New York: Atheneum.
   1970. *The Social Contract: A Personal Enquiry into the Evolutionary Sources of Order & Disorder*. New York: Atheneum.
Beach, Frank A. 1950. "The Snark was a Boojum," *American Psychologist*, 5:115–24.
   1955. "The descent of instinct," *Psychological Review*, 62:401–10.
   1978. "Sociobiology and interspecific comparisons of behaviour," in Gregory, Michael S., Anita Silvers, and Diane Sutch (eds.), *Sociobiology and Human Nature: An Interdisciplinary Critique and Defence*, pp. 116–35. San Francisco: Jossey-Bass.
Breland, Keller and Marian Breland. 1961. "The misbehaviour of organisms," *American Psychologist*, 15:681–4.
Burkhardt, Richard W., Jr. 1997. "The founders of ethology and the problem of animal subjective experience," in Dol, Marcel, Soemini Kasanmoentalib, Susanne Lijmbach, Esteban Rivas, and Ruud van den Bos (eds.), *Animal Consciousness and Animal Ethics: Perspectives from the Netherlands*, pp. 1–13. Assen: Van Gorcum.

2005. *Patterns of Behaviour: Konrad Lorenz, Niko Tinbergen, and the Founding of Ethology.* Chicago: University of Chicago Press.

Canty, Nora and James L. Gould. 1995. "The hawk/goose experiment: sources of variability," *Animal Behaviour,* 50:1091–5.

Education Development Center, Inc. [1970a] 1983. *Man: A Course of Study. Talks to Teachers. 1983 Edition.* Washington, DC: Curriculum Development Associates.

1970b. *Man: A Course of Study. Teacher's Guide Number 4: Herring Gulls.* Washington, DC: Curriculum Development Associates.

1970c. *Herring Gulls.* Washington, DC: Curriculum Development Associates.

1976. *Man: A Course of Study. A Guide to the Course. 1976 Edition.* Washington, DC: Curriculum Development Associates.

Föger, Benedikt and Klaus Taschwer. 2001. *Die andere Seite des Spiegels: Konrad Lorenz und Nationalsozialismus.* Vienna: Czernin Verlag.

Goethe, Friedrich. 1937. "Beobachtungen und Erfahrungen bei der Aufzucht von deutschem Auerwild," *Deutsche Jagd,* 6 and 7.

Gray, Philip Howard. 1966. "Historical notes on the aerial predator reaction and the Tinbergen hypothesis," *Journal of the History of the Behavioural Sciences,* 2:330–4.

Green, Ronald, W. J. Carr, and Marsha Green. 1968. "The hawk-goose phenomenon: further confirmation and a search for the releaser," *Journal of Psychology,* 69:271–6.

Heinroth, Oskar. 1910. "Beiträge zur Biologie: namentlich Ethologie und Psychologie der Anatiden," in Schalow, Herman (ed.), *Verhandlungen des 5. Internationalen Ornithologen-Kongresses in Berlin, 30 Mai bis 4. Juni 1910,* pp. 589–702. Berlin: Deutsche Ornithologische Gesellschaft.

Heinroth, Oskar and Konrad Lorenz. 1988. *Wozu aber hat das Vieh diesen Schnabel? Briefe aus der frühen Verhaltensforschung, 1930–1940,* Koenig, Otto (ed.). Munich: Piper.

Heinroth, Oskar and Magdalena Heinroth. 1924–1934. *Die Vögel Mitteleuropas in allen Lebens- und Entwicklungsstufen photographisch aufgenommen und in ihrem Seelenleben bei der Aufzucht vom Ei ab beobachtet,* 4 vols. Berlin: H. Bermühler.

Hinde, R. A. 1990. "Nikolaas Tinbergen," *Biographical Memoirs of Fellows of the Royal Society,* 36:547–65.

Hirsch, Jerry. 1957. "Careful reporting and experimental analysis – a comment," *Journal of Comparative and Physiological Psychology,* 50:415.

Hirsch, Jerry, R. H. Lindley, and E. C.Tolman. 1955. "An experimental test of an alleged innate sign stimulus," *Journal of Comparative and Physiological Psychology,* 48:278–80.

Lehrman, Daniel S. 1953. "A Critique of Konrad Lorenz's theory of instinctive behaviour," *Quarterly Review of Biology,* 298:337–63.

Lockard, Robert B. 1971. "Reflections on the fall of comparative psychology: is there a message for us all?" *American Psychologist,* 26:168–79.

Lorenz, Konrad Z. 1935. "Moral und Waffen der Tiere," *Neues Wiener Tagblatt,* November 15, 1935.

1939. "Vergleichende Verhaltensforschung," *Verhandlungen der Deutschen Zoologischen Gesellschaft, Zoologischer Anzeiger,* 12(supp.):69–102.

1950. "The comparative method in studying innate behaviour patterns," *Symposia of the Society for Experimental Biology,* 4:221–68.

1952. *King Solomon's Ring: New Light on Animal Ways.* London: Methuen.

1965. *Evolution and Modification of Behaviour*. Chicago: University of Chicago Press.

1966. *On Aggression*. New York: Harcourt Brace and World.

[1932] 1970a. "A consideration of methods of identification of species-specific instinctive behaviour patterns in birds," in Lorenz, Konrad, *Studies in Animal and Human Behaviour*, vol. I, pp. 57–100. Cambridge, MA: Harvard University Press.

[1935] 1970b. "Companions as factors in the bird's environment," in Lorenz, Konrad, *Studies in Animal and Human Behaviour*, vol. I, pp. 101–258. Cambridge, MA: Harvard University Press.

[1937] 1970c. "The establishment of the instinct concept," in Lorenz, Konrad, *Studies in Animal and Human Behaviour*, vol. I, pp. 259–315. Cambridge, MA: Harvard University Press.

1978. *Vergleichende Verhaltensforschung: Grundlagen der Ethologie*. Vienna and New York: Springer-Verlag.

Lorenz, Konrad Z., and N. Tinbergen. 1938. "Taxis und Instinkthandlung in der Eirollbewegung der Graugans," *Zeitschrift für Tierpsychologie*, 2:1–29.

Manning, Aubrey. 1967. *An Introduction to Animal Behaviour*. Reading, MA: Addison-Wesley.

Marler, Peter and William J. Hamilton, III. 1966. *Mechanisms of Animal Behaviour*. New York: Wiley.

Morris, Desmond. 1967. *The Naked Ape*. London: Jonathan Cape.

Mueller, Helmut C. and Patricia G. Parker. 1980. "Naive ducklings show different cardiac responses to hawk than goose models," *Behaviour*, 74:101–13.

Nelkin, Dorothy. 1977. *Science Textbook Controversies and the Politics of Equal Time*. Cambridge, MA: MIT Press.

Schleidt, Wolfgang M. 1961. "Reaktionen von Truthühnern auf fliegende Raubvögel und Versuche zur Analyse ihres AAM's," *Zeitschrift für Tierpsychologie*, 18:534–60.

Schleidt, Wolfgang M., Michael D. Shalter, and Humberto Moura-Neto. (Forthcoming) "The Hawk/Goose story: the classical ethological experiments of Lorenz and Tinbergen, revisited," *Journal of Comparative Psychology*.

Storr, Anthony. 1968. *Human Aggression*. New York: Atheneum.

Taschwer, Klaus and Benedikt Föger. 2003. *Konrad Lorenz: Biographie*. Vienna: Paul Zsolnay Verlag.

Tiger, Lionel and Robin Fox. 1971. *The Imperial Animal*. New York: Holt, Rinehart and Winston.

Tinbergen, N. 1939. "Why do birds behave as they do? Part II," *Bird Lore*, 41:23–30.

1942. "An objectivistic study of the innate behaviour of animals," *Bibliotheca Biotheoretica*, 1:39–98.

1948. "Social releasers and the experimental method required for their study," *Wilson Bulletin*, 60:6–51.

1951. *The Study of Instinct*. Oxford: Clarendon Press.

1953a. *The Herring Gull's World: A Study of the Social Behaviour of Birds*, London: Collins.

1953b. *Social Behaviour in Animals*. London: Methuen.

1957. "On anti-predator responses in certain birds – a reply," *Journal of comparative and physiological psychology*, 50:412–14.

1958. *Curious Naturalists*. London: Country Life Limited.

1963. "On aims and methods of zoology," *Zeitschrift für Tierpsychologie*, 20:410–33.

1965. *Animal Behaviour*. New York: Time-Life Books.

1968. "On war and peace in animals and man: an ethologist's approach to the biology of aggression," *Science*, 160:1411–8.

Wilson, Edward O. 1975. *Sociobiology: The New Synthesis*. Cambridge, MA: Harvard University Press.

EIGHT

TRAVELLING FACTS ABOUT CROWDED RATS: RODENT EXPERIMENTATION AND THE HUMAN SCIENCES

EDMUND RAMSDEN

1. Introduction

It is a commonplace that the findings of laboratory experiments will come to do work outside the laboratory walls. Less often explored is how that process is actually effected – more so in the case of the social and behavioural sciences. In what follows, we will see how facts generated by one particular series of experiments with rodents came to stand as evidence for social scientists, planners, architects, environmentalists and population activists concerned with human problems. How did it happen that these specific claims were received by such a broad audience and became evidence for diverse claims? The laboratory experiments were crowding studies conducted by the animal ecologist and psychologist John B. Calhoun at the National Institute of Mental Health (NIMH) from 1954 until 1986.[1] They explored the detrimental effects of high population density or crowding among various strains of laboratory rats and mice. Calhoun's first paper documenting his results – "Population Density and Social Pathology," published in *Scientific American* in 1962 – rapidly became one of the most widely referenced in psychology and in studies of urban populations. It even became a source of information for the design of buildings, such as hospitals, prisons and college dormitories, by architects, planners and psychologists. Although concern with crowding was not new, as one psychologist

---

[1] After training as an ecologist, and then being employed as such at Ohio State University and Johns Hopkins University, he then went on to redefine himself as a psychologist after joining the NIMH at the National Institutes of Health (NIH), until his retirement in 1986, describing himself as having "rejected, though not lost sight of, my "father," the discipline of animal ecology" (Calhoun 1972a).

recollected: "[I]t was a study by Calhoun with rodents that stimulated social and environmental psychologists into action" (Paulus 1988, p. 1).[2]

However, for the sociologist Amos Hawley, this interest in Calhoun's work within the human sciences was a "curious phenomenon" (1972, p. 522). Certainly, Calhoun had presented his rats and mice as models for man: He argued that with increased population density, social animals became overloaded by unwanted interactions. This, in turn, resulted in social disorganisation and psychological, even physiological, breakdown. He graphically illustrated a variety of behavioural pathologies that emerged: aggression, withdrawal and sexual deviance, pathologies that surely resonated with concerns surrounding human population growth and urbanisation. Nevertheless, social scientists had long been careful in circumscribing the boundary between human and animal. Indeed, the interest in Calhoun's findings from "people who have so studiously held their work aloof from any comparison with findings of biological researches," was, Hawley observed, "rather ironic" (1972, p. 522). In spite of common interest in the ecology of human and animal populations, relations between social and biological scientists were strained and superficial, borrowings metaphorical (Gaziano 1996). Although Gregg Mitman (1992, p. 1) describes ecology as "the borderland between the social and the biological sciences through the study of the interrelationships between and among individual organisms and their environment," this was a border that was carefully policed. Many in the social sciences felt a great deal of distrust towards biological modes of explanation – fears that were hardly allayed by the claims of biologists such as Raymond Pearl that Malthusian problems were better studied through "lower forms of life in the laboratory, under physically and chemically controlled conditions, than from any manipulation of never quite satisfactory demographic statistics" (Pearl 1925, p. 5).

The aim of this paper is not only to establish how, but how *well*, the facts of crowding pathology, generated in the rodent laboratories of NIMH,

---

[2] The study of the crowd had been central to the emergence of social psychology and urban sociology in the late nineteenth and early twentieth centuries, the work of Gustav Le Bon (2002 [1895]) in France and Wilfred Trotter (1919) in Britain stimulating the development of group psychology (Freud 1922; Pick 1995), and the ecological sociology of Robert Ezra Park (1904) that would become central to the so-called "Chicago School" that rose to prominence in the interwar era (McPhail 1991). Interest in crowding processes had, however, waned by the 1940s and 1950s. Crowd psychology was not only considered to lie outside of the realm of the experimentalist, but to be politically dubious – the study of the malevolent and atavistic "group mind" by early "folk" psychologists such as Le Bon and Sir Martin Conway having an unhealthy association with social Darwinist, eugenicist and totalitarian ideals (cf. Conway 1915; Ash 2005).

travelled to an alternative setting: the cities, institutions and buildings of the social and behavioural scientist. I will be focussing particular attention on the field of environmental psychology, which grew out of social psychology in the late 1960s and was concerned explicitly with the effects of the physical environment on human behaviour. Although, as we shall see, Calhoun has come to be credited with the more general (and overly simplistic) notion that high levels of population density lead to social pathology, his experiments contained a cluster of facts: a range of pathologies determined by a number of social processes such as group size, interaction and hierarchy. These processes were, in turn, influenced by the physical elements of space and numbers, elements that could be controlled to achieve a variety of behavioural responses, some negative and some positive. For Calhoun, not only were these facts interdependent, but within this interdependence lay their relevance for mankind – as normal social relations became disrupted in crowded (and thus poorly structured and designed) environments, abnormal and destructive behaviours would emerge, culminating in physiological breakdown.

In transferring Calhoun's studies to the human setting, environmental psychologists interpreted his experiments in different ways, privileging some facts over others. These different interpretations were then reinforced by alternative experimental methods: some turning to the laboratory, seeking to provide controlled and objective studies of the effects of density on behaviour; others turning to the field, studying crowded conditions in prisons, hospitals, schools and colleges, conditions they considered more faithful to those experienced by Calhoun's rats and mice. I will be focussing on two leading examples of these alternative approaches used by psychologists. The first is a series of human experiments in the laboratory pioneered by Jonathan Freedman, the second, a series of field studies carried out in the college dormitory by Andrew Baum and Stuart Valins.[3]

Both approaches began with a similar set of assumptions – that crowding would result in behavioural pathologies such as aggression; both reached a similar set of conclusions – that crowding pathology could be ameliorated through the more effective design of social and physical environments. Both, however, came to these conclusions in different ways, their alternative experimental practices building upon different interpretations of

---

[3] Freedman was based first at Stanford University and then Columbia; Baum was first at Trinity College, Connecticut, and later, the School of Medicine of the Uniformed Services University, while Stuart Valins was based at the State University of New York at Stony Brook. Baum, in particular, would dedicate much of his career as an environmental psychologist to the problems of crowding and stress.

Calhoun's facts and their uses to environmental psychology as a discipline. For Freedman, Calhoun's experiments provided the psychologist with the opportunity of an objective measure of the effects of the physical environment – in this case, density, or number of individuals per square unit area – on social behaviour. Yet in Calhoun's flawed experimental design, Freedman also believed that one could see the naïveté and pessimism of the biologist: that increased physical density or crowding would result directly and inevitably in social pathology.

For Baum and his colleagues, Calhoun was not simply concerned with density, but with the more complicated processes of social interaction. They also noted how Calhoun's work allowed for optimism: Through designing buildings, cities and networks that controlled social interaction more effectively, it was possible to adapt to increased population density. As we shall see, the varying degrees of influence of these interpretations of what Calhoun's most significant facts were would determine the perception of his work by a generation of social and behavioural scientists.

2. The Ecology of Crowding: From Rodent Utopia to Urban Hell

In 1940, Raymond Pearl encouraged Calhoun to explore the growth of populations in confined environments while he was working for his PhD in zoology at Northwestern (Calhoun 1977). After a number of temporary appointments, Calhoun finally found the opportunity to do so when employed on the Rodent Ecology Project (1946–1949) at Johns Hopkins University School of Public Health and Hygiene.[4] The project increasingly sought to control Baltimore's burgeoning rat population by ecological means – through the control of space. Under the direction of the ecologists John T. Emlen and David E. Davis, the project succeeded in reducing rat populations through restricting the access of animals to nesting sites and food sources. Increased competition for resources reduced the number of rats that could survive in a given area.

The scientists were, however, left with a puzzle. Why was it that rat populations stabilised at a certain level? And why did blocks of the same size have different-sized populations? These would remain stable through time,

---

[4] The project was established during World War II, supported first by the Office of Scientific Research and Development and the City of Baltimore, and later, by the Rockefeller Foundation and the United States Public Health Service. In war-time Britain and the United States, rodents were considered a greater problem due to the losses of food-stuffs and the spread of disease. See Chitty (1996) and Keiner (2005).

even when there was an increased availability of food. Even after a successful period of poisoning, rat populations returned to their original pre-poisoning level (Calhoun 1963, pp. 1–3). Seeking to understand the underlying social and biological forces that functioned to regulate population size, Calhoun and his colleague, John J. Christian, turned to the laboratory. In his first experiments, Calhoun enclosed a population of wild Norway rats in a quarter-acre pen near his home in Towson, Maryland, and let them breed. The population grew rapidly, protected as they were from predation and supplied with ample food and water. Extrapolating from the size of an individual laboratory cage, Calhoun calculated that the size of the pen could, theoretically, support up to 5,000 adults. After two years, however, their numbers had never exceeded 200 and stabilised at 150.[5] Once employed at the NIMH from 1954, his experiments became increasingly sophisticated – first in a rented barn near Rockville, Maryland, and later in a specifically constructed animal research centre near Poolesville. He now designed numerous "rodent universes" – room-sized pens that could be viewed from the attic above via windows cut through the ceiling. Providing a variety of strains of rats and mice with all their physical needs, he described his experimental universes as a "rat utopia" or "mouse paradise."[6] The one resource lacking was space. As the populations grew, this became increasingly problematic. Rats and mice achieve sexual maturity within five weeks of birth, and are in oestrus every four or five days thereafter. Gestation is not usually more than twenty-two days, and litters can number twelve pups. Within months, the pens heaved with animals. As one of Calhoun's assistants put it, rodent "utopia" had rapidly become "hell" (Marsden 1972, p. 9).

It was a hell characterised by a series of pathologies. Most striking was the dramatic increase in levels of aggression and violence. These behaviours are hardly alien to rat populations, dominant males using violence to establish control over territory and females. This would prove important to Calhoun's experiments, as we can see in the panel from his *Scientific American* article published in 1962. Calhoun had divided the universe into four sections, connected by ramps. Each pen was constructed to support twelve adults, which Calhoun believed to be the maximum, optimal group

---

[5] Although Calhoun did publish a few research papers on these experiments (see Calhoun, 1952), these were short pieces in ecological journals, not conducive to travelling into the social and behavioural sciences. The publication of his text on the subject, *The Ecology and Sociology of the Norway Rat*, was delayed until 1963.

[6] Extensive materials relating to the design of these experiments (including some physical remnants of the model environments themselves) can be found among the John B. Calhoun Papers, National Library of Medicine (NLM), NIH, Bethesda, MD.

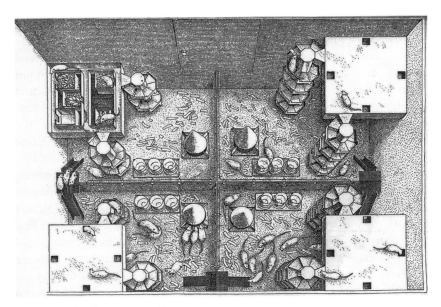

Figure 8.1. "Rodent universes."
From: Calhoun, 1962, pp. 140–1, © Scientific American.

size for this domesticated strain of Norway rat, and many other social animals, including man. Dominant male rats were able to defend the pens with only one ramp to guard. These so-called "despots" lived in relative comfort with a harem of eight to ten females. As Calhoun allowed this population to grow, to eighty in the first instance, this meant that the other rats were crowded into the other two pens. Here the males were unable to establish any kind of meaningful territory or hierarchy. As a result, they became trapped in cycles of perpetual violence, Calhoun describing them as "going berserk, attacking females, juveniles and the less active males, and showing a particular predilection – which rats do not normally display – for biting other animals on the tail" (1962, p. 146), see Figure 8.1.

The aggression of these male "probers" extended into their sexual behaviour. Increasingly hypersexual, they relentlessly pursued oestrous females. More subordinate males became "pansexual," even "homosexual." Females began neglecting their young – failing to build proper nests and, in some cases, attacking their own offspring. As a consequence, mortality ran at 96 per cent in the crowded sections of the pens, the dead pups cannibalised by the adults. With such levels of violence and depravity, many rats and mice simply withdrew, ensuring their physical existence but at an immense psychological cost. These animals looked like healthy rats and mice – well fed with luxuriant coats – but this was because they no longer behaved as rats and mice. The males among them did not compete for territory, status, food or mates, preferring to huddle together as a large vacant mass on the floor, moving only to eat, drink and groom themselves excessively. With an ever-increasing proportion of the population existing in this withdrawn, asexual state, the population was destined for extinction.

As macabre as the experiments may seem, Calhoun was one of a number of biologists who had dedicated themselves to the study of crowding in animal populations over several decades. As the experiments of John J. Christian showed repeatedly, population density resulted in increased levels of interaction and competition, which then led to social disorganisation or social "strife," and finally, to a breakdown in bodily systems (Christian 1950, 1961). While Calhoun focused on the behavioural pathologies associated with social strife, Christian focused on the physiological effects of stress as identified by the physiologist Hans Selye (1936, 1950). In a series of experiments beginning in the 1930s, Selye's laboratory rats were drowned, traumatised in treadmills and injected with a variety of noxious agents, all resulting in the same physiological pathologies: adrenal hypertrophy, atrophy of lymphatic structures and the ulceration of stomach and duodenum. Following a period of study for the Naval Research Academy in the 1950s, Christian joined a number of pathologists, endocrinologists and ecologists such as Robert L. Snyder and Herbert Ratcliffe at the Penrose Research Laboratory at Philadelphia Zoo. Here attention was focussed on the problems of density and stress in a number of species such as chickens, hamsters, voles, deer, monkeys and, of course, rats and mice (Ratcliffe 1968).

Those such as Christian and Snyder were seeking to address a question fundamental to modern population ecology. From the 1920s, many had become increasingly concerned with identifying the various density-dependent mechanisms that ensured that a species rarely, if ever, outstripped their food supply, a process that would seem to be illogical when viewed from an evolutionary perspective (Mitman 1992; Kingsland 1995). For some species, the mechanisms were more clumsy and destructive than others. Periods of population growth were followed by population crashes – none more spectacular than the cycles of boom and bust among the Norwegian lemming. While the famed British ecologist, Charles Elton, had sought explanations in cosmic events, others identified more traditional Malthusian factors, such as the limits of food sources or the spread of disease (Erickson and Mitman 2007). Yet the argument that density was itself an inhibitory factor was becoming increasingly popular. Through the study of population dynamics in the laboratory, biologists had sought to identify that various social, physiological or behavioural pathologies that emerged under conditions of extreme density could function as population control mechanisms – such as reproductive dysfunction in mice or cannibalism among flour beetles (McAtee 1936; Southwick 1971; Chitty 1996). The idea that stress served a homeostatic regulatory function was an important addition to this mode of thinking.

Although concerned primarily with animal ecology, these scientists did see their work as relevant to human population problems. For Ratcliffe, Philadelphia Zoo could "serve as a continuing experiment in which many kinds of animals share the environmental defects with which urban man surrounds himself" (1968, p. 243). Yet they were often cautious in extrapolating directly from animal to man in terms of *behavioural* similarities, preferring to ever more carefully document the *physiological* effects of crowding stress, following Selye's cue (Ratcliffe focused on hypertension and heart disease in chickens). These effects, they supposed, were shared by all animals, including man; yet it was up to others to establish scientifically the cause and effect of stress among human populations.

Calhoun, in contrast, sought to transgress directly the boundaries between the social and the biological, animal and man.[7] His first foray yielded little reward, however. In a paper delivered to the 1957 Cold Spring Harbor Symposium on Quantitative Biology, he used the concept of "social welfare" to make direct connections between the ecology of animal populations and urban environments. With the help of the mathematical biophysicists Murray Eden and Nikolas Rashevsky, he formulated a "mechanics of social interaction which might be equally applicable to mouse or man" (Calhoun 1957, p. 349). Applying this complicated model to human beings, he argued that the growth of economies and populations had been mutually reinforcing. Unfortunately, optimal population size for economic advance was greater than that required for social well-being. As a consequence, growth had contributed social, psychological and physiological stress. His approach had precedent: Hadn't sociologists such as Georg Simmel (1950 [1903]) and Louis Wirth (1938) famously identified crowded urban spaces as sources of conflict, frustration and psychological breakdown?[8] Wasn't

---

[7] For an analysis of the negotiation over the animal–human boundary in ethology, see Burkhardt, this volume.

[8] For Simmel and Wirth (Park's respective tutor in Germany and colleague at Chicago), individuals were overloaded by social contacts and stimuli in the city, leading to fleeting social relations and personality disorders, such as anomie, loneliness and social withdrawal.

Amos Hawley (1950) attempting to resuscitate an interest in ecology that had proven so inspirational to sociologists of earlier generations?

Nevertheless, response to the paper was muted and (where present) critical. Calhoun's 1962 *Scientific American* article was, in contrast, a runaway success. Why the turn-around? Most obviously, these were very different forms of publication. The annual Cold Spring Harbor meetings were enormously influential among population scientists, his invitation to attend, prestigious. Contributions were generally sober and considered attempts at community building within and between the biological sciences (broadly conceived) through theory and method. In this regard, the 1957 meeting between animal ecologists and human demographers was an abject failure. Social scientists reacted unfavourably to the ecologist Lamont Cole's description of demography as "a branch of general biology," and to his crediting of Pearl as the man who had moved "populations into the laboratory where they can be studied under controlled, or at least specifiable, conditions" (Cole 1957, pp. 5–6). In response, the sociologist and demographer Frank Lorimer (1957, p. 17) criticised the tendency of biologists such as Pearl to apply complicated and abstract mathematics to conditions that were contingent on varying, and fundamentally social, processes. The boundaries between human and rodent were not going to be traversed through opaque mathematical formulations.

The *Scientific American* was a more popular venue, one that allowed Calhoun to communicate with a broad range of scientific expertise. Although the publication now targets the general public, in the 1960s, it was still a press where the latest, most cutting-edge and often controversial research was published for scientists interested in issues outside their own areas of expertise. It was an important site of knowledge transfer, or what Peter Howlett (2008) describes as a "communication space" – a perfect vehicle for Calhoun to transmit his facts about crowded rats beyond the confines of animal psychology, ecology and ethology to the social and medical sciences and the design and planning professions more generally.

In his 1962 article, Calhoun simply relied on the descriptive power of his experiments. He told a story – and a dramatic one at that. He provided a careful and detailed account of the range of pathologies on display, from parental neglect, hypersexual and homosexual behaviour, withdrawn "somnambulists," to animals "going berserk" (Calhoun 1962 pp. 144 & 146). He included a number of images, most notably an overview of the pen (see Figure 8.1) as he and his collaborators saw it from the 3-by-5-foot glass window cut in the floor of the hayloft above the rodent universe (Hall 1966, p. 25). Readers were therefore invited to witness the various pathologies present in the "lower" pens: two rats fighting, probers relentlessly pursuing a female over a ramp, a group of dishevelled males waiting on another ramp for the opportunity for a raid on an adjacent pen once the dominant male fell asleep.

Calhoun (1962, p.144) also coined a very influential phrase, the "behavioral sink." This described a process of pathological "togetherness" by which animals were drawn to the more crowded food hoppers and water bottles, the result of animals becoming so used to contact when eating and drinking that they begin to associate these processes with the presence of others. Calhoun (1962, p. 144) noted that the "unhealthy connotations of the term are not accidental." Neither was his choice of words – the anthropologist Edward Hall describing the connotations of the word "sink" to "mean a receptacle of foul or waste things" (Hall 1966, p. 26).[9] Hall, a leader in the study of man's negotiation of physical space and interaction, took up the term in his own writing, as did many scientists, writers and public commentators concerned with urban degeneration. Indeed, the attractiveness of Calhoun's writing can be gauged by the success of his term – behavioural sink – in travelling to the work of writers and journalists of the day. Through Hall, Tom Wolfe was introduced to Calhoun's work, dedicating a chapter of his *The Pump House Gang* (1968) to the problems of crowding, titled: "Oh Rotten Gotham! Sliding Down into the Behavioral Sink." Calhoun's terminology was designed to travel well and to carry his facts with it, Hunter S. Thompson described the phrase as "a word jewel," "a flat-out winner, no question about it."[10]

In the context of urban living, these were trying times for the American population. The 1960s had witnessed rioting across a number of cities, the most famous being the Watts Riots in Los Angeles, where, over a period of six days in the summer of 1965, 34 people died, 1,032 were injured and 3,952 arrested (Boskin 1969). Seeking an explanation, many turned to the problem of crowding. Density not only led to social discontent by intensifying

---

[9] Building upon Calhoun's work and that of the Swiss zoo psychologist Heini Hediger, Hall established the field of "proxemics": the study of personal space and interaction. Through books such as *The Hidden Dimension*, Hall was a leading figure in establishing the idea that the majority of communication between humans was physical rather than verbal. Although the measurable distances between people as they interact differed in various cultures – northern Europeans having larger personal space requirements than the southern – he grounded these spatial needs in biology, common as they were across species.

[10] Thompson to Wolfe, 21 April 1968, in Thompson (2001). For a detailed analysis of Calhoun's cultural impact among writers and journalists, see Ramsden and Adams (2009). The relationship between literature and travelling facts is also explored by Adams and Schell, both this volume.

poverty and lowering living standards, but could lead directly to an increase in violent behaviour just as in Calhoun's rats and mice. Coupled to concerns with the so-called population "explosion" or population "bomb," the riots signalled a future of increased urban, perhaps even global, unrest.[11] As the Canadian demographer and population-control advocate Nathan Keyfitz (1966, p. 873), declared:

> [f]ood riots in Bombay, and civil riots in Newark, Memphis, and even Washington, D. C. This ultimate manifestation of population density, which colors the social history of all continents, is a challenge that can no longer be deferred... It will not cease until population control is a fact.

The 1960s and 1970s also saw the rapid rise of social movements concerned with the protection of the environment, urban ecology and the threat of nuclear war – all of which were connected to the problems of population growth (see Connelly 2008). Indeed, for those concerned with population control, Calhoun's rats proved a most useful tool in demanding increased support from governmental and non-governmental agencies for the promotion of family planning technology. When accepting the Nobel Peace Prize for his contribution to the "Green Revolution," Norman Borlaug (1999, p. 477) advocated population control. The influence of Calhoun's work was evident when he spoke of Malthus's failure to foresee:

> ... the disturbing and destructive physical and mental consequences of the grotesque concentration of human beings into the poisoned and clangorous environment of pathologically hypertrophied megalopoles. Can human beings endure the strain? Abnormal stresses and strains tend to accentuate man's animal instincts and provoke irrational and socially disruptive behavior among the less stable individuals in the maddening crowd.[12]

While others worked on the problems of crowding among animals, Calhoun's experiments were, by far, the most influential, and as the psychologist Irwin Altman observed, the "most dramatic" (Altman 1975, p. 170). Calhoun had made the study of crowding *behaviour* his own, and further, had identified a wide range of pathologies – violence, sexual depravity, withdrawal – that resonated with the range of problematic behaviours associated with city life. Reflecting this breadth, Lewis Mumford, the noted critic of urban planning argued:

---

[11] The "population bomb" is also discussed by Adams, this volume.

[12] Borlaug received the Nobel Prize in 1970, and was concerned that even if the Green Revolution rid human societies of famine, they would have to cope with the social and psychological problems that resulted from increased population density.

No small part of this ugly urban barbarization has been due to sheer physical congestion: a diagnosis now partly confirmed with scientific experiments with rats – for when they are placed in equally congested quarters, they exhibit the same symptoms of stress, alienation, hostility, sexual perversion, parental incompetence, and rabid violence that we now find in Megalopolis (Mumford 1968, p. 210).

Through the concept of the "behavioural sink," Calhoun's work even suggested why people continued to be drawn to high-density living in spite of the pathology. In one experimental universe, so many of man's social ills seemed confirmed and explained in relation to crowding. Calhoun was complicit in this spread, writing in his 1962 paper of the potential relevance to man:

> In time, refinement of experimental procedures and of the interpretation of these studies may advance our understanding to the point where they may contribute to the making of value judgments about analogous problems confronting the human species. (Calhoun 1962, p. 148)

Although this statement was relatively cautious, it jarred with the language he began to adopt in later publications and in his own research notes: The aggressive males became "delinquents," the behaviour of the mothers to their young akin to "infant abuse" and "battered child syndrome," the healthy yet withdrawn individuals became the "beautiful ones," or occasionally "social misfits" and "dropouts": "Autistic-like creatures, capable only of the most simple behaviours compatible with physiological survival... Their spirit has died" (Calhoun 1973a, p. 86).[13] In a telling interview, a journalist from *Newsweek* asked if the "phenomenon of the beautiful ones" was evident in "the dropout, drug culture?" Calhoun stated that he could give "no scientifically provable reply..., but rather to my surprise," wrote the reporter, "he did not think the question ridiculous."[14] Calhoun was clearly determined that people recognise, and act upon, the problems of population density

---

[13] For examples of Calhoun's vivid use of language, see Calhoun (1972b, 1973b and 1973c, 1976) and also "Universal Autism: Extinction Resulting from Failure to Develop Relationships," March 25, 1986, Calhoun Papers, Box 18, NLM. Also enlightening is his anthropomorphic use of language in his written notes when observing his rodent universes.

[14] Stewart Alsop, "Dr. Calhoun's Horrible Mousery," *Newsweek*, August 17, 1970, p. 96. Many examples of how such journalistic pieces led to further media and academic interest can be found in Calhoun's papers. One Stanford Friedman, professor of pediatrics and psychiatry at the School of Medicine and Dentistry, University of Rochester, asked for further reprints from having read the *Newsweek* article (letter, August 13, 1970, Box 1, Folder: Reprint Request 1970, NLM). In response to the *Newsweek* editorial, *The Washington Post* carried its own article on Calhoun's work (Box 49, Folder: 8 narrative H items [1968–91], NLM).

among human populations – problems that had become evident through his studies of rats and mice.

3. Studies of Human Crowding

Important to the success of Calhoun's rat facts travelling so far and wide was their intersection with growing scientific interest in questions of space and numbers in the 1960s and 1970s. Coming from the biological sciences, theories of "territoriality" were particularly influential. Ethologists such as Konrad Lorenz (1963) argued that all animals (humans included) required a degree of personal space and territory. With increased density, territorial boundaries were more frequently transgressed – leading inevitably to aggressive response. Therefore, it was necessary, argued popular-ethologists Robert Ardrey and Desmond Morris, to design urban environments in accordance with these basic biological needs (Ardrey 1966; Morris 1967). Such ideas extended into architecture where leading figures such as Ian McHarg at the University of Pennsylvania drew from ethology and Calhoun specifically, in his demands that man design with, rather than against, the laws of the natural world (McHarg 1964, 1969).

In the social sciences, too, Calhoun's work was used in demands for the physical to be included in the study of the social. Pessimism in the failings of American society was tempered by perceived opportunities for change. Sociologists saw the ideas of founding fathers reflected in Calhoun's work and thought to engage once again with the study of density would allow them to realise an interdisciplinary ecology, an opportunity that earlier generations had missed. William Michelson, for example, argued that previous sociologists had used ecology as a mere method, dividing the urban landscape into convenient areas to measure relationships between variables that were social and economic, not spatial or physical:

> ... [T]hese rat experiments gave some clue as to the social concomitants of the physical pathologies observed in previous behavioral sinks... With these findings as incentive, one might well imagine a groundswell of activity among human ecologists to apply this perspective to human life... Yet... pioneers of ecology have left the study of these crucial phenomena behind them in the dust. (Michelson 1970, p. 199)

Some had taken interest, however. They accepted Calhoun's facts about crowded rats, and now sought to replicate his work: "We... take the animal studies as a serious model for human populations" (Galle, Gove and MacPherson 1972, p. 23). Through identifying Calhoun's crowding pathologies among human populations, psychologists and sociologists believed that they would open up a new area of research, often discussed but rarely studied effectively: the effects of the physical environment on social behaviour.

Clearly, however, a student of human populations faced a problem. One cannot place human beings in a laboratory, however "utopian" its surroundings, and breed them to extinction. Therefore, social and behavioural scientists adopted a number of research strategies that they believed capable of uncovering the processes of crowding in human beings in ways that were comparable to Calhoun's studies with rats and mice.

Sociologists turned to statistical data gathered in censuses and surveys as a means of correlating density, measured by persons-per-acre or persons-per-room, to "a series of variables similar to or suggested by Calhoun's work" (Winsborough 1965, p. 123). Sexual deviance was measured through cases of sexual assault, and violence through crime statistics. Most controversial was the decision to measure the breakdown in maternal behaviour in Calhoun's laboratory by families on welfare (Schmitt 1963, 1966; Galle et al. 1972; Gillis 1974).[15]

Calhoun's work was not only connecting with existing social and scientific trends, it was also inspiring new approaches. It was particularly influential among a new generation of social psychologists – a generation disaffected with the failure of the previous generation to address the problems of space and numbers (Altman 1978, pp. 7–8). The study of density offered this emerging field of environmental psychology a seemingly objective, quantifiable and transferrable measure – number of individuals per square unit area – that would have clear and important behavioural consequences. It is to these adaptations of Calhoun's approach to the study of human beings in laboratory and field that we will now turn.

Jonathan Freedman began his studies of crowding among human beings as a recent PhD in psychology at Stanford University in the late 1960s, studies that he continued through the 1970s when appointed as professor of psychology at Columbia.[16] His approach reflected the traditional interest

---

[15] For a detailed and very critical appraisal of the use of such measures, see Harvey Choldin (1978).

[16] Freedman, a recent PhD, had been encouraged to look at the relations between crowding and behaviour by his colleague at Stanford, the biologist and population control activist, Paul Ehrlich, with whom he collaborated. For Ehrlich, evidence that crowding caused stress and that it was, therefore, detrimental to mental and physical health and social order would be a useful weapon in promoting population control policy. See also, Freedman et al (1971) and also, Ehrlich and Holdren (1972).

of the social psychologist in the laboratory, seemingly making Calhoun's approach easily transferrable to the needs of the social scientist. In a series of experiments, Freedman and various collaborators recruited high school and university students and placed them in different-sized rooms, ranging from large – one hundred sixty square feet, to small – thirty-five square feet, while keeping the size of the groups, ranging from four to ten people, constant. By doing so, they could arrive at an objective measure of density. They then assigned these individuals a number of tasks – object-uses, word-association, concentration, coordination and public speaking, and games of competitiveness, such as the prisoner's dilemma. Combining density with such exercises, they intended to establish its effects in terms of stress, discomfort, aggression, competitiveness and general "unpleasantness."

Freedman claimed to have started his research "with the familiar naive assumption that crowding was 'a bad thing' and would have negative effects on people's behavior" (1975, p. 78). He then went on to describe having been "startled" by the results. As he reported in a series of influential papers and a book, *Crowding and Behavior*, density seemed to have no appreciable negative effect (Freedman 1971, 1975; Freedman et al. 1971, 1972, 1975). This was further supported by a statistical study of density, delinquency and mental illness in New York (Freedman 1972). Similar to other statistical studies in sociology, the results were inconsistent: Although some variables were positively correlated to density, others were not, and some researchers even identified an inverse relationship (Schmitt 1966; Mitchell 1971, 1974; Cassel 1972; Factor and Waldron 1973; Booth 1976).

Inconsistency encouraged criticism, the following statement in a review of the field by two environmental psychologists being typical:

> Everybody assumed that lack of space would be harmful. When in 1975 Freedman announced that he could find no evidence that this was so, there was a collective gasp, similar to the one that must have occurred when a child pointed out that the emperor had no clothes. (Russell and Ward 1982, p. 673)

Early optimism in uncovering a direct relationship between crowding and pathology began to dissipate. The focus now shifted away from establishing the pathological consequences of density toward factors that mediated its effect. In discussions of human crowding, the accounts taken from animal studies were increasingly simplified. Calhoun's precise delineation of physical space that had become the defining feature of his experiments was now reduced to the simple causal claim – increased density leads to increased pathology. The complexity of Calhoun's approach and his findings were ignored: Although increased density *could* result in pathology, this also relied on a series of subsidiary facts – the degree to which an area was defensible, the control an individual had in limiting unwanted interaction, the numbers in (as opposed to the size of) a physical space and so on. Calhoun's studies had now been stylised, even caricatured. There was now a distinction to be made between density, a crude physical measure as envisaged by the animal ecologists, and crowding, a subjective response – an individual could only *feel* crowded (Stokols 1972). This feeling was mediated through a range of social and psychological factors, such as an individual's personal history and their relations with others present. Factors such as space and numbers were important, but they now functioned as one of many antecedents. One could feel crowded when alone, and not at a football match. Perception was everything, and density no longer a primary, nor even a necessary, explanatory variable. Crowding was becoming an individualised and amorphous concept, one of many environmental stressors in social psychology that could be relieved by adequate coping mechanisms. Although Calhoun's rodents may well have suffered from their experience of crowding, human beings were capable of adapting to high density through their higher intelligence and culture.

In making this distinction between density and crowding, psychologists were also making a division between physical and mental processes. They accepted that stress caused physical illness, even death, as had been shown from animal studies such as Calhoun and Christian's – a problem that required the attention of the human psychologist. However, when it came to the psychological *causes* of stress, these were not so simply comparable across species. Human beings did not "go berserk" in crowded rooms, as they were capable of controlling and adapting to situations. Thus, as Calhoun had long complained, when it came to the transfer of *physiological* facts from animal to human, this was considered acceptable; the transfer of *behavioural* facts were, in contrast, more frequently denied:

> We are still weighed down by the heritage of Descartes dualism of mind and body. We are prone to accept the similarities of body, of structure and physiological function, shared by man and other animals, but hold out for a human uniqueness of mind, that is to say, awareness and consciousness (Calhoun 1973b, p. 93).

4. Rescuing Calhoun's Facts

Although Freedman's work encouraged criticism of the density-pathology equation, and with it a shift *away* from the spatial and the physical, he rejected the growing distinction made between density and crowding. Retaining crowding as a physical variable, as "amount of space per person," was, he argued

> ... more basic and more interesting. Once people feel crowded, virtually by definition this is a negative state, and, presumably, they will respond in ways people always respond to negative states. As far as I can tell, there is nothing that distinguishes this negative emotional state from others, and, therefore, it is not of special interest to the environmental psychologist. (Freedman 1979, p. 168)

The view that he was attempting to establish a boundary between animal and human was also mistaken. Freedman was not attempting to disassociate psychology from Calhoun's rats. He was attempting to enrol them. For Freedman, his own psychological experiments that had revealed no relation between crowding and pathology were correct, and further, they were consistent with Calhoun's studies. The pathology identified in Calhoun's study had resulted not from density, but, once again, from excessive social interaction. Not all of Calhoun's rats, Freedman noted, had gone berserk. Those that managed to retain some control over space had relatively normal lives. Yet he did not credit Calhoun with this insight:

> Something more is needed. *I* would propose that the effects are due at least in part to problems involved in the social interaction among the animals. Much of the evidence indicates that the number of animals may be more important than the amount of space per animal. (Freedman 1975, p. 35, emphasis added)

In Freedman's view, Calhoun's pathologies had resulted from a poorly designed rodent universe. He had not created a rodent utopia. From the very initiation of his experiments, he had created a rodent hell. The problem with Calhoun's research was not that he had tampered with natural processes, but that he had not tampered enough.

Freedman concluded his book with a chapter "in praise of cities" in which he extolled the benefits of high density, once planners, advised by psychologists, designed environments that maintained privacy while building for community. This was in contrast to the existing situation where "many inhabitants of high-rise, low-cost housing are like caged animals, making forays from their nests to get food or clothing and then returning to 'safety'" (Freedman 1975, p. 124). Once having rescued Calhoun's rats from density-determined pathology, Freedman was able to conclude by describing human responses to poorly designed environments as "remarkably similar to those of other animals" (1980, p. 205).

Others, while supportive of Freedman's emphasis on the importance of social interaction, were nevertheless critical of his experimental approach as embodying the naiveté of much of social psychology. Behaviour needed

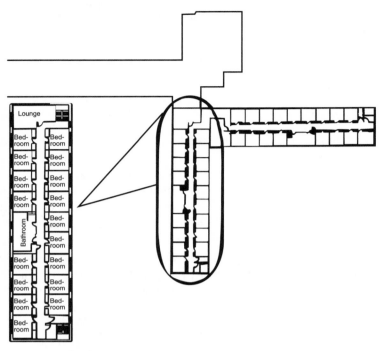

Figure 8.2. Corridor design.
From: Baum and Valins, 1977, p. 21.

to be studied in a natural setting, rather than the artificial environment of the laboratory. Freedman had only confined his subjects to a room for a short period, often for only one hour, four at the most. They were few in number and were willing, paid participants. Andrew Baum and Stuart Valins argued that such studies had, therefore, failed to study conditions analogous to Calhoun's rats – conditions that confined people against their wishes in crowded spaces for considerable periods (Baum and Valins 1977).

They turned to the college dormitory, where they carried out detailed studies using a variety of tools – statistical data, institutional records, interviews, observation and physical examination (palmar sweating seen as an indicator of stress). Unlike Freedman, they drew a distinction between density and crowding. The former was a physical condition; the latter, a psychological response mediated by the degree to which individuals had control over the frequency and quality of their interactions with others. By comparing the results of a study of two different environments, one large corridor

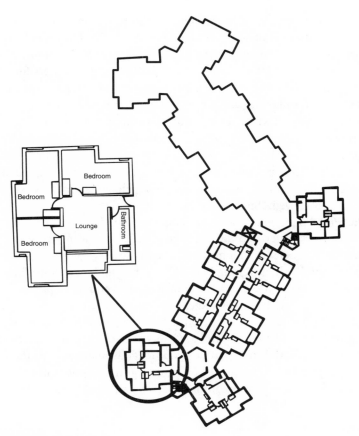

Figure 8.3. Suite-style design.
From: Baum and Valins, 1977, p. 22.

dormitory (Figure 8.2) and the other, a "suite-style" that was sub-divided into smaller communities (Figure 8.3), they provided evidence of both pathology and the potential for its amelioration. In the corridor, students perceived the environment as crowded and exhibited increased stress levels, which affected health and academic success. In the latter, designers had successfully reduced unwanted social interaction without significantly decreasing density.

These insights had a direct influence on dormitory design. In a later experiment, Baum and his colleague, Glenn Davis, were allowed to alter the design of the original corridor dormitory that they had used in their earlier studies. They replaced three bedrooms with lounge areas and provided unlocked doors bisecting the corridor. In transforming the physical structure of the corridor from long to short, they were again able to ameliorate crowding stress. This reduced the overall numbers of inhabitants on the floor from forty-three to thirty-nine, and it only cost the university $4,000 a year in lost revenues (Baum and Davis 1980, p. 480).[17] Baum would later join others focused on the study of prison environments at the University of Texas at Arlington, and again Calhoun's influence would be direct, both the researchers and Calhoun serving as key expert witnesses in a number of landmark court rulings against overcrowded prisons in the 1970s and 1980s (Paulus 1988; Schaeffer et al. 1988). Once again, the key was not simply to provide standard guidelines for space allocation, but to understand how architecture determined interaction, and consequently, health, well-being and rehabilitation.[18] As Paul Paulus (1988, p. 5) observed: "Space can't "bite," but people can."[19]

Like Freedman, Baum and his colleagues had drawn strongly upon Calhoun's work. They had identified pathologies resulting from increased and unwanted social interaction (admittedly milder forms of withdrawal, rather than individuals "going berserk").[20] They had also allowed for optimism, emphasising the importance of design over size of physical space. Yet they did not use this to challenge Calhoun's conclusions. They recognised, as had Calhoun, that behavioural pathologies would not be identical among rodents and man. Granted, Calhoun had seen the increased violence and withdrawal in urban environments as analogous to the behaviours emerging as a consequence of crowding in his rodent universes. Yet, more significantly, what was common between different species in crowded

---

[17] The housing office had originally considered reducing density through simply reducing the occupancy of each room from two to one. However, such a strategy would have been costly.

[18] Calhoun also reflected on how, having been so closely involved with psychologists, sociologists and behavioural scientists: "More recently, we have come to interact very closely with architects and city planners." Letter to Norman R. Brown, November 6, 1973. Calhoun Papers, Box 2 B, NLM.

[19] Army prison standards defined living space to be seventy-two square feet per person, dropping to fifty-five if necessary, and never to be below forty square feet in any event (Sommer 1974, p. 36). For environmental psychologists such as Paul Paulus, it was often more desirable to ensure privacy through single cells, even at the cost of space for inmates, as enforcing standards of space per person could mean that inmates could be placed in large open-plan dormitories in which they could exert little control over social interaction. For further discussion of how small changes in the details of architectural styles can make a significant difference to the usage of buildings, see Schneider, this volume.

[20] In this, they again went against a trend of associating Calhoun's rats with extreme aggression and then seeking analogous behaviours among human populations. Withdrawal was, they argued, a defence mechanism against crowding stress. If prolonged, it could become pathological, as Calhoun had argued. It is also notable that, perhaps unsurprisingly, there were few studies of crowding and sexual deviance among human populations.

environments was the breakdown at the meta-level, in higher and more complex forms of social organisation – the successful breeding and rearing of young among rodents or the intricacies of social engagement and the planning of family, education, career and life-style among human beings. Furthermore, they noted how Calhoun had also recognised that his own experiments were not simply about density, but about social interaction. Indeed, their use of Calhoun's conceptual framework is indicative of this alternative interpretation. While it is for the term "behavioural sink" that Calhoun is best remembered, he considered his concept of "social velocity" to be equally, if not more, important (Calhoun 1977, p. 30). This was a measure of social interaction by which one ranked animals: the more confident and dominant an animal, the more active; the more subordinate, the more withdrawn. Baum and Valins were amongst the few researchers to apply "social velocity" in their study of college dormitories – the more active individuals were those with the power to control the degree and quality of social interaction; those experiencing crowding stress were more likely to avoid interacting with their peers.

Calhoun had also recognised that one could control social velocity through design, and had encouraged Baum and Valins to do likewise – ensuring spaces for privacy and community could improve the quality of life among both rodents and man. This was the subject of his later rodent experiments in the 1970s, where he focused his attention on the *amelioration* of crowding pathology. Through the effective design of space and the control of interaction, determining which of the rats and mice could eat, sleep, live with whom, he sought to design ever more intelligent and collaborative rodent communities, capable of withstanding ever greater degrees of density. Calhoun had shown how to create madness; now he wanted to show how to avoid it. There are, however, fewer references to this later work, particularly among environmental psychologists, and even fewer uses of his concept of "social velocity." Although both Baum and Paulus had been explicit in their acknowledgement of Calhoun,[21] it was the simple yet powerful facts of association between population density and social pathology, along with vivid images of rats "going berserk" in the "behavioural sink," that tended to travel into the publications of social and behavioural scientists – only to be denied in the context of human populations. This wasn't the legacy he had hoped for.

---

[21] In their preface, Baum and Valins state that they are "especially indebted to John B. Calhoun," the influence of his "conceptualizations and comments" being evident throughout the text (Baum and Valins 1977: ix).

5. Good News for People Who Love Bad News

On one level, Calhoun's facts travelled very well into the social and behavioural sciences – they captured the imagination of a generation concerned with the problems of population growth and urbanisation, while offering a deceptively simple, and measurable, cause for the pathologies associated with these processes. However, as we have seen, Calhoun's experiments embodied more than a single, headline-grabbing fact – that increased density led to increased pathology. They involved many facts – not only in terms of pathologies, but of the behavioural processes and designs that determined them. Some of these facts fared well, travelling from Calhoun's rodent laboratories into the psychology of human behaviour; others less so.

This is reflected in the two alternative approaches to the study of environmental psychology explored in this chapter. Both recognised the significance of social interaction to the formation of pathological behaviour among Calhoun's rats and mice, both focused attention on to the psychology of physical space and both sought to provide a more positive assessment of mankind's future in a crowded world. In their choice of laboratory or the field, however, they interpreted and adapted Calhoun's insights in markedly different ways.

For Freedman, through the carefully designed laboratory experiments of the social psychologist, it was possible to gain a basic and objective understanding of crowding among human beings that could contribute to debates over population growth and urbanisation, debates in which he was actively involved. Like many social scientists working in these areas, he was increasingly concerned with countering what he saw as undue pessimism of biologists such as Desmond Morris and Konrad Lorenz, who, he believed, suggested social pathology was inevitable without aggressive policies to "uncrowd."[22] For Freedman, population growth was indeed a problem, yet cities of high density could provide a solution to the ecological destruction wreaked by suburban sprawl:

> The cities of the world are not doomed. They are not necessarily condemned to high crime rates, riots, and violence… the race will not destroy itself simply because it will be crowded. There is some hope for the cities and for mankind – if,

---

[22] Although Calhoun had no association with right-wing politics Lorenz did have, as Richard Burkhardt argues in his contribution to this volume. It was not uncommon for critics of Calhoun's influence to follow the discussion of his crowded rats with a critique of Ardrey, Morris or Lorenz's "killer apes," therefore tainting Calhoun's facts through association. For examples, see Freedman (1975, p. 42) and also Fischer (1978, p. 132).

instead of taking a fatalistic view or concentrating entirely on problems of population, the world turns its attention to the problems that something can be done about. (Freedman 1975, pp. 107–8)

Freedman settled, albeit theoretically, on an argument that crowding intensified existing psychological states, both negative *and positive*. In so doing, he embedded it within inequality – the main source of urban social ills. The biological perspective of Calhoun, in focusing solely upon density, would, he feared, focus attention away from this problem. He interpreted the behaviour of Calhoun's rats in this light: Those controlling space had comfortable lives in direct contrast to their subordinates.

Andrew Baum, in contrast, was involved directly in public health and the design of environments such as the college dormitory and prison. Baum and his colleagues were concerned with showing how specific architectural interventions could reduce the stress of high-density situations. In making a distinction between density and crowding, Baum retained a focus on crowding as a pathological response to the stress of unwanted social interaction. Yet his use of Calhoun was not limited to the creation of pathology, but its amelioration. Baum, along with numerous architects and planners, saw a very different message embodied in Calhoun's experimental findings – the power of design in determining, and improving, human behaviour, health and well-being. They did not cut Calhoun out of this story, but recognised that his organisation and interpretation of his experiments reflected a degree of optimism. Calhoun's facts were, therefore, not only directly relevant to the study of crowding pathology, but to providing a cure for the problems of population density. In turn, Calhoun grew closer to these behavioural scientists and planners concerned with environmental design, so much so that the high degree to which he was referenced in these fields contrasted to the declining frequency with which he was referred to in the field where he began his career – animal ecology.[23]

In contrast, Freedman's determination to provide insights of more general applicability, while countering pessimism, led him to reject the distinction between density and crowding. In so doing, he had attempted to retain a strong link between the study of rodents and human beings. Yet he had also appropriated and redefined Calhoun's facts regarding the importance of group size and social interaction. This tendency to attribute the more

---

[23] While those such as Christian and David E. Davis continued to reference him as having designed the first "rat city," it was increasingly in passing – rarely building on his new ideas. When Robert Snyder (1968), Christian's colleague at Philadelphia Zoo, completed a detailed review of the history of research on crowded animals, Calhoun's studies were not even mentioned.

positive interpretation to the social and environmental psychologist, as opposed to Calhoun, continues:

> It has been suggested, for example, that part of the reason for the behavioural sink in Calhoun's study was the use of solid food pellets that forced animals to remain in the central chamber while eating. It has also been suggested that the effects were more to do with group size then density per se. (Halpern 1995, p. 107)

There is nothing particularly problematic about this statement – what is missing is the fact that Calhoun *himself* made both points in his 1962 publication as a means of emphasising the power of environmental design. Although many of the facts that Calhoun identified were travelling, and travelling well, this was not through his publications – they were travelling through the work of others. As facts travel, they lose paternity, and also, in this case, their qualifications, complexity and detail.[24] The very success of Calhoun's 1962 publication, the exemplar of crowding studies among laboratory animals, resulted in the simplification of Calhoun's facts in the discussions of social scientists. In this way, Calhoun served a useful purpose – a dramatic means of focussing attention on the issues of density, space and numbers as important problems. Their study and control in the human context was, however, to be left to the human scientist. Thus, it was not Calhoun's interpretation of his experiments that were proving so influential, but interpretation of those experiments by psychologists.

We have seen how there were both optimistic and pessimistic perspectives contained in Calhoun's work. Although Calhoun liked to emphasise the former, ultimately, the pessimistic interpretation of his work won out.[25] Man may have shared many stress-related pathologies with animals, yet when it came to resolving problems, man was a very different beast. This was, in no doubt, reinforced by the common cultural view of the rat itself: intelligent and adaptable, yes, but always to be associated with urban degeneration, not its amelioration.[26] Indeed, in the words of one reviewer, Freedman was to be congratulated for "clearing out the rat's nest" from the study of human populations (Loyd 1977, p. 55). For those focusing on the design of institutions, Calhoun's facts continued to be used to challenge the treatment of inmates as "caged animals." For the social and

---

[24] Oreskes and Schell, both this volume, also note on the phenomenon of facts travelling without their parents.

[25] See also Oreskes, this volume, on travelling facts and their relationship to good news/bad news stories.

[26] With my colleague, Jon Adams, I have further explored the cultural significance of the "rat" in Calhoun's experiments (Ramsden and Adams 2009).

psychological sciences more generally, however, it seems that simply associating Calhoun's rodent universes with pathology, rather than its amelioration, was an opportunity considered too attractive, or perhaps too convenient, to miss.

Acknowledgements

This work was supported by the ESRC/Leverhulme Trust project, "The Nature of Evidence: How Well Do 'Facts' Travel?" (grant F/07004/Z) at the Department of Economic History, LSE. I would like to thank David Cantor, John Rees and Johnny Ho for their help and advice during numerous visits to the National Library of Medicine, NIH. I would also like to thank my colleagues at LSE for their many useful comments and insights at various stages of this paper's development, Jon Adams, Sabina Leonelli, Mary Morgan and Simona Valeriani in particular; and also Rachel Ankeny, Richard Burkhardt, Thomas Gieryn, Martina Merz and other contributors to the various "facts" project workshops.

Bibliography

Altman, Irwin. 1975. *The Environment and Social Behavior*. Monterey CA: Brooks-Cole.

　　1978. Crowding: Historical and Contemporary Trends in Crowding Research. In Andrew Baum and Yakov M. Epstein (eds.) *Human Response to Crowding*. Hillsdale NJ: Laurence Erlbaum.

Ardrey, Robert. 1966. *The Territorial Imperative*. New York: Athenaeum Press.

Ash, Mitchell G. 2005. "The Uses and Usefulness of Psychology," *Annals of the American Academy of Political and Social Science*, 600:99–114.

Baum, Andrew and Glenn E. Davis. 1980. "Reducing the Stress of High-Density Living: An Architectural Intervention," *Journal of Personality and Social Psychology*, 38:471–81.

Baum, Andrew, and Stuart Valins. 1977. *Architecture and Social Behavior: Psychological Studies of Social Density*. Hillsdale NJ: Lawrence Erlbaum.

Baum, A. and Valins, S. 1979. Architectural Mediation of Residential Density and Control: Crowding and the Regulation of Social Contact. In Leonard Berkowitz (ed.) *Advances in Experimental Social Psychology, Volume 12*. New York: Academic Press, 1979.

Boskin, Joseph. 1969. *Urban Racial Violence in the Twentieth Century*. London: The Glencoe Press.

Booth, A. 1976. *Urban Crowding and Its Consequences*. New York: Praeger.

Borlaug, Norman. 1999 [1972]. The Green Revolution, Peace, and Humanity. In Frederick W. Haberman (ed.) *Nobel Lectures, Peace 1951–1970*. Singapore: World Scientific Publishing.

Calhoun, J. B. 1952. "The Social Aspects of Population Dynamics," *Journal of Mammalogy*, 33:139–50.

1957. Social Welfare as a Variable in Population Dynamics. In *Cold Spring Harbor symposia on quantitative biology, 1957, volume 22, Population studies: Animal ecology and demography*. New York: Cold Spring Harbor.

1962. "Population Density and Social Pathology," *Scientific American*, 306:139–48.

1963. *The Ecology and Sociology of the Norway Rat*. Bethesda MD: U.S. Department of Health, Education and Welfare.

1972a. "The Population Crisis Leading to the Compassionate Revolution and Environmental Design," *World Journal of Psychosynthesis*, 4:21–8.

1972b. "Disruption of Behavioral States as a Cause of Aggression," *Nebraska Symposium on Motivation*, 20:183–260.

1973a. "Death Squared: The Explosive Growth and Demise of a Mouse Population," *Proceedings of the Royal Society of Medicine*, 66:80–9.

1973b. "From Mice to Men," *Transaction and Studies of the College of Physicians of Philadelphia*, 41:92–118.

1973c. "What sort of box?" *Man-Environment Systems*, 3:3–30.

1976. "Scientific Quest for a Path to the Future," *Populi*, 3:19–27.

1977. Looking Backward from 'The Beautiful Ones.' In W. R. Klemm (ed.) *Discovery Processes in Modern Biology*. Melbourne FL: Krieger.

Cassel, J. 1972. Health Consequences of Population Density and Crowding. In R. Gutman (ed.) *People and Buildings*. New York: Basic Books.

Chitty, Dennis. 1996. *Do Lemmings Commit Suicide? Beautiful Hypotheses and Ugly Facts*. New York and Oxford: Oxford University Press.

Choldin, Harvey M. 1978. "Urban Density and Social Pathology," *Annual Review of Sociology*, 4:91–113.

Christian, John J. 1950. "The Adreno-Pituitary System and Population Cycles in Mammals," *Journal of Mammalogy*, 31:247–59.

1961. "Phenomena associated with population density," *Proceedings of the National Academy of Sciences of the United States of America*, 47:428–49.

Cole, Lamont C. 1957. Sketches of general and comparative demography. In *Cold Spring Harbor symposia on quantitative biology, 1957, volume 22, Population studies: Animal ecology and demography*. New York: Cold Spring Harbor.

Connelly, Mathew, 2008. *Fatal Misconceptions*. Cambridge MA: Harvard University Press.

Conway, Martin. 1915. *The Crowd in Peace and War*. London: Longmans, Green and Company.

Ehrlich, P. R. and J. P. Holdren, 1972. Impact of Population Growth. In R. G. Ridker (ed.) *Research Reports, Vol. III: Population Resources and the Environment*. Washington, DC: Commission on Population Growth and the American Future.

Erickson, Paul and Gregg Mitman. 2007. "When Rabbits Became Human (and Humans, Rabbits): Stability, Order, and History in the Study of Populations," *Working Papers on The Nature of Evidence: How Well Do "Facts" Travel? 19/07*. Department of Economic History, LSE.

Factor, R. and Waldron, I. 1973. "Contemporary Population Densities and Human Health," *Nature*, 243:381–4.

Fischer, Claude. 1978. Sociological Comments on Psychological Approaches to Urban Life. In A. Baum, J. M. Singer, and S. Valins (eds.) *Advances in Environmental Psychology, Volume 1, The Urban Environment*. Hillsdale NJ: Lawrence Erlbaum.

Freedman, Jonathan L. 1971. "The Crowd: Maybe Not so Madding After All," *Psychology Today*, 5:58–61.

Freedman, J. L. 1972. Population Density, Juvenile Delinquency and Mental Illness in New York City. In Sara M. Maize (ed.) *Population Distribution and Policy*. Washington, DC: Gov Printing Office.

1975. *Crowding and Behavior*. San Francisco: W. H. Freeman.

1979. Current Status of Work on Crowding and Suggestions for Housing Design. In J. R. Aiello and A. Baum (eds.) *Residential Crowding and Design*. New York and London: Plenum Press.

1980. Human Reactions to Population Density. In M. Cohen, R. S. Malpass, H. G. Klein (eds.) *Biosocial Mechanisms of Population Regulation*. New Haven and London: Yale University Press.

Freedman, J. L. Klevansky, S., Ehrlich, Paul R. 1971. "The Effect of Crowding on Human Task Performance," *Journal of Applied Social Psychology*, 1:7–25.

Freedman, J. L., Levy, A. S., Buchanan, R. W., and Price, J. 1972. "Crowding and Human Aggressiveness," *Journal of Experimental Social Psychology*, 8:528–48.

Freedman, Jonathan L., Heshka, Stanley, and Levy, Alan, 1975. "Population Density and Pathology: Is There a Relationship?" *Journal of Experimental Social Psychology*, 11:539–52.

Freud, Sigmund. 1922. *Group Psychology and the Analysis of the Ego*. London: Hogarth Press.

Galle, O. R. Gove, W. R. and McPherson, J. M. 1972. "Population Density and Pathology: What Are the Relations for Man?" *Science*, 176:23–60.

Gaziano, Emanuel. 1996. "Ecological Metaphors as Scientific Boundary Work: Innovation and Authority in Interwar Sociology and Biology," *American Journal of Sociology*, 101:874–907.

Gillis, A. R. 1974. "Population Density and Social Pathology: The Case of Building Type, Social Allowance and Juvenile Delinquency," *Social Forces*, 53:306–14.

Hall, Edward T. 1966. *The Hidden Dimension*. New York: Doubleday.

Halpern, D. 1995. *Mental Health and the Built Environment: More than Bricks and Mortar?* London: Taylor and Francis.

Hawley, Amos H. 1950. *Human Ecology: A Theory of Community Structure*. New York: Ronald Press.

1972. "Population Density and the City," *Demography*, 9:521–9.

Howlett, Peter. 2008. "Travelling in the Social Science Community: Assessing the Impact of the Indian Green Revolution Across Disciplines," *Working Paper on The Nature of Evidence: How Well Do "Facts" Travel?* Department of Economic History, LSE.

Keiner, Christine. 2005. "Wartime Rat Control, Rodent Ecology, and the Rise and Fall of Chemical Rodenticides," *Endeavour*, 29:119–25.

Keyfitz, Nathan. 1966. "Population Density and the Style of Social Life," *Bioscience*, 16:868–73.

Kingsland, Sharon. 1995. *Modeling Nature: Episodes in the History of Population Ecology*, 2nd ed. Chicago: University of Chicago Press.

Le Bon, Gustave. 2002 [1895]. *The Crowd: A Study of the Popular Mind*. New York: Courier Dover.

Lorenz, Konrad. 1963. *On Aggression*. New York: Harcourt, Brace and World.

Lorimer, Frank. 1957. Human Populations: Historical Study, Introductory Remarks of the Chairman. In *Cold Spring Harbor Symposia on Quantitative Biology, 1957, Volume 22, Population Studies: Animal Ecology and Demography*. New York: Cold Spring Harbor.

Loyd, Bonnie. 1977. "Clearing out the Rat's Nest," *Growth and Change*, 8:54–5.

Marsden, H. M. 1972. Crowding and Animal Behavior. In J. F. Wohlwill and D. Carson (eds.) *Environment and the Social Sciences: Perspectives and Applications*. Washington, DC: American Psychological Association.

McAtee, W. L. 1936. "The Malthusian Principle in Nature," *The Scientific Monthly*, 42:444–56.

McHarg, Ian L. 1964. "The Place of Nature in the City of Man," *Annals of the American Academy of Political and Social Science*, 352:1–12.

  1969. *Design with Nature*. New York: Natural History Press.

McPhail, Clark. 1991. *The Myth of the Madding Crowd*. Hawthorne NY: Aldine.

Michelson, W. 1970. *Man and His Urban Environment*. Reading MA: Addison-Wesley.

Mitchell, R. E. 1971. "Some Social Implications of High Density Housing," *American Sociological Review*, 36:18–29.

  1974. "Misconceptions about Man-Made Space: In Partial Defense of High Density Housing," *The Family Coordinator*, 23:51–6.

Mitman, G. 1992. *The State of Nature: Ecology, Community, and American Social thought, 1900–1950*. Chicago: University of Chicago Press.

Morris, D. 1967. *The Naked Ape*. London: Jonathan Cape.

Mumford, L. 1968. *The Urban Prospect*. New York: Harcourt, Brace and World.

Park, Robert E. 1904. *Masse und Publikum* (Ph.D. thesis). Bern: Lack and Grunau.

Paulus, Paul. 1988. *Prison Crowding: A Psychological Perspective*. New York: Springer-Verlag.

Pearl, R. 1925. *The Biology of Population Growth*. Baltimore: The Williams and Wilkins Company.

Pick, Daniel. 1995. "Freud's *Group Psychology* and the History of the Crowd," *History Workshop Journal*, 40:39–61.

Ramsden, Edmund and Adams, Jon, 2009. "The Rodent Experiments of John B. Calhoun and Their Cultural Influence," *Journal of Social History*, 42:761–92.

Ratcliffe, Herbert L. (1968) "Contribution of a Zoo to an Ecology of Disease," *Proceedings of the American Philosophical Society*, 112:235–44.

Russell, J. A. and L. M. Ward. 1982. "Environmental Psychology," *Annual Review of Psychology*, 33:651–88.

Schaeffer, Marc A., Andrew Baum, Paul B. Paulus and Gerald G. Gaes. 1988. "Architecturally Mediated Effects of Social Density in Prison," *Environment and Behavior*, 20:3–20.

Schmitt, R. C. 1963. "Implications of Density in Hong Kong," *Journal of American Institute of Planners*, 29:210–7.

  1966. "Density, Health, and Social Disorganization," *Journal of the American Planning Association*, 32:38–40.

Selye, H. 1936. "A Syndrome Produced by Diverse Nocuous Agents," *Nature*, 138:32.

  1950. *Life Stress and Bodily Disease*. Baltimore: The Williams and Wilkins Company.

Simmel, Georg. 1950 [1903]. The Metropolis and Mental Life. Kurt Wolff (Trans.) *The Sociology of Georg Simmel*. New York: Free Press.

Snyder, Robert L. 1961. "Evolution and Integration of Mechanisms that Regulate Population Growth," *Proceedings of the National Academy of Sciences of the United States of America*, 47:449–55.

1968. Reproduction and Population Pressures. In Eliot Steller and James M. Sprague (eds.) *Progress in Physiological Psychology, Volume 2*. New York and London: Academic Press.

Sommer, Robert. 1974. *Tight Spaces: Hard Architecture and How to Humanize It*. Englewood Cliffs NJ: Prentice Hall.

Southwick, C. H. 1971. "The Biology and Psychology of Crowding in Man and Animals," *The Ohio Journal of Science*, 71:65–72.

Stokols, Daniel. 1972. "A Social-psychological Model of Human Crowding," *Journal of the American Institute of Planners*, 38:72–84.

Thompson, Hunter S. 2001. *Fear and Loathing in America: The Brutal Odyssey of an Outlaw Journalist 1968–1976*. New York: Simon and Schuster.

Trotter, Wilfred. 1919. *Instincts of the Herd in Peace and War*. New York: MacMillan.

Winsborough, H. H. 1965. "The Social Consequences of High Population Density," *Law and Contemporary Problems*, 10:120–6.

Wirth, L. 1938. "Urbanism as a Way of Life," *American Journal of Sociology*, 4:1–24.

Wolfe, Tom. 1968. *The Pump House Gang*. New York: Bantam.

NINE

USING CASES TO ESTABLISH NOVEL DIAGNOSES: CREATING GENERIC FACTS BY MAKING PARTICULAR FACTS TRAVEL TOGETHER

RACHEL A. ANKENY

1. Introduction

Although some critics have questioned the epistemological value of cases, the case report continues to be exceedingly popular within medicine.[1] It is estimated that 40,000 new case report publications are entered into the Medline database each year, with the core one hundred twenty clinical journals, on average, having 13.5 per cent of their references devoted to case reports (Rosselli and Otero 2002). In most of these journals, there are specific guidelines for what must be presented in a case report and what warrants reporting. *The Lancet* has a long history of publishing case reports, and began a peer-reviewed section in 1995 aimed at allowing clinicians an outlet for publication, with a particular focus on reports that have a 'striking message' (Bignall and Horton 1995). The *New England Journal of Medicine* includes brief case reports, which usually describe one to three patients or a single family, as well as case records from Massachusetts General Hospital. Many case reports begin as notifications published in the *Mortality and Morbidity Weekly Report* (*MMWR*) of the U.S. Centers for Disease Control (CDC), particularly if they are describing infectious or other types of diseases with serious public health implications. In general, the overwhelming majority of case reports depict complaints arising in specialty or subspecialty settings,

---

[1] A brief note on terminology: In this paper, I use the term 'case' to refer collectively to published, individual case reports or studies as well as case series presented as a unified whole. I include consideration of all material associated with the published case to which a reader would have access, such as introductory and concluding discussions as well as editor's notes.

and describe uncommon or even 'unique' clinical occurrences (McCarthy and Reilly 2000). A recent review of case studies noted that many cases report rare conditions for which trials of various types of therapies (particularly randomised or controlled trials) are not feasible due to low patient numbers or other issues, but that some cases are well received and can influence research as well as clinical practice (Albrecht et al. 2005). In contemporary medicine, cases may offer what is considered to be fairly definitive evidence in modern scientific terms, especially, for instance, with regard to unusual or unexpected occurrences such as adverse drug reactions (see, e.g., Aronson and Hauben 2006; Glasziou et al. 2007; Hauben and Aronson 2007). It is claimed that a good case study 'begets awareness, jogs the memory and aids understanding' (Morgan 1985, p. 353), a description that indicates the mixture of educative and epistemologic goals inherent in cases.

However, it remains unclear how we should understand the epistemic role of cases and how it is that they bring together evidence, or what here we might term 'facts.' On the one hand, case studies and reports, and at the extreme, so-called syndrome letters or pedagogical anecdotes (see Hunter 1986, 1990) are essential ways of providing information about particular clinical phenomena, usually as observed in a single or a few individuals under uncontrolled circumstances (Simpson and Griggs 1985). They allow practitioners to recognise similar patterns as new patients present themselves, and to expand their background knowledge beyond their experiences of the typical or the usual in the clinic (on the genre of cases more generally, see Hurwitz 2006). On the other hand, single cases are seen by some as problematic in as much as they are deviations even from the norm of what is abnormal, as it were.[2] They capture exceptions rather than rules, and heighten practitioners' awareness that their field is, in fact, a 'science of particulars' (to use the term coined in Gorovitz and MacIntyre 1976), or even, as often claimed, an art rather than a science. They highlight the importance of clinicians having diagnostic 'puzzle-solving' skills and not only scientific knowledge, a fact that has been well recognised in popular culture in Oliver Sacks's books and, more recently, the television series *House, MD*.[3]

Despite the notes of scepticism detailed previously, their ubiquitousness alone seems to indicate that cases can play crucial roles in various

---

[2] An interesting example of a single case and problematic travelling facts is presented by David Haycock, this volume.

[3] For a discussion of Sacks's work in a different context, see Jon Adams's contribution to this volume.

stages in the practice of medicine and medical science. This paper focuses on one of the earliest stages of medical practice, namely the development of hypotheses and questions for research in the process of establishing new or modified diagnostic categories (either because they are truly newly created or occurring, or are novel to conventional Western medicine). It presents two examples of cases from the latter half of the twentieth century that helped to establish novel diagnostic categories, and traces the evolution of each of these cases and the disposition of the facts they brought together and transported in order to explore how cases that start as a collection of discrete or isolated facts become a form of evidence that is epistemologically powerful. Two additional cases are briefly discussed to illustrate how, once cases are established, they can be used as the basis for diagnosis.

Suppose a patient presents to a physician or to hospital with several symptoms that are individually identifiable. It may be that these symptoms have not previously been seen in this combination and have not been recognised as together constituting a discrete syndrome or disease condition. In other instances, this pattern or cluster of symptoms may not 'make sense' as a singular disease condition in this type of patient – for instance, due to his or her baseline health condition or personal history. The pattern being sought is a diagnostic category describing a syndrome or disease that can be used to make decisions about the provision of therapies or prognoses. Otherwise, if no pattern is detectable, the symptoms may need to be treated individually or else multiple diagnoses will need to be made.

As will be seen in the examples to follow, the case in its original form typically serves as a sort of vehicle that holds what at the beginning are individual facts travelling separately. For a (loose) metaphor, picture a train carriage (representing the initial case, including the index patient described by it) with facts sitting in the various compartments. As in a mystery novel, we do not know at the outset what the relationship is, if any, between any of these facts (the assorted symptoms and other information thought to be relevant). In all but the most trivial sense of happening to be riding in the same train carriage (i.e., occurring in the same patient), these facts are not yet in any deeper sense 'travelling together,' since medical practitioners have not yet identified a recognizable pattern that captures the real relationship between them. In the process of working up a case relating to such a patient, there are often 'red herrings,' which are defined in this context as facts that are not false, incorrect or misconstrued when taken by themselves in isolation, but which seem to point in a direction that proves later to be misleading (and hence, eventually alight from the train).

Practitioners then engage in a process of gathering additional information using a variety of mechanisms, such as case reporting via the CDC or comparison to other cases in the published literature (so new facts get into the carriage at various stops along the way, if you will). By organizing (and re-organizing) the various facts, medical practitioners can propose a diagnostic category that establishes which facts are relevant and how they are interrelated. Those facts that are most essential will be upgraded to travel together in a first class cabin and given a place of prominence within the case as the main symptoms associated with the diagnostic category; other facts that are relevant but less central will stay in the carriage, that is, remain part of the diagnostic category captured by the case, but de-prioritised and hence relegated to second class. Some will join the existing cluster of facts as late arrivals, and some will be pushed off the train (out of the case) altogether.

This paper illustrates different types of travel than many of the other authors in this volume explore, inasmuch as the facts at issue here do not travel across disciplines, geographic space or even time. Instead, they travel in two different senses: first, as the case is refined, which facts 'ride together' often changes as their interrelation is investigated and altered once additional evidence is obtained and some facts are eliminated from carriage altogether. Second, the facts grouped together travel from what might be considered to be their original site, an index patient located in a particular clinical setting, to another type of site, a diagnostic category that can be used outside its original context.[4] Publication in a case format 'smoothes out' the particular facts associated with the individual index patient and renders them in a condition that makes them more likely to be generalisable for future use, assuming the train reaches its destination, that is, once the diagnosis has become established as a recognised disease condition. Once a particular set of facts has been recognised as travelling together in a significant way in the case (whether or not it has been labelled as a disease condition), the case report can be found again by others in the field looking for that particular set of characteristics: In this way, the facts captured in the case become generic.[5] Thus, the case can be made the basis for a diagnosis applied to other patients in other places, and can potentially provide enough information to allow efficacious treatment and/or accurate prognosis.

[4] The importance of labelling as well as of de-contextualisation to allow the travel of facts is discussed by Sabina Leonelli in this volume.

[5] For a related discussion of the difficulties of 'diagnosing' the nature of something based on the evidence or facts contained in them, see Alison Wylie's discussion of mound sites in this volume.

2. Toxic Shock Syndrome

The first example describes a case that can be seen as permeable and open to entry of facts from the outside not contained in the original, as well as an instance where a case has been used to refine a disease category. The original case report typically associated with toxic shock syndrome (TSS) (CDC 1997) contained details of what was claimed to be a new, severe and acute disease process associated with the presence of a toxin-producing strain of *Staphylococcus* (Todd et al. 1978; example noted originally in Morris 1989). It described seven paediatric patients between the ages of eight and seventeen seen with the syndrome, one of whom had died, and provided a detailed case of one, a young woman of around sixteen years of age. It ruled out a number of possible causal factors, including toxins or other vectors, as well as drug overdose. The authors noted that similar case reports of illnesses that had been thought to be *Staphylococcal* scarlet fever had been noted as far back as 1927. The only common demographic characteristics explicitly described by the paediatric specialists reporting this case were that the syndrome had been only observed in older children.

The CDC, the U.S. federal agency charged with monitoring epidemiology, began tracking cases of the syndrome soon thereafter, particularly given a growing number of deaths, and issued an alert in their *MMWR* in May 1980 (CDC 1980a), which showed a positive correlation between the syndrome and tampon usage. Of the cases reported to the CDC in 1979 to 1980, 95 per cent were in women; a menstrual history was obtained in 80 per cent of these, of which 95 per cent had the onset of illness within the five-day period following onset of menses. This evidence was felt to be strong enough for the editors to state that the syndrome 'affects primarily young women of child-bearing age who have been previously healthy' (CDC 1980a, 229). Two case control studies were published later that year (Davis et al. 1980; Shands et al. 1980), which showed the same positive correlation of this form of syndrome with tampon usage. The case control studies also clearly demonstrated that use of various brands and styles of tampons was, by far, the most important risk factor for menstrual TSS: As the CDC editors noted in an historical retrospective, '[T]he most plausible explanation for the "emergence" of menstrual TSS in the late 1970s was the manufacture and widespread use of more absorbent tampons made of a variety of materials not previously used in tampons'(1997, p. 495) rather than some sort of superbug form of *Staphylococcus*. Although it took considerable time to develop a full understanding of the toxic processes associated with the syndrome, by the end of 1980, the case report and observational control studies provided enough evidence to allow promulgation of public health recommendations (CDC 1980b, 1980c), which led to a substantial reduction in menstrual TSS.

Several issues arise from the history of this case: First, what is striking in retrospect is that there is no mention in the original case of how many of the patients were female and what their menstrual status was, given the clear association later noted between many cases of TSS and tampon use. The lead author later clarified that three of the four girls were menstruating when the illness developed, noting that 'it should have been obvious that the group of [ill] young women with "vaginitis" were of menstruating age...but we missed completely the possibility of any connection with tampon use. Fortunately, other pediatricians took up the slack' (Todd 1981, p. 922; cf. MacLure 1998). Despite this conspicuous gap, the original case still served as the entry point to closer investigation of the syndrome, and it is cited in all of the major publications on TSS that followed.[6] In addition, the very name for the syndrome as given in this original case is the one that has remained with it to this day, even though it is probably a bit too general and overlooks key causative factors. Clearly researchers in this field view this case as a sort of vehicle that carried some essential facts even though it is lacking other facts that later proved critical to the processes under examination. In part, this is because as the original case was constructed, other types of facts that proved to be relevant (the sex and tampon usage status of the patients) were not in any way explicitly ruled out or contradicted by the information contained in the original case, and in this sense it was permeable to their later entrance. In addition, it did not contain any incorrect or misleading facts, nor any suggestions of or associations with facts that were actually incorrect; instead, the key issue was which facts were truly relevant.

Second, once these additional facts 'entered' the case, the syndrome began to be subdivided, and the category of menstrual TSS became the focus of attention. Hence, the original case served as a call for closer tracking and reporting of this newly recognised syndrome, despite the fact that it was missing some critically important facts relating to what became its central features. Although cases of TSS in men also occurred during this time at a low and stable rate (5 per cent of the CDC cases mentioned earlier), the later case control studies caused investigators to focus on this subset of TSS cases and possible risk factors associated with the syndrome.

---

[6] A search for cited references to Todd et al. 1978 using ISI Web of Knowledge generates 780 hits, most of which (40 per cent) occurred between 1980 and 1985.

Finally, the epistemological structure of the initial TSS case is highly bio-
logically descriptive and rather open-ended, with few suggestions or speculations about mechanisms of disease or causality, in contrast to the next
case to follow, that of AIDS. The investigators' focus in the case is less at a
population level or on public health measures to prevent the syndrome, and
more on the etiological link to *Staphylococcus* and the activity of staphylococcal toxins. However, it does list a series of demographic factors for the
affected patients that leave open the possibility of more such categories and
facts within them entering the case. Hence, although cases often might be
in some sense highly speculative when first published, one advantage of a
case report is that it can serve as a springboard for future investigations,
particularly when it contains gaps in the facts but can remain relatively
static in rough outline as new information emerges (cf. Todd 1981).

3. Recognizing and Defining Aids

Perhaps the most famous example of the use of a case to establish a novel
diagnostic category was the publication in 1981 of a series of case reports
that is widely recognised as the first published observations of the disease
now known as acquired immunodeficiency syndrome or AIDS.[7] Although
the human immunodeficiency virus (HIV) associated with AIDS obviously
existed prior to this time (except according to those who are non-realists
about disease conditions or certain African politicians), the disease was
noticed by clinicians in California, who in late 1979 started seeing patients
with symptoms of what appeared to be a mononucleosis-like syndrome,
including fever, weight loss and swollen lymph nodes (Grmek 1990; see also
Shilts 1987). The patients also had thrush and diarrhoea, which together
with the fact that they did not show the typical improvement associated
with mononucleosis, ruled out this diagnosis. These symptoms taken
together suggested an immunodeficiency syndrome of some kind. In addition, all of the patients were young, homosexual men. Meanwhile, in New
York City during the end of 1979, an aggressive form of skin cancer, Kaposi
sarcoma, was observed in a number of homosexual, young men with friends
in common; this cancer had previously only been seen in elderly men of
Mediterranean descent.

---

[7] Unlike the first example, narratives associated with the recognition of AIDS abound, and
hence I must reiterate that my aim in this paper is to examine the logic behind the use of
published cases, rather than to establish a definitive historical narrative relating to any of
these diseases. Hence, my account is necessarily selective in order to permit focus on the
epistemological considerations that are central to this paper.

When an immunologist, Michael Gottlieb, asked one of his immunology fellows to look for interesting teaching cases, he was told about a young, homosexual man with what seemed to be a severely damaged immune system (Fee and Brown 2006).[8] The patient who was to become the index patient for the original case report had sought medical attention due to unexplained fevers and severe and increased weight loss, and was found to have severe oral thrush (via infection by the yeast *Candida albicans*): '[I]n the medical universe of 1980, it was a distinctly unusual event for doctors to be confronted with a patient in previous excellent health who had this yeast infection, one that in adults was virtually always seen in patients already known to be immune deficient' (Gottlieb 1998, p. 365).

To decide that this patient, in fact, was a new 'case,' that is had a previously un-described disease condition, the clinician ruled out the known causes of this sort of condition using a series of facts: chemotherapy for cancer, drugs used for immune suppression post-transplantation or an inborn autoimmune disorder (Gottlieb 1998). Candidiasis was known to be typically associated with a deficiency in T-lymphocytes, and blood tests revealed a marked decline in the number and function of this type of lymphocyte. A case report discussion among the teaching hospital's immunology post-doctoral fellows and internal medicine residents resulted in the use of an experimental technique to identify sub-classes of lymphocytes, and in this case, a severe depletion of CD4+ helper cells was identified. The patient was discharged without a diagnosis, but re-admitted one week later with what proved to be a rare, life-threatening form of pneumonia, *Pneumocystis carnii*, usually only found in immunocompromised patients. Soon thereafter, several additional patients with the same symptoms who were also homosexual men were referred, and it was found that they had the same form of pneumonia and similar blood results, as well as infection with cytomegalovirus (CMV).

Gottlieb retrospectively describes himself as realizing that this discovery could have major public health implications: When he telephoned the editor of the *New England Journal of Medicine*, he declared that he had an article that was 'possibly a bigger story than Legionnaire's disease...[which] now seems a colossal understatement' (Gottlieb 2001, p. 1788). He was advised that peer-reviewed publication would take a minimum of three months, and that he should first submit a brief article to CDC's *MMWR* in order to

---

[8] I mention the origins of this case for purposes only of historical accuracy, as the topic of the use of cases for pedagogical purposes raises a range of separate issues that I do not investigate in this paper: on cases as pedagogical devices, see, for example, Hunter 1986, 1996.

both 'stake his claim' and alert public health officials, as there were considerable possible public health implications, since the disease processes had rapidly advanced and homosexual men were known to have a high incidence of sexually transmitted diseases.

The case published in June 1981 presented a series of five cases of young, homosexual men with *P. carnii*, CMV infection and severe oral thrush caused by *Candida*, two of whom had died (CDC 1981a). Originally, this case report was generally overlooked according to Gottlieb and medical historians (Fee and Brown 2006), as few physicians bothered to read the *MMWR*. However, once a second report in early July of a spike in cases of Kaposi sarcoma, as well as *Pneumocystis* pneumonia among homosexual men, was published (CDC 1981b) and the popular media began its coverage (Kinsella 1989, p. 115), the public as well as the medical community began to take notice. By December 1981, there was enough evidence to warrant publication of several articles related to this disease condition (which had yet to be named) in the *New England Journal of Medicine*, which appeared as research articles but the content of which still was largely case-based (Gottlieb et al. 1981; Masur et al. 1981; Siegal et al. 1981). By 1982, the literature on what came to be known as AIDS began to grow, and the second CDC article (CDC 1981b) was frequently cited over the next four years (more than two hundred citations per year during this period). Within about one year, cases with similar symptoms emerged in different populations, including Haitian migrants, heterosexual men with haemophilia and other blood transfusion recipients without known risk factors, as well as women, which helped to further refine necessary and sufficient conditions for the diagnostic category. To return to the metaphor, additional facts got on the train and caused re-evaluation of the relationships and assumptions about the existing facts that already were on board, particularly the association of the disease solely with homosexual men.

It has recently been claimed by the editors of the *American Journal of Public Health* that the original *MMWR* case report (CDC 1981a) 'has become a classic, a reference point for medical and public recognition of the start of an epidemic that is perhaps the defining public health issues of our times' (Anonymous 2006, p. 981). However, in strictly quantitative terms, the original case report had an extremely low impact even once the AIDS crisis ensued: An analysis using ISI Web of Knowledge reveals a total of three citations in the medical literature in the 1980s, and only a total of eighteen to the present day. How can its reputed impact be explained? The original case (CDC 1981a) can be viewed as providing the initial impetus for the gathering of additional, similar cases that summarised a series of key facts, and the eventual identification of the underlying causative agent, for two reasons. The first is pragmatic: Publication in the *MMWR* provides a mechanism for reporting and tracking similar cases. Indeed, within six months, one hundred fifty-nine cases had been registered by the CDC. By the beginning of 1982, the count had exceeded two hundred cases, seemingly arising from three geographic locations: New York City, Los Angeles and San Francisco (Grmek 1990, p. 11), which began to give clues about its mode of transmission.

The second reason why the case (likely together with a longer, more formal publication of the same material in Gottlieb et al., 1981) worked in this manner was that its epistemological structure reflected selectivity as to the relevant information and a specific form of organization and emphasis on particular facts as essential, which signalled its authors' suspicions about the nature and mode of transmission of the disease. Within the first two lines, it is stated that the patients were young: Among other points, this fact indicates the very need for an explanation beyond the diagnosis of pneumonia (or thrush, or any of the other conditions present in these patients, which are, hence, viewed merely as signs of a more serious underlying condition). The fact that the patients are active homosexuals also is made explicit: This inclusion makes it clear that the authors view this information as central to the description of the patients that follows, even if they cannot yet explain its relevance (or even justify inclusion of this information). Although the total number of patients described is relatively small (five), the authors plainly think it can be no coincidence that these were homosexual men. The fact that they are sexually active points to the hypothesis that the underlying disease condition causing the various symptoms is likely to be infectious in nature. As the editors' note included at the end of the case indicates, this form of pneumonia in previously healthy individuals is 'unusual…[and] the fact that these patients were homosexuals suggests an association between some aspect of a homosexual lifestyle or disease acquired through sexual contact and *Pneumocystis* pneumonia in this population' (CDC 1981a, p. 250). Hence, in this example, it is clear that in this domain facts are typically not disputed as facts; in other words, their truth status is not questioned, but more importantly, their relevance and whether they are essential (or just incidental) becomes a key matter for debate.

The opening paragraph introducing the case series also contains a fact that later would be proven to be a 'red herring,' namely that all were infected with cytomegalovirus (CMV). As Gottlieb later put it, '[W]e rushed to judgment, overlooking the fact that CMV is a common opportunistic infection in immune compromised organ transplant recipients' (1998, p. 367).

The editors' comments clearly sound a cautious note about the possible role of CMV infection, noting that the 'observations suggest a possibility of a cellular-immune dysfunction related to a common exposure that pre-disposes individuals to opportunistic infections' (CDC 1981a, p. 252). This exchange underscores the potential fruitfulness of cases that contain facts that later prove to be irrelevant (or even wrong): If the balance of the facts within a case proves to be relevant, the case can be used to guide other practitioners' investigations despite inclusion of misleading facts or later de-prioritisation of what were initially thought to be essential facts. Such misleading facts can be identified and weeded out with the addition of more data, which can show that these are mere correlations or coincidences (as in the case of CMV infection, which proved not to be directly relevant to the aetiology of AIDS) or that a particular attribute does not apply beyond the small population described in the case at hand and hence is an ines-sential fact. In a similar way, the fact that many who had the disease were homosexuals was later determined to be relevant but not essential, which in part led to the abandonment of one of the original names for the disease, GRID (gay-related immunodeficiency disease). Thus, cases often serve as permeable vehicles for facts, allowing them to alight and disembark, and it is arguable that the cases most likely to withstand scrutiny over time will be those that can carry facts in a loose grouping but which can also release those facts that later prove not to be relevant, asking them to leave the train, as it were, while still allowing the hard core of information to be retained in a form that remains identifiable as the 'same' case.

The closing paragraph lists a series of additional facts on which neither the authors nor editors comment – for instance, that the patients did not know each other and had no known common contacts, that they were not aware of sexual partners with similar illnesses and that all five patients reported using nitrite inhalant drugs ('poppers'). Hence, another facet of the epistemological value of cases (vs., for instance, fully detailed causative theories generated through formal research studies) is that they can carry facts without necessarily requiring the logic of their inclusion to be made explicit. The facts need only be grouped together to be carried within the case. One can surmise that the facts about sexual contacts and popper use were included as a series of promissory notes about possible hypotheses about the aetiology of the disease (namely that the disease was infectious and transmissible through blood and/or sexual contact). Explicit notice of the facts that the patients lacked known common contacts or ill sexual partners hinted at facts that would later emerge when tracing the disease's epidemiology, namely that transmission was occurring in part through movement of individuals with numerous and often casual sexual partners and also that the disease was not affecting all people who were carrying HIV in the same way.

4. Using Cases to Diagnose

In this section, two examples are presented where published cases have been used to diagnose new patients, in other words, where a set of facts has been previously recognised as travelling together in a significant way and captured by a case report. First, a twenty-six-year-old young woman was admitted to an American hospital with acute confusion; she had developed a severe headache followed by increasing amounts of paranoia and other psychiatric symptoms, despite having no history (Sanders 2008).[9] After developing a fever and having a seizure while under treatment in a psychiatric facility, she went into a coma and was transferred to a specialist teaching hospital where she remained unresponsive even to pain stimuli. After an extensive workup, including blood tests, numerous scans and examination by various specialists, the only finding of potential relevance was a small cyst on her left ovary on a computed tomography (CT) scan. An attending resident suggested that a teratoma (one more unusual type of ovarian cyst, which is a type of germ cell tumour that may contain several different types of tissue, including mature elements such as teeth, hair, muscle and bone) could cause the symptoms the patient was experiencing, even though teratomas usually cause no difficulties and are only removed if they cause pain or become large.

We can reconstruct the typical process of diagnostic reasoning in which the clinician assigned to the case would have engaged in order to assess this patient and to determine the most likely diagnosis (and hence, the most appropriate treatment). First, particularly given the lack of a definitive test to detect a disease condition (unlike, for example, diagnosing a condition that causes a specific change in blood or other laboratory results), she would need to search the available literature for patients with similar histories and symptoms to the patient under examination, in other words, attempt to locate a parallel case study. In this example, the lack of a psychiatric history, injury or insult to which the patient's current condition could be attributed would be essential facts that would serve as the basis for such a search, and

---

[9] I first came across this example through a column in the *New York Times Magazine*; I am grateful to Mary Morgan for drawing my attention to it and to Lisa Sanders, who wrote the column, for references relating to the story recounted therein.

the chance finding of an ovarian teratoma would also be included as a potentially relevant fact along with any other abnormalities that had been noted.

Such a search would have resulted in locating several articles in the literature describing female patients with similar symptoms (headaches and psychiatric symptoms) that culminated in severe neurological symptoms, including becoming comatose (see Okamura, Oomori and Uchitomi 1997; Nokura et al. 1997 and Dalmau et al. 2007). In both cases, treatment of the psychiatric symptoms and other types of treatments did not result in improvement, but removal of the teratoma resulted in elimination of most major symptoms within a few days following surgery, with a limited number of sequelae. This sort of disease condition is known as a paraneoplastic syndrome, which is a general term used to describe a disease that is the consequence of the presence of cancer in the body at some distance from the site of the symptoms, but is not due to the local presence of cancer cells. Paraneoplastic syndromes are typically found only among middle-aged to elderly patients with lung cancer (Alamowitch et al. 1997). Hence, this diagnosis would not have been at all obvious for several reasons: In the first instance, paraneoplastic syndromes are unusual in their manifestations, given the distance and lack of obvious connection between the causative factor (in this case, the ovarian teratoma) and the type of symptoms (psychiatric and neurological). Second, the patient's profile did not fit that of a typical sufferer of a paraneoplastic syndrome, inasmuch as her tumour was not in the lung and she was very young.

However, the publication of the case studies describing patients with this condition brought information (e.g., facts relating more generally to paraneoplastic syndromes that were previously available primarily in specialist textbooks relating to cancer) to a more general audience. The clinician trying to diagnose the patient did not need to recognise the fact that she had a teratoma was essential to the diagnosis; she merely needed to include that finding in her search criteria along with the more striking neurological and psychiatric symptoms. But for the case studies, if she had included facts about the patient's age or the site of her tumour as part of the description of the patient's condition, she would not have easily uncovered this diagnosis based on the existing literature, which reviews the more typical or usual diseases and syndromes. The case studies capture an extremely rare condition (with only ten published reports from 1997–2008; for a review of all published literature, see Van Altena et al. 2008), and allow the similarities between the new patient and those reported in the literature to be recognised more readily, and for the hypothesis about the diagnosis to be tested by removing the tumour.

Often, diagnosis is less than clear in patients with complicated histories, and thus case reports that detail disentangling more than one syndrome are not uncommon, as seen in the second example (e.g., Neira et al. 2006). Some of these cases are fairly straightforward exercises in recognition of something unusual or a correlation not likely otherwise to be considered (see also Hunter 1996), often by consulting already available case reports. For instance, a brief case report described a young woman who came into the emergency room after swallowing a toothbrush after falling on the wet bathroom floor while brushing her teeth (Faust and Schreiner 2001). On examination, the patient's physical exam, including her weight and general health, were otherwise normal. However, an x-ray revealed that the toothbrush was lodged in the reverse direction (handle down) to how it should have been if the patient had been accurately describing the accident.

The patient admitted that she was trying to induce vomiting with the handle of the toothbrush, and proceeded to report other symptoms associated with bulimia, including a previous incident of a swallowed toothbrush removed by endoscope. A Medline search revealed that this is a rarely reported event (only forty cases in the published literature over the previous ten years), but critical to assess correctly, given the underlying nature of the disease conditions. The authors conclude that 'finding an unusual foreign body in the oesophagus or stomach should make the attending physician suspicious of bulimia and anorexia nervosa' (Faust and Schreiner 2001, p. 357). As one group of practitioners writes, 'Such cases are a bit like recognising a camel – once you have seen one you know what it looks like' (Morice, Ojoo and Kastelik 2003). A fact that might not be easily detectable (that the woman had bulimia) helps to explain an unusual fact (the ingestion of the toothbrush with the brush in the upright position) once other cases are identified, which bring these facts together and point toward a diagnosis. The publication of such cases is an essential part of the process of building a store of medical knowledge, particularly of the unusual, complicated or confusing.

5. From Patient to Diagnosis: How Cases Turn Particular Facts into Generic Ones

So in summary, how do cases serve as vehicles for facts and how does this structure allow them to do 'work'? In the examples explored earlier, the cases seem to function as useful in part because they contradict one or more implicit or explicit background assumptions. Such a background assumption has been termed a 'mental control group,' which with the publication of a case (and later responses to it) becomes more explicitly articulated (Vandenbroucke 2001). A published case study usually highlights something novel, unforeseen or surprising, something that a practitioner has been unable either to map onto existing disease categories, or something unusual that undermines existing theories or practices. For example, it was unexpected that severe immunosuppression without any obvious cause would develop in young patients who were otherwise healthy, as in the case of AIDS, nor was it usual for neurological symptoms to result from what would seem to be an asymptomatic ovarian cyst. Hence, certain facts within cases are essential and ground the case: They may seem to be relatively minor details, but for practitioners of these cases, they are the major, motivating facts, the surprise as it were, without which there would be little reason to publish the case study. The facts provided in a case highlight what is thought to be most essential to understanding this case as distinct and at the same time for identifying other similar cases that might occur in the future.

In addition, cases can be useful forms of knowledge because they serve as vehicles for loosely gathering facts in one place and putting them in contact with each other, as it were.[10] This function is in part due to their epistemic structure, which allows otherwise disparate, highly descriptive, observationally based facts to be brought together without imposing the rigors that are implicit in many types of medical research, which require larger samples or case-controlled studies. As I have argued elsewhere with regard to the biological sciences (Ankeny 2000, 2005), these types of descriptions can serve as resource materials for future hypothesis testing, and are typically not motivated (at least initially or immediately) by their future potential explanatory value.[11] In a similar way, medical cases typically do not provide proven causal mechanisms, explanations or even hypotheses for why something has occurred in a patient, but instead only that it has (such as in the case of TSS). Cases thus can allow recognition of new instances of the condition described while the diagnostic category is in the process of refinement, which is crucial for clinical practice. Furthermore, in that they

---

[10] There are obvious parallels with many of the other chapters in this volume with regard to the behaviour of facts as described here; see, for example, Simona Valeriani's chapter in this volume, which discuses artefacts as a vehicle for carrying a bundle of facts.

[11] But, of course, to the extent that a clinical case reflects (in a complex manner) physical processes and events, it always carries potential explanatory power, for instance, that could be used to justify proposed clinical interventions, and perhaps may even have stronger explanatory power if and when the observed patterns recur and underlying mechanisms can be identified. I am grateful to Brian Hurwitz for emphasising this point to me.

are largely descriptive, many cases are left open-ended or rather vague; in other words, facts newly recognised or determined to be relevant can later join the existing cluster of facts captured by the case. A more formalised research report would require more detail about underlying disease mechanisms or processes, which if included in the original case might have caused it to become invalid or otherwise not useful in the future, particularly as practitioners' understanding of the syndrome was systematised and refined. These qualities of openness to the entrance of new facts, as well as the re-organization and re-prioritisation of the existing facts, are precisely what give cases epistemic strength.

Further, the case format assumes that this collection of facts can somehow be projected from the typically small group of patients, or even the individual patient explored in the original case, onto a new patient. In other words, the publication of a case is in some sense a proposal or promise that the particular will be able to be made generic through a process of systematizing the original facts, smoothing out their particularity and refining the original case until it is applicable beyond an individual patient. This type of reasoning is convincing and valid because the patients described in cases need not be representative in any broad sense: Much as biological scientists reason from one model organism to 'higher' level ones (Ankeny 2000), so, too, can reasoning occur via cases, so long as the limitations of such reasoning are acknowledged. These limitations arise from the variability of individuals, and knowing how to assess what are the relevant facts and what is merely incidental variability when making a diagnosis. For instance, over time, clinicians learned to be cautious about taking a careful history before eliminating a diagnosis of AIDS, given the recognition that the disease did not only affect homosexual men.

Finally, cases involve the transformation of facts about individual patients, which were initially viewed as surprising or abnormal, into new norms. Instead of continuing to view future patients with these or similar conditions as abnormal, the publication and subsequent refinement of a case allows a new diagnostic category to be established, and this then comes to define a norm in the process of translating the facts from being particulars about individual patients into generic facts that are generalisable beyond their initial boundaries. In some sense, then, the refined case (or, more precisely, the diagnostic category that is ultimately generated following refinements to facts within the original case or cases observed) becomes like a field guide, with a set of identifying characteristics (essential facts) against which a patient who newly presents is compared in order to determine whether he or she fits into the category. Hence, even in their refined and smoothed state, cases that become diagnostic categories often are still a form of vehicle that carries a complex set of necessary and sufficient conditions, inasmuch as any individual patient might not have all of the attributes associated with the disease condition.

Paying attention to cases allows us to develop a more descriptively accurate understanding of medical practice and research, inasmuch as what often has been described as serendipity or 'basic science' is more accurately viewed as knowledge resulting from careful recording and reporting of cases, grounded by the selective gathering, prioritisation and organisation of facts, as the examples detailed previously have shown. Developing a richer view of cases as essential to the early, descriptive stages of medical research and practice allows us to more adequately capture the typical characterisation of medicine as both a prudential undertaking and a scientific enterprise: Cases involve judgements and cautious gathering of facts, the elimination of red herrings (misleading) or inessential facts and the organisation and prioritisation of such facts in a manner that can communicate the necessary message to a clinician-reader.

This chapter has illustrated the way in which cases can have an epistemically interesting role in the establishment and refinement of emerging disease categories, and argued for a range of types of purposes they serve in this process. Why then are cases often ignored as a valid component of medical reasoning? In part, cases have a bad reputation, as it were, because there often is conflation (even among those who support case reports) between their educative purposes (for instance, serving as a story or anecdote, which allows knowledge to be more easily organised and recollected)[12] and their role in medical research, discovery and progress (e.g., Bignall and Horton 1995). The emphasis on the unusual and provocative, and a general preference in the field for cases that 'appeal to the emotions' (Nathan 1967), often causes critics of cases to overlook their power as the starting point for research questions that can be rigorously scientifically assessed within a case. In addition, the requirements that journals impose on the publication of cases in a sense are contradictory: A case needs to be both novel (by which the editors seem to mean unique or highly unusual) and credible to be published, qualities that are typically inversely correlated (Morgan 1985).

However, as has been noted in different ways by commentators on medical reasoning, the underlying epistemology of medical diagnosis and treatment

---

[12] The power of narrative as a vehicle for travelling facts is discussed by Adams in this volume.

is less like that of the experimental sciences and more similar to that of the social sciences, including history, in that hypotheses (if they are ever explicitly formulated) typically come primarily from observations or what we would term 'facts' (Hunter 1990; see also Blois 1988, and Cole 1994 on similar practices in the history of statistics). These facts then are smoothed out and made to fit together into larger patterns, rendering generic what was particular and specific. Thus, the starting point for additional investigations of such patterns, whether or not they can be later subjected to more formalised trials or larger-scale studies, oftentimes must be a rich, descriptive case report on an individual patient or small series of patients, even if the canonical epistemological structure that governs cases and the rules for their use are rarely explicitly recognised by medical practitioners.

Acknowledgments

I am grateful to all of the participants in the Facts workshops for their feedback, but particularly Harro Maas and Ed Ramsden for their useful commentaries (and cartoons, in Harro's case) on the first version of this paper at the final workshop. A version of this paper was also presented at the History of Science Society Annual Meeting in 2008 where attendees' questions were helpful in preparation of the final draft of this paper. Brian Hurwitz provided invaluable feedback on a final draft. Most of all, my gratitude is due to Peter Howlett and especially Mary Morgan for inviting me to join the Facts project, and to both of them as well as the postdoctoral scholars for the constant challenges and intellectual stimulation; the project has been a model of collaboration, which has been extremely important in furthering my scholarship over the past five years. None of this would have been possible without the grant funding from the Leverhulme Trust and the ESRC (grant F/07004/Z), which enabled me to spend extended periods working with the Facts group at LSE.

Bibliography

Albrecht J., Meves A., and Bigby M. Case reports and case series from *Lancet* had significant impact on medical literature. *Journal of Clinical Epidemiology* 58:1227–32, 2005.

Alamowitch S., Graus F., Uchuya M. et al. Limbic encephalitis and small cell lung cancer: clinical and immunological features. *Brain* 120:923–8, 1997.

Anonymous. Editor's note to Voices from the past. *American Journal of Public Health* 96:981, 2006.

Ankeny R. A. Fashioning descriptive models in biology: of worms and wiring diagrams. *Philosophy of Science* 67:S260–72, 2000.

Case-based reasoning in the biomedical and human sciences: lessons from model organisms, in P. Hajek, L. Valdés-Villanueva, and D. Westerståhl (eds.) *Logic, Methodology and Philosophy of Science: Proceedings of the Twelfth International Congress.* London: King's College Press, 229–42, 2005.

Aronson J. K. and Hauben M. Anecdotes that provide definitive evidence. *British Medical Journal* 333:1267–9, 2006.

Bignall J. and Horton R. Learning from stories – *The Lancet's* case reports. *The Lancet* 346:1246, 1995.

Blois M. S. Medicine and the nature of vertical reasoning. *New England Journal of Medicine* 318:847–51, 1988.

Centers for Disease Control (CDC) [Chesney P. J., Chesney R. W., Purdy W. et al.]. Toxic-shock syndrome – United States. *Mortality & Morbidity Weekly Report* 29:229–30, 1980a.

Centers for Disease Control (CDC). Follow-up on toxic-shock syndrome – United States. *Mortality & Morbidity Weekly Report* 29:297–9, 1980b.

Follow-up on toxic-shock syndrome. *Mortality & Morbidity Weekly Report* 29:441–5, 1980c.

Centers for Disease Control (CDC) [Gottlieb M. S., Schanker H. M., Fan P. T. et al.]. *Pneumocystis* pneumonia – Los Angeles. *Mortality & Morbidity Weekly Report* 30:250, 1981a.

Centers for Disease Control (CDC). Kaposi's sarcoma and pneumocystis among homosexual men – New York City and California. *Mortality & Morbidity Weekly Report* 30:305, 1981b.

Centers for Disease Control (CDC) [Reingold A. L., Matthews G. W. and Broome C. V.]. Editorial note to Epidemiologic notes and reports toxic-shock syndrome–United States (historic reprint). *Mortality & Morbidity Weekly Report* 46:492–5, 1997.

Cole J. The chaos of particular facts: statistics, medicine and the social body in early 19th-century France. *History of the Human Sciences* 7:1–27, 1994.

Dalmau J., Tüzün E., Wu H., et al. Paraneoplastic anti-N-methyl-D-aspartate receptor encephalitis associated with ovarian teratoma. *Annals of Neurology* 61:25–36, 2007.

Davis J. P., Chesney P. J., Wand P. J. et al. Toxic-shock syndrome: epidemiologic features, recurrence, risk factors, and prevention. *New England Journal of Medicine* 303:1429–35, 1980.

Faust J. and Schriener O. A swallowed toothbrush (case report). *The Lancet* 357:1012, 2001.

Fee E. and Brown T. M. Michael S. Gottlieb and the identification of AIDS. *American Journal of Public Health* 96:982–3, 2006.

Glasziou P. et al. When are randomised trials unnecessary? Picking signal from noise. *British Medical Journal* 334:349–51, 2007.

Gorovitz S. and MacIntyre A. Toward a theory of medical fallibility. *Journal of Medicine and Philosophy* 1:51–71, 1976.

Gottlieb M. S., Schroff R., Schanker H. M., et al. Pneumocystis carinii pneumonia and mucosal candidiasis in previously healthy homosexual men: evidence of a new acquired cellular immunodeficiency. *New England Journal of Medicine* 305:1425–31, 1981.

Gottlieb M. S. Discovering AIDS. *Epidemiology* 9:365–7, 1998.

AIDS–past and future. *New England Journal of Medicine* 344:1788–91, 2001.

Grmek M. D. *History of AIDS: Emergence and Origin of a Modern Pandemic.* Trans. R. C. Maulitz and J. Duffin. Princeton: Princeton University Press, 1990.

Hauben M. and Aronson J. K. Gold standards in pharmacovigilance: the use of definitive anecdotal reports of adverse drug reactions as pure gold and high grade ore. *Drug Safety* 30:645–55, 2007.

Hunter K. M. 'There was this one guy…': the uses of anecdotes in medicine. *Perspectives in Biology and Medicine* 29:619–30, 1986.

An n of 1: syndrome letters in *The New England Journal of Medicine. Perspectives in Biology and Medicine* 33:237–51, 1990.

'Don't think zebras': uncertainty, interpretation, and the place of paradox in clinical education. *Theoretical Medicine* 17:225–41, 1996.

Hurwitz B. Form and representation in clinical case reports. *Literature and Medicine* 25:216–40, 2006.

Kinsella J. *Covering the Plague: AIDS and the American Media.* New Brunswick, NJ: Rutgers University Press, 1989.

McCarthy L. H. and Reilly K. E. How to write a case report. *Family Medicine* 32:190–5, 2000.

MacLure M. Inventing the AIDS virus hypothesis: an illustration of scientific vs. unscientific induction. *Epidemiology* 9:467–73, 1998.

Masur H., Michelis M. A., Greene J. B. et al. An outbreak of community-acquired Pneumocystis carinii pneumonia: initial manifestation of cellular immune dysfunction. *New England Journal of Medicine* 305:1431–8, 1981.

Morgan P. P. Why case reports? (editorial). *Canadian Medical Association Journal* 133:353, 1985.

Morice A. H., Ojoo J. C. and Kastelik J. A. Authors' reply to Weinberger M. Disabling cough: habit disorder or tic syndrome? (letter to the editor). *The Lancet* 361:1991, 2003.

Morris B. A. The importance of case reports (letter to the editor). *Canadian Medical Association Journal* 141:875–6, 1989.

Nathan P. W. What is an anecdote? *Lancet* 2:607, 1967.

Neira M. I., Sánchez J., Moreno I. et al. Occam can be wrong: a young man with lumbar pain and acute weakness of the legs (case report). *The Lancet* 367:540, 2006.

Nokura K., Yamoamot H., Okawara Y. et al. Reversible limbic encephalitis caused by ovarian teratoma. *Acta Neurologica Scandinavica* 95:367–73, 1997.

Okamura H, Oomori N. and Uchitomi Y. An acutely confused 15-year-old girl. *The Lancet* 350:488, 1997.

Rosselli D. and Otero A. The case report is far from dead (letter). *The Lancet* 359:84, 2002.

Sanders L. Brain drain. *The New York Times Magazine*, 9 November 2008, 22, 24.

Shands K. N., Schmid G. P., Dan B. B. et al. Toxic-shock syndrome in menstruating women: association with tampon use and Staphylococcus aureus and clinical features in 52 cases. *New England Journal of Medicine* 303:1436–42, 1980.

Shilts R. *And the Band Played On: Politics, People, and the AIDS Epidemic.* New York: St. Martin's Press, 1987.

Siegal F. P., Lopez C. M., Hammer G. S. et al. Severe acquired immunodeficiency in male homosexuals, manifested by chronic perianal ulcerative herpes simplex lesions. *New England Journal of Medicine* 305:1439–44, 1981.

Simpson R. J. and Griggs T. R. Case reports and medical progress. *Perspectives in Biology and Medicine* 28:402–6, 1985.

Todd J., Fishaut M., Kapral F. et al. Toxic-shock syndrome associated with phage-group-I-Staphylococci. *Lancet* 2:1116–8, 1978.

Todd J. Toxic shock syndrome – scientific uncertainty and the public media. *Pediatrics* 67(6):921–3, 1981.

VanAltena A. M., Wijnberg G. J., Kolwijck E. et al. A patient with bilateral immature ovarian teratoma presenting with paraneoplastic encephalitis. *Gynecologic Oncology* 108:445–8, 2008.

Vandenbroucke J. P. In defense of case reports and case series. *Annals of Internal Medicine* 134:330–4, 2001.

TEN

TECHNOLOGY TRANSFER AND TRAVELLING FACTS: A PERSPECTIVE FROM INDIAN AGRICULTURE

PETER HOWLETT AND AASHISH VELKAR

1. Introduction

This chapter is concerned with technology transfer in Indian agriculture. The existing literature on technology transfer has tended to focus on spatial and temporal diffusion, emphasising socio-economic factors such as the role of social networks and social learning.[1] However, we are interested in the facts that travel in association with the technology's adoption rather than in the transfer of technology per se. From our perspective, different facts travel during the different stages of technology transfer through different spaces. Although such travel involves many different types of facts (technical, procedural, scientific, etc.), we argue that the central fact that needs to travel to potential users in any technological transfer project is that adoption of the new technology will deliver noticeable benefits. The integrity of this fact is dependent on a package of related facts, including those related to the financial costs of adoption, the knowledge needed to implement the technology successfully, access to necessary inputs, the benefits of the technology and the ability of the user to capture those benefits in the form of increased income or profitability.

Technology transfer is a staged process, and one of the first stages involves issues such as how do potential users learn about the benefits of the technology and how do facts travel from the technological or scientific domain

---

[1] On diffusion see Rogers (1962), Feder et al. (1985) and Geroski (2000); on social networks and learning, see Foster and Rosenzweig (1995), Bandiera and Rasul (2006) and Warner (2008). Interestingly, Warner draws on Latour's notion of the circulatory system of scientific facts (Latour 1999).

to the user domain? Are some vehicles of transmission better than others – and if so, what are the characteristics of those vehicles? Also, this process of travel may not be a linear process – facts may travel in multiple directions, influence the transmission mechanisms and interact with other facts spanning different points in time and space.

Subsequent stages of adoption and implementation of technology typically involve facts that are more technical and experiential in nature, that is, commentator's facts as distinct from facts associated with the technology per se. Here, we are concerned with adoption and late adoption, with access to inputs and knowledge required to implement the technology, with people who have designed the technology and others who have used the technology and with how the technology actually works in practice. How well the facts travel reflects a variety of institutional and socio-economic considerations in addition to the technical and technological considerations. Such travel also reflects the expertise of various groups: the scientists and technicians who design the technology, the users who have their own expertise and experiences regarding the context in which the technology is received, policy makers who encourage the diffusion of technologies and so on. Transfer may also depend upon the degree to which the different expertise, and the facts they may generate, contest or complement each other.

Beyond adoption, facts about the technology may continue to travel, and travel well. Demonstration effects from early adopters may enable facts about the technology, its benefits, how it works and so on, to travel further, reflecting in part the efforts of those who design the technology and promote it. Further travel also reflects the interaction between adopters and non-adopters and what facts travel from the former to the latter. Here, too, the institutional and socio-economic considerations play an important role compared to the purely technical considerations: Is the technology reliable? Are the people promoting it trustworthy? Transmission vehicles reflect complex relationships between users and potential users, involving formal as well as informal mechanisms.

The simple schematic presented earlier is not meant to offer an exhaustive or definitive account of technology transfer. It illustrates how thinking about travelling facts might help us to better understand the transfer processes. We show that the travelling of facts associated with technology transfer was far from being a linear process – that it was a dynamic process where feedback mechanisms were important and where trust (in facts and individuals) played a significant role.

The issue of how *well* facts travel lends another dimension to thinking about the technology transfer process. We consider the wellness of travel from multiple perspectives: the extent of travel (i.e., from early adopters to late adopters, crossing domain spaces, etc.), institutional developments that aid travel (i.e., emergence of cooperative associations), the integrity with which a set of facts travels (i.e., technical facts travelling along with experiential facts) and the consequences or impact of travel (i.e., enhancing skill levels, improving living standards, etc.).

2. The Precision Farming Project

We study the Tamil Nadu Precision Farming Project (TNPFP), an extension education programme sponsored by Government of Tamil Nadu and Tamil Nadu Agricultural University (TNAU) between 2004 and 2007.[2] The programme involved scientists from TNAU (led by the director of extension education, Dr. Vadivel – see Figure 10.1a) providing four hundred farmers from Dharmapuri and Krishnagiri in the northern part of Tamil Nadu (see Figure 10.2), with precision farming technologies, plus instruction on how best to use the technology.[3] The rationale for selecting these two districts was primarily their socio-economic status – they were backward, impoverished and water-scarce areas dominated by traditional agricultural practices. The new technologies revolved around fertigation, that is, water- and labour-saving drip irrigation methods involving the use of water-soluble fertilizers (WSF). We study how farmers gained facts about precision farming, how well facts about precision farming travelled within the TNPFP and how well such facts travelled beyond the TNPFP.[4]

TNPFP had two objectives. First, promote high-tech horticulture through the use of precision technology and transfer latest cultivation and post-harvest technologies to the farmers. Second, promote market-led horticulture by encouraging farmers' forums and associations, and increase the overall value accruing to the farmers.[5] The project was concentrated around clusters, with two hundred farmers being selected in each of the two districts. The absorption of farmers into the scheme happened progressively in three stages: By the end of the first stage, one hundred farmers were recruited, by the second, an additional two hundred farmers were part of the scheme and

[2] The TNPFP is part of the state promotion of modern technology in agriculture that intensified from the mid-1960s in India and that is often referred to as the Green Revolution (Farmer 1977; Byres 1983).

[3] Although we study 'precision farming technology,' the meaning of the term 'precision' is here restricted to the actor's usage of the term, that is, precise application of the technology, rather than understanding precision in terms of 'sameness' of usage post-adoption. Thus, we do not consider wellness in terms of the precision of facts during travel, where precision refers to objectification or quantification or both (Wise 1995, pp. 6–7).

[4] A more detailed discussion of some of the points raised in this chapter can be found in Howlett and Velkar (2008).

[5] Vadivel (2006, p. 1).

Figure 10.1a. Dr. Vadivel, director of extension education, TNAU, addressing a meeting of precision farmers.
From: © A. Velkar.

Figure 10.1b. Precision farmers show TNAU scientist Dr. S. Annadurai and Aashish Velkar (second and third from left) their fertigation tank.
From: © A. Velkar.

Figure 10.1c. Aashish Velkar interviews a precision farmer in his field – note the drip irrigation laterals in front of them.
From: © A. Velkar.

Figure 10.1d. Peter Howlett (second from right) and Aashish Velkar (second from left) outside the Erode Precision Farm Producer Company shop.
From: © A. Velkar.

Figure 10.2. Map showing location of TNPFP project districts.
From: PlaneMad/Wikipedia: http://en.wikipedia.org/wiki/File:India_Tamil_Nadu_
locator_map.svg.

by the third year, the remaining one hundred farmers were recruited into
the scheme. Farmers were selected on the basis of several criteria, including
minimum land area, ability to provide for a minimum quantity of water,
nature of the soil, location of the farm in relation to the cluster, and so on, as
well as other 'soft' criteria, such as willingness to participate in associations,
willingness to conform to practices recommended by TNAU and so on.[6]

The TNAU scientists offered precision farming as a 'package' of different
technologies to the beneficiary farmers. These could be broadly classified

---

[6] Vadivel (2006, pp. 3–4) for details.

as *fertigation technology* and *other farm technologies*. The fertigation technology used drip irrigation systems and fertigation units along with WSF and fertigation schedules developed for each soil and crop type by the scientists (Figure 10.1b shows a fertigation tank). The other farm technologies included cultivation techniques, pest-management techniques, grading and sorting of produce and several others that guide cultivation and post-harvest activities.[7]

The precision farming technologies were made available to the beneficiary farmers through financial assistance that included the cost of the fertigation equipment and the cost of installation (including the installation of water channels called laterals – see Figure 10.1c) and the cost of cultivation (including the cost of WSF). Depending upon the crops grown and other factors, the cost of 'converting' to precision farming ranged between Rs. 75,000 and Rs. 150,000. TNPFP assumed a standard conversion cost of Rs. 115,000 and structured the subsidy amount on a progressively reducing scale over the years as follows: Year 1 – 100 per cent of the standard conversion cost; Year 2 – 90 per cent of the cost, Year 3 – 80 per cent of the cost. It is important to note that each farmer received financial assistance only for the first year in which they joined the project. Cultivation expenses for the subsequent years (including the cost of the WSF) were to be the responsibility of the farmer.

The main evidence for this study were in-depth interviews with a sample of fifty-two farmers, including those who were in the TNPFP (beneficiary farmers) and some who were not in the project (non-beneficiary farmers). The interviews were conducted in August 2007 in both Dharmapuri (twenty-one farmers) and Krishnagiri (thirty-one farmers). The beneficiary farmers interviewed (thirty-four) were spread over all the three years of joining: nine joined the TNPFP in the first year, seventeen in the second year and eight in the third year. The non-beneficiary farmers (eighteen) included both those who had applied to be part of the TNPFP but had been rejected (eight) and those who had not applied (ten). The interviews were based on a set of common questions; however, the discussion was essentially free-flowing and conversational. These interviews were conducted on a one-on-one basis and reflect the opinions of individual farmers, rather than a collective group. In the text we cite the farmers using only their initials. Data from the fieldwork are supplemented by additional information from published statistics and other primary sources such as harvest records maintained by the beneficiary farmers.

---

[7] Vadivel (2006) provides details of the individual precision farming technologies.

3. What is Travelling?: Facts and Domains

Many different facts were travelling within the TNPFP, and for functional reasons we distinguish between four types. We term facts associated with certain physical objects or artefacts to be technological facts.[8] We also distinguish between facts derived from scientific claims (i.e., claims arising from scientific knowledge and experimentation) and experiential claims (i.e., claims arising from experience or use of the technology). In essence, both types of claims lead to facts. For the purpose of this paper, we limit the use of the term experiential facts just to those claims that arise out of use or experience of the technology and technical facts to be those based on scientific claims. Such technical facts include scientific advice, or facts that carry instructions based on such advice, for the user. We do not assume that scientific knowledge resides only with the scientists or that experiential knowledge resides only with the farmers. On the contrary, we discover that farmers accumulate expertise through experience as well as experimentation.[9] Consequently, the direction in which facts travel is not self-evident or straightforward. Some facts – both experiential as well as technical – travel from the farmers to the scientists, just as they travel from the scientists to the farmers.[10] We also identify several institutional facts travelling within the TNPFP. Many of the facts we study in this article were combinations of two or more types.

Finally, we differentiate the travelling process in terms of three spaces, or domains: the fertigation domain, the associated technologies domain and the enabling institutions domain. These domains are overlapping and interactive. The *fertigation domain* encompassed *fertigation technologies* that include the drip irrigation and fertigation equipment, WSF and the fertigation schedule. The first two are technological facts (in economic terms, one is a fixed cost and the other a variable cost), whilst the latter is a technical fact (it provides a procedural schedule that embodies scientific research). All three of these elements were non-negotiable, and monitored carefully

---

[8] For an example of physical objects as important vehicles and embodiments of travelling facts, see Valeriani (2006).

[9] See Collins and Evans (2007, pp. 13–14) whose 'periodic table of expertise' also discusses how specialist expertise goes beyond just knowing facts and requires specialist tacit knowledge (p. 13–14).

[10] For a discussion of experiential knowledge, see Epstein (2005, pp. 3–4); for an Indian agricultural technology context, see Foster and Rosenzweig (1995). More generally on the issue of the relationship between producers and users of technology, see Oudshoorn and Pinch (2005).

by the scientists in the first year a farmer was part of the project. The financial subsidy ensured that the drip irrigation and fertigation equipment did travel. As we shall see, it was highly unlikely that a farmer would abandon this technology, as, at a most basic level, it ensured a better delivery of water and fertiliser than had previously been possible. The other two elements – the use of WSF and the fertigation schedule – can only be judged to have travelled successfully if farmers continued to use them after the first year. This in turn depended upon claims made by scientists and how well these claims travelled. Given the water and income constraints faced by the farmers, two such claims were that the precision farming technology would deliver significant water savings and significant improvements in income. Other important claims were that precision farming would deliver labour savings and lead to improved crop yield and health. The farmers would only believe these claims if their own experience of using the fertigation technologies matched the claims. In many ways, these experiential facts were the most important for the TNAU scientists – that precision farming could provide significant benefits in terms of water saving and income even for small-scale farmers. TNAU wanted to ensure that *these* facts travelled widely across various audiences, not just to the participating farmers.

The *associated technologies domain* encompassed *other farm technologies* and scientific advice provided by the scientists regarding cultivation, farm management and post-harvest management techniques, excluding fertigation. At times, some of the technology in this domain involved only technical facts travelling, for example, crop spacing. However, much of it involved combinations of facts travelling. For example, the use of pesticides involved both the travelling of a technological fact (the technology embodied in the pesticide itself) and a technical fact (the advice the scientists gave to the farmer about how to use the pesticide). Similarly, post-harvest techniques included technical facts (sorting and grading) as well as institutional facts (that sorting and grading reflected the norms expected by modern terminal commodity markets or buyers). These technologies and scientific advice were offered as part of the general extension education effort, and their implementation by farmers was not monitored by the scientists. In this space, TNAU encouraged the travel of facts concerning the management of modern farms and the operation of modern commodity markets. They were keen to promote the fact that a better understanding of these processes would enable the farmers to earn better incomes.

The *enabling institutions domain* encompassed at least two important institutions that help to explain the wellness of travel. They are the TNAU's Directorate of Extension Education (DEE) and the farmer associations formed as part of the TNPFP. As we will show, these institutions enabled the facts to travel between and within the various groups (beneficiary and non-beneficiary farmers, scientists, etc.). For instance, the generous financial subsidy delivered to the farmers, and channelled through the DEE, was vital for the small-scale farmers to adopt the fertigation technologies – and thus for the travel of key technological facts. But the fact that the full benefit of the subsidy reached the farmers also helped the fact that the scientists were trustworthy and reliable to travel. The institutional facts regarding the integrity and trustworthiness of the scientists were also reinforced by the particular model of extension education adopted by the DEE. The intensity of the extension efforts, the close supervision that the scientists provided and the personal involvement they showed in solving various issues (both at the individual level of the farmer as well as at the collective level of the farmer associations) helped many of the different facts to travel. Similarly, the farmer associations enabled several technical, experiential and institutional facts to travel. For instance, TNAU was keen to promote the fact that cooperation and coordinated efforts amongst farmers would translate into better crop yields, stronger negotiating position in the market, better access to scientific and experiential knowledge (of other farmers) and so on.

4. Prior Knowledge About Precision Farming and TNPFP

The central fact about precision farming (PF) technologies that the scientists sought to communicate was that the adoption of PF would result in better and more consistent quality of produce and an overall increase in profit margins. However, before the benefits of PF could accrue to the farmers, several deficits needed to be overcome: a knowledge deficit (i.e., prior knowledge about precision farming and its benefits), a skills deficit (i.e., how do farmers learn to apply precision farming techniques successfully) and a financial deficit (i.e., how do farmers raise the money necessary to acquire the technology?). How did farmers obtain facts about PF, and what were the sources of these facts? There are two possibilities: Farmers were either aware of precision farming before they heard of TNPFP, or they heard of precision farming at the same time that they heard of TNPFP. In the latter case, we tried to establish if farmers heard of TNPFP through contact with TNAU scientists or prior to meeting with them, that is, through other sources. The significance of this was to ascertain the extent to which facts about TNPFP and precision farming travelled through more indirect sources, such as other farmers, media and so on.

About one-third of the farmers interviewed had knowledge of precision farming before they heard about the TNPFP. Just over half of these gained their knowledge from existing demonstration schemes in other states. Newspapers and the television were another important source of knowledge for these farmers.[11] In our sample, three farmers were using drip irrigation before the arrival of the TNPFP, including one who grew mulberry and had adopted drip irrigation (without the use of WSF) because of a subsidy provided by the Silk Board. In addition to these farmers, if we include farmers who heard about precision farming at the same time as they heard about TNPFP, then the proportion of farmers with prior knowledge in our sample increases from one-third to 75 per cent.[12] All of these additional farmers gained their knowledge about precision farming from other farmers. Therefore, an overwhelming proportion of our sample had gained their knowledge about precision farming by observing demonstration schemes, including the TNPFP.[13] How did farmers gain their knowledge about the TNPFP? Half of the fifty-two farmers interviewed gained their knowledge from existing beneficiary farmers,[14] whilst more than a third first heard about the scheme at meetings organised by TNAU; only one farmer cited the Horticulture Department as their source of knowledge. There was, however, a significant difference between the beneficiary and non-beneficiary farmers in the sample: Sixteen of the eighteen non-beneficiary farmers gained their knowledge about TNPFP from neighbours who were beneficiary farmers, whilst half of the beneficiary farmers learnt about the project by attending a TNAU meeting.

What facts about precision farming or TNPFP travelled prior to contact or adoption? This could be gauged from motivations for adopting precision farming techniques. It was clear from the general tenor of the interviews that most, if not all, beneficiary farmers would not have joined the project without the generous subsidy. Hence, it was important that the facts that precision farming techniques came bundled with a generous subsidy and that the full benefit of the subsidy would reach the farmers did travel to the farmers.

[11] According to a 2003 national survey, 29.3 per cent of farmer households in India accessed information on modern agricultural technology via media sources (Birner and Anderson 2007, p. 7).

[12] This is a proportion of all the farmers (beneficiary and non-beneficiary) interviewed.

[13] This seems to contradict the 2003 survey of Indian agriculture, which reported that only 2 per cent of farmer households gained their knowledge about modern agricultural technology from government demonstration schemes, although another 16.7 per cent gained their knowledge from 'other progressive farmers' and 5.7 per cent from extension workers (Birner and Anderson 2007, p. 7).

[14] On the importance of social learning more generally, see Munshi (2004).

Two-thirds of the beneficiary farmers also explicitly mentioned other factors that influenced their reason to join the TNPFP, and by far the most important of these were water saving (mentioned by sixteen farmers) and labour saving (eleven farmers). Other reasons cited included increased yield or crop growth (six farmers), using advanced or high-tech technology (two farmers) and the fact that the technology seemed to be convenient and easy to use. One farmer said that joining TNPFP allowed him to watch television whilst the fields were irrigated! In terms of successful travelling, such kinds of reasons should not be underestimated. We observe that simple facts will find it easier to travel in this sort of environment than more complex facts.

Thus, there is evidence that demonstration effects helped experiential facts about precision farming to travel to a wider community. It is perhaps necessary to reflect upon whether post-adoption experiential facts (judging from farmers' responses) validated and reinforced the scientific claims made by the scientists, and in turn helped facts (about precision farming in general) to travel at large, both within the farming community as well as reflecting back to the scientists. We consider the effect of demonstrations on the travel of technological and technical facts in the following sections.

5. Adoption of Fertigation Technologies

One indication of the successful travel of technological facts is that there was no reported case of a beneficiary farmer leaving the TNPFP or abandoning the use of the drip and fertigation equipment subsequent to the first year. As far as the use of WSF and the fertigation schedule is concerned, inspection of farm records suggests that in the first year of participation the beneficiary farmers closely followed the fertigation schedule recommended by the scientists.[15] All but two of the interviewed beneficiary farmers claimed that they continued to follow the recommended fertigation schedule in subsequent years without any deviation. One exception was a farmer who claimed to have made 'mini changes,' based on his experience and judgement, to the fertigation schedule according to the nutrient content of the soil. Another was a farmer who claimed that he used both soluble and non-soluble fertiliser, without consulting the scientists, to save on cost (WSF is more expensive than conventional fertilisers). Further, he claimed that some other farmers were also following this practice. If this was true, then the deviation from the scientists' recommendation may be more widespread than our interviews implied. Nevertheless, most farmers

---

[15] Individual farm records were maintained only for the first year of joining TNPFP.

interviewed stated that although WSF was more expensive, following the fertigation schedule meant that they used *less* fertiliser than previously and that the result was that their overall fertiliser costs had declined.

One of the major reasons why there was little or no deviation in the fertigation practices appears to be the fact that scientists would be present during the mixing of WSF, at least in the first year that the farmer was part of TNPFP. This ensured that the correct dose was applied during fertigation, which also had a demonstration effect on the beneficiary farmers who could observe and learn the proper methods of mixing and applying the WSF. Furthermore, the fertigation schedules were not static. The scientists reviewed them annually and made minor refinements based on their increased experience of the region and the crops, and on evidence from the precision farms. This made the schedules specific and relevant.

An important aspect for the successful travel of the fertigation technology was that the farmers could see that it delivered on the claims made by the scientists. Almost all of the farmers interviewed reported a significant labour and water savings as a result of adopting precision farming techniques. Often, this was cited as one of the most important aspects or benefit of precision farming technology. Although the extent of labour and water saved on individual farms was not assessed quantitatively, most farmers agreed that they were using at least half of both water and labour than previously. These savings are manifested not only in the reduced quantum of labour or water used, but also the reduced effort applied for irrigation, weeding and other soil preparation activities. Reasons given for labour-saving effects were fairly unanimous: Drip irrigation reduced labour needs both for irrigation itself and for weeding, whilst the more porous soil engendered by drip irrigation made it easier to work and to plough, again reducing labour needs. Similar savings were experienced in the case of water requirements: '[F]or the same amount of water I needed previously to irrigate 1 acre, I can now irrigate 3 acres' [V.]. Another aspect of the core technology that impressed several farmers was its impact on soil aeration: R. explained how older methods, such as channel and flood irrigation, left the soil hard, while fertigation left the soil loose, which in turn promoted growth through better root condition and better yield.

In terms of the impact of precision farming on yield, the quantitative evidence that is available is unambiguous: Precision farming increased output obtained by the farmers by several-fold, and the average yield obtained by the beneficiary farmers was also several times that of the national average.[16]

---

[16] The sample of farmers used varied by crop, and these samples are not the same as the interview samples. We do not report the data in detail here, but it is available upon request.

For example, a survey of nine crops grown by TNPFP farmers showed that all crops yielded multiple harvests, implying lengthened crop duration and increased harvest periods through precision farming techniques. Furthermore, the average tonnage obtained in one season was also considerable: The average TNPFP yields for tomato, brinjal (aubergine) and banana were at least three to twelve times higher than the national average. Such findings were corroborated through conversations with individual farmers, the majority of whom were willing to testify to the positive impact that precision farming had upon the yield and quality of their crop and the income they derived from this.

Apart from yield, the quality and consistency of produce obtained was also reported to be high. For example, D. claimed that his tomatoes had 'good personality and were very attractive,' another said he now produced 'shiny tomatoes,' whilst a cabbage farmer reported that the average size of cabbage produced had increased from 2.5 kg to 3.5 kg. Consistency was seen to be linked directly with fertigation: L. reported that due to fertigation, the 'quality and size of the product was maintained, and there was uniformity in yield.' He also compared the results of the precision farming techniques to older methods: He said that for tomatoes, in precision farming, every four to five days there is equal application of fertilizers, leading to even growth through the life of the crop; he also claimed that this resulted in an extended shelf life of the product, 'sometimes up to 15 days, where ordinary products would be 4–5 days.' One result of better quality was that farmers received a better price. Examples of these better prices quoted by different farmers included Rs. 20 extra per crate of tomatoes for the same weight or volume; about Rs. 5 more per kg for tomatoes due to improved quality; a doubling in the income derived from the sugarcane crop; for the same crop and hybrids, whereas under non-precision farming they had received Rs. 180 per crate, under precision farming they received Rs. 200–210 per crate. Consistency also had a price benefit: According to S., the uniformity of water and fertiliser delivery meant that even on the last harvest he received the same price, whereas previously, later harvests would get a lower price. Perhaps the excitement that precision farming has given farmers is best captured by C.R. who told us that before he adopted precision farming he found it difficult to sell but now buyers fight among themselves to get his product!

The impact of savings on inputs, higher yields and better and more consistent quality was that the income of the beneficiary farmers increased, often quite considerably so. The impact on income is captured by the following comments from farmers: 'income has doubled'; 'for cabbage I get twice as much income as before'; 'before I was earning Rs. 1 lakh but now I am earning 2–3 lakhs'; 'from the first precision farming crop on 1.25 acres I made 3.5 lakhs, which was 80,000 rupees more than I made before precision farming, and my precision farming expenses are 20% less than previous farming techniques.'[17] In some cases the impact on the life of the farmer and his family has been extraordinary: One farmer said that before being part of the TNPFP he was finding it difficult to educate his children, but now because of his increased income he could fund them at college, and his son is studying for an MBA in Indiana in the United States.

On the whole, the interviews suggest that the farmers had a strong incentive not to stray from the precision farming techniques recommended by the scientists, as the improvements in yield and market value were directly attributable to the precision farming practices and fertigation. This vindication of claims made by the scientists ensured that the farmers believed in the science behind the project and in turn enhanced the reputations of the scientists and TNAU generally. This can be illustrated with statements from two farmers. D. said that when the project ended he wanted the '[university] people to continue to stay with us' as new diseases and pests would be encountered, and he also said that 'we will find it difficult if the university people leave us [because] department people [i.e., state extension education officials] do not give such good advice compared to university people.' K. illustrated the importance of the overall package provided by TNAU when he stated that 'from sowing to market, the university guided me.'

Thus, in the fertigation domain, technological and technical facts travelled successfully from the scientists to the farmers. Almost all of the experiential facts that emerged post-adoption were in consonance with the technical (scientific) facts that the scientists were promoting. This helped to validate the claims made by the scientists, and at least on one occasion – the fertigation schedules – helped to qualify or refine them. In this particular instance, experiential facts travelled well from the farmers to the scientists. The validation of technological and technical facts by the experiential facts also appears to make the central fact about precision farming travel well – that its adoption results in better quality crops and higher profit margins. This can be gauged by the increase in the number of applications to participate in the TNPFP in the second and third years of its operation. So successfully did this central fact travel that TNAU had to reject a larger proportion of applications in the second and third years when they had at times struggled to recruit farmers in some clusters in the first year.

---

[17] Rs. 1 lakh = Rs. 100,000.

6. The Associated Technologies Domain

Our initial assumption was that many of the cultivation, farm management and post-harvest management technologies operating in the associated technologies domain were less of a black box to the farmers. Unlike the fertigation domain, the farmers had existing experiential knowledge, and the interesting question is how this was reconciled with the scientific advice they received from TNAU scientists. Their facts based on scientific claims were likely to be subject to greater scrutiny by the farmers, which is what we found. The farmers were keener to experiment and innovate within the *associated* technologies domain as compared to the fertigation technologies domain. For instance, several farmers mentioned that they experimented with the spacing between the crops to ascertain the 'optimal' distance and crop density. This was a deviation from the 'standard' distance recommended by the scientists for each crop.[18] This was often done without consultation with the scientists. For instance, P. experimented with 6-foot and 3-foot spacing for sugarcane, instead of following the 5 feet recommended by TNAU. He told us that he discovered that 3 feet was a more optimal distance than 5 feet for sugarcane.

Facts related to post-harvest management and marketing issues also formed a part of this domain, such as technical facts that improved existing methods of preparing and transporting produce to buyers or commodity markets. Evidence that facts travelled well in this domain comes from changes in the marketing practices of the beneficiary farmers. Precision farming improved, or in many cases introduced, a system of grading using fairly simple methods, such as sorting produce into crates indicating different product quality. Sorting potentially helped to obtain better prices for produce. Information from terminal market operators suggests that precision farms were beginning to establish a marketing network, were gaining a recognition in the market and that they could access various markets beyond their local ones.[19] This, coupled with improved post-harvest practices such as transport of produce in crates or improved packing practices, meant that beneficiary farmers could decide on the best markets to sell. Beneficiary farmers either learnt about or improved their existing knowledge about the broad institutional aspects of modern commodity markets and how they functioned.

---

[18] We see another example of this in the way that nineteenth-century American builders experimented with certain elements of the imported classical style of architecture but left essential elements intact (Schneider, this volume).

[19] This information is based upon an interview conducted with a senior executive of the SAFAL Market, a commodity exchange and terminal market for horticulture crops in Bangalore.

Thus, in this domain, we observe that in some instances technical facts that travelled to the farmers were scrutinized, questioned and compared with experiential facts before they were accepted. Also, given the scrutiny and experimentation, farmers were more willing to modify or adapt facts they received from TNPFP than they were with facts in the fertigation domain. Perhaps this was an important aspect of the overall project – a domain wherein farmers could exercise their own judgement, a domain where their expertise was allowed the same, if not more, validity as that of the scientists. However, it is difficult to establish whether such flexibility within this domain was incidental or through design.

The other important observation in this associated domain is the travel of institutional facts that promote the importance of applying technologies in a community setting – technologies that promote and encourage cooperative behaviour. This is somewhat distinct from the nature of facts (and technologies) in the fertigation domain, which are individualistic in nature. Although the creation of community nurseries for seedlings is one example of such institutional facts travelling, the other example is the creation of a brand name for the products grown under the TNPFP. The branding of precision farming products encouraged market recognition: '[T]he brand name helps buyers to identify the source of the product and therefore that it is of better quality (because they are associated with precision farming techniques) and this helps with price' [A. R.].

Such observations lead us to consider that the travel of facts in the fertigation domain – and indeed the central precision farming fact – might have been aided by the nature of travel within the associated technologies domain. While this may be so, we also identify some enabling institutions that helped the travel of various technological, technical, experiential and institutional facts.

7. The Enabling Institutions Domain

In this domain, we have focussed on the role of two institutions in particular: the DEE – especially the extension model they adopted for TNPFP – and the farmer associations that were formed as part of the project. Historically, extension models in India follow the researcher-to-development officials-to-farmer mode, a three step process usually mediated by state officials or by non-governmental organisations (NGOs).[20] For the TNPFP, the DEE

---

[20] This is the standard model of extension, practiced in other counties such as the United States, in which there is a clear distinction between the laboratory and the field, with extension agents as the link between the two (Röling and Engel 1990).

followed a simpler and more direct model: that of researcher-to-farmer. In addition to two project scientists, seventeen field scientists were assigned in each district to the project on a full-time basis. The field scientists had regular contact with the farmers and provided direct assistance to them on technical and farm management issues. Thus, the level of human resources dedicated to the project by the DEE was considerable. This reduced the chance for error in the transmission of technology, skills and knowledge, as it ensured that scientists talked directly to farmers; it also ensured that scientists could quickly respond to queries from farmers, receive the experiential facts generated at the farm level and foster a cooperative dialogue such that farmers felt that their concerns and input were relevant to the project.[21] In one sense, the field scientists acted as good companions to the facts, but the interactive dialogue meant that they were not passive but proactive companions – they not only conveyed facts, but also listened and responded to the farmers.[22] Such a close relationship obviously had high costs for the scientists, but it is probably one of the most important innovations of the TNPFP and undoubtedly played an important part in the successful travel of facts. Such an institutional set-up helped all types of facts to travel well – technological, technical (scientific), experiential and institutional.[23]

For instance, the two most compelling reasons for non-adoption of precision farming (even though facts about precision farming had travelled well to non-adopters) were the lack of adequate financing and a knowledge deficit about how precision farming actually worked. Four farmers, who had seen precision farming in Andhra Pradesh, felt that they lacked the necessary knowledge to implement the techniques on their own – they needed more detailed technical facts and required expert guidance. Two of the farmers also explicitly cited concerns about the quality of extension programs. N. R. said that other state-sponsored extension programs usually supplied farmers with poor quality equipment and did not provide proper training or support. The state officials would 'promise something, bring something else.' He also stated that under previous government extension

---

[21] In listening to farmers and taking their local knowledge seriously, the attitude of the TNAU scientists was a marked improvement on what B. Harriss (1977) had observed in Tamil Nadu's North district more than three decades before.

[22] By engaging proactively with the farmers and making them a central part of the project, the scientists were, in effect, engaging in a form of participatory extension education (see Chambers et al 1989; Pretty 1995; Röling and Wagemakers 1998; Bonny et al 2005).

[23] The role of scientists in making their facts travel here can be contrasted with the activities of climate scientists in Oreskes, this volume.

programs he would not receive the full extent of the financial subsidy and that the officials would, in most cases, 'take their share.' This issue also came up in another interview. G. mentioned that during an initial TNAU meeting, whilst most farmers accepted the scientific claims, many were unwilling to join the project because they did not believe that the entire subsidy would reach them.[24]

The institutional set-up of the extension model ensured that the farmers had access to both knowledge as well as money to receive and use the best of the technology promised to them. Farmers received the full extent of the subsidy, were delivered with the best drip irrigation and fertigation equipment available and were able to interact closely with the scientists. The scientists not only monitored the fertigation activities regularly, but also adapted their advice according to farmers' experience in the fields. The integrity and hands-on approach of the scientists helped important experiential and institutional facts to travel, both from the scientists to the farmers as well as from the farmers to the scientists.

Similarly, the farmer associations (FA) performed vital roles in the adoption of precision farming and the travel of facts associated with it. FA served as nodes for exchanging knowledge and information, and helped farmers obtain better value in the markets. FA also supported the extension efforts of TNAU by functioning as information nodes. According to the president of the Moliyanur Precision Farmers Association, they would hold regular monthly meetings to discuss marketing and other issues on the second day of every month. Often, TNAU scientists attended these meetings and were able to offer expert advice, but even in their absence local issues were raised and resolved multi-laterally. Many such association meetings were attended by non-beneficiary farmers, which not only raised the profile of the TNPFP, but also substituted for the lack of direct scientist-to-non-beneficiary farmer interactions. The associations acted as vehicles for a variety of different kinds of facts (experiential, technical, institutional, etc.) about precision farming.[25]

The FA helped in the travel of facts in both the fertigation as well as the associated technologies domain. V. felt that associations were very helpful

---

[24] For a discussion of the weaknesses of the traditional extension education system, including its vulnerability to corruption, see Chambers and Wickremanayake (1977, pp. 160–3) and Birner and Anderson (2007, pp. 14–16).

[25] Only one farmer we interviewed felt he did not need the associations. Although he did discuss issues with members of his association, he felt most of his learning was 'from his own fields.' This view was unusual, and most of the farmers we talked to viewed the associations positively.

in disseminating information about the use of various technologies. For example, he was better informed about fertigation as well as about plant protection measures through the associations. D. felt that the benefit of association was the regular meetings 'on how to improve individual farms, use technology and discuss marketing.' R., a non-beneficiary farmer who attended local association meetings, said that he learnt a lot about precision farming methods in these meetings. He said that he 'got to know how the drip irrigation system can save water' through regular interactions with precision farmers at such meetings. He further said that the farmer meetings and discussions 'have taught [me] about plant protection measures, what chemicals to use and how much to spray.' Thus, when 'representatives of pesticide companies visited me, I was able to make up my own mind about what is [good] for my crops.'

FA also helped beneficiary farmers obtain better value by improving their negotiating position vis-à-vis buyers and input providers: They 'bring unity among farmers [and] give them better bargaining power.' Organized markets in the region increasingly prefer to deal with farmer associations, as it helps to eliminate risks of delivery failure while providing a greater assurance of quality. This was beneficial to the TNPFP farmers, as it helped them to secure better value by costing out of the revenue delivery failures and in-transit damage to produce. M. said that several precision farmers from his cluster would collectively send about forty to fifty crates of produce each to the market. Such practices meant that it led to 'sharing cost of transportation between farmers and saving of time for all.' L. explained how large buyers approached the FA with large volume requirements, which then coordinated the delivery details. By assuring minimum quality through proper grading and sorting, FA helped farmers obtain better average prices than comparable produce sold without the associations' involvement. The associations also helped the farmers in many other ways: to negotiate better prices for inputs such as fertilizers, pesticides, seeds and so on, by guaranteeing minimum quantity; to negotiate for or arranging *timely* supply of inputs; and by helping to pool together resources to transport produce to the market, saving time and effort, and guaranteeing delivery (Figure 10.1d shows an association shop in Erode).

The travel of this institutional fact – that farmer associations helped farmers to get the greater benefits from precision farming – culminated in the formation of a private limited company, Dharmapuri Precision Farmers Agro Services, Limited (DPFAS) by the two hundred Dharmapuri beneficiary farmers, with encouragement from TNAU. DPFAS represents a significant institutional development, as it acts as the distributor or dealer for several agri-product corporations and sells input materials such as seeds, fertilizers, plant protection materials and other agriculture inputs. The products sold are of better quality and at cheaper rates. Furthermore, it has become a dealer for Jain Irrigation Systems. Ltd., the supplier of the fertigation tanks and drip irrigation equipment used in the TNPFP, and in an intriguing twist has begun selling fertigation equipment to other farmers in the district. This is evidence of technological facts about precision farming travelling beyond TNPFP.[26]

In summary, we observe a greater degree of scrutiny of facts by farmers in this space (i.e., the benefit of farmer associations) and a greater degree of multi-directional travel (i.e., farmer to farmer, farmers to scientists, scientists to farmers, etc.).

8. Facts Travelling Beyond TNPFP

Considering adoption of fertigation technologies to be the ultimate sign of successful travel, we asked two important questions. First, did beneficiary farmers extend precision farming techniques to their non-TNPFP land? Second, did non-beneficiary farmers adopt precision farming?

More than half of the beneficiary farmers we interviewed had already extended or were in the process of extending precision farming techniques to their non-TNPFP land at their own cost, even where it involved buying another fertigation tank. Three of the farmers had or were about to convert all their land to precision farming, while P. K. intended to purchase four more acres for precision farming. The largest extension mentioned was by G., who said that he and some other farmers had bought equipment that would allow them to extend precision farming to another twenty acres. But would beneficiary farmers continue with precision farming after the project, and hence after the subsidy and TNAU support, ended in 2008? Overwhelmingly the beneficiary farmers responded that they would continue because of the impressive benefits precision farming delivered. Of course, we do not know how many have since followed through, but the creation of the DPFAS at the very least is a support and commitment mechanism that suggests precision farming in the Dharmapuri district will be a long-term phenomenon.

It became evident during the interviews that several non-beneficiary farmers had adopted some of the precision farming techniques. This was

---

[26] This degree of travelling well provides a good case of the way that people 'act on facts' – see Morgan, this volume.

primarily due to their interaction with beneficiary farmers, which confirms the signalling effect of demonstration projects and their importance in the travel of facts.[27] The discussion of the associations has already shown that facts were travelling from the beneficiary farmers, and the scientists, to non-beneficiary farmers through that particular node. Experiential facts were travelling, and the non-beneficiary farmers could see the advantages of precision farming for the beneficiary farmers. Non-beneficiary farmers would also informally talk to neighbouring beneficiary farmers about precision farming or visit the precision farms. For example, A. claimed that he had received many visitors and that his field has become an 'exhibition plot,' and that several visitors have started growing the same hybrid bananas that he does. Similarly, D. claimed that he had 'plenty of visitors' and that most of them were generally aware of the TNPFP. He claimed that 'after seeing [his] results, they regretted not having applied for the scheme.'

Nine non-beneficiary farmers interviewed had officially applied to the TNPFP and had been rejected. Of the remaining, three had not heard about the project until quite recently, one was waiting to see how his brother (a beneficiary farmer) did and the remaining four had wanted to join but were prevented from doing so, each for a different reason (lack of adequate water or electricity or land and, in one case, incapacitation – he was in hospital).[28] We then asked them if, having seen the benefits of fertigation, they would invest in this technology. Only one stated that he had already done so, whilst three others said they probably, or possibly, would do so using their own money or by getting a loan. Three other non-beneficiary farmers responded that they would adopt drip irrigation in the future only if they were successful in receiving state assistance through subsidy schemes (minimum 50 per cent assistance) or extension programs. Two farmers said they were unlikely to adopt drip irrigation despite knowing the benefits it could provide; they said they were constrained by either basic water supply problems or finance problems. Although claims that non-beneficiary farmers had adopted the core technology cannot be easily verified, the increased sale of fertigation equipment to non-beneficiary farmers in Dharmapuri district can be considered direct evidence of the impact of precision farming on the larger community. One DPFAS member claimed that as distributors of Jain Irrigation in Dharmapuri, DPFAS had supplied fertigation equipment to

---

[27] TNAU did not collect any information on whether their precision farming techniques did travel beyond the beneficiary farmers, and so our data about such travelling is based primarily on our interviews and to that extent may be subject to selection biases.

[28] The answer by one farmer to this question was confused and so has been ignored here.

about eighty-two farmers without any subsidy involved in the eight months beginning January 2007. During the same period, about fifty beneficiary farmers of the PFP were supplied with fertigation equipment to extend the existing area under fertigation.[29]

Although travel of the technology to non-beneficiary farmers was limited, travel of technical facts – that is, use of hybrid seeds, plant protection measures, field preparation methods and so on – has been comparatively greater.[30] However, the non-beneficiary farmers interviewed felt that in order to gain the full benefits from precision farming they would have to adopt the fertigation technologies and not only other improved cultivation and farm management methods. Thus, these experiential facts travelled well amongst the non-adopters.

Non-adoption of precision farming was also affected by existing attitudes. F. J. suggested that there was an overall reluctance to adopt precision farming because 'people in this area are not well educated and are very tied to traditional crops and methods.' Nevertheless, some cases were reported where such a conservative attitude was overcome.

A. R. reported that 'some people with 50% subsidy have adopted precision farming although they have only adopted drip irrigation and not water soluble fertilisers.' P. K., a floriculturist, told us that he tried the 'open system' of Dutch rose cultivation on advice on university staff and that as a result of its impressive yield and high income, fifty farmers in his area (only five of whom are in the TNPFP) abandoned their traditional system of cultivation and had adopted the new technique. On the whole, most beneficiary farmers felt that non-adopters were deterred from adopting precision farming without the benefit of some kind of financial assistance. Given the potential of significant financial returns generated by precision farming, this suggests a very conservative attitude to risk both by farmers as well as by those who could extend credit to them.[31]

We observe that although non-beneficiary farmers received key technical facts, on the whole, their attitude towards risk and uncertainty remained unchanged. We cannot help wondering if this is a function of institutional

[29] We have been unable to verify these figures with the equipment supplier.
[30] Given that the fertigation technologies represent the main fixed-cost part of the TNPFP package, this would seem to confirm the conclusion of Feder and O'Mara (1981, p. 73): '[R]isk aversion can be argued to be a deterrent to innovation adoption by small farmers only to the extent that adoption entails fixed costs.'
[31] Access to credit is a well-known major problem facing many small-scale farmers in India. See, for example, J. Harriss (1977) and Binswanger et al (1993). The reasons for slow adoption and the role of risk and uncertainty has been noted by, among others, classic articles by Ryan and Goss (1943) and Griliches (1957); for a Tamil Nadu perspective, see B. Harriss (1977).

changes required to enable technology adoption rather than how poorly certain facts travel in spaces beyond the TNPFP project. For example, we noticed that though non-beneficiary farmers accepted many of the facts regarding the fertigation technologies and its benefits (increased yield, water and labour savings, etc.), they were unwilling to make the initial capital investment. This may be dependant upon the general expectation by the farmers that the state should extend financial assistance for adoption of new technologies (i.e., in the form of subsidies) rather than poor travel of facts associated with the technology in question. Such social issues, in the context of our study, may be a limiting factor – a barrier – to the travel of key technological facts – drip irrigation and fertigation technologies. In other words, social and institutional factors may act as barriers for 'expensive' facts such as technologies, even though other facts – technical, experiential and so on – may travel across the same boundaries.

9. Concluding Remarks

This study shows how a successful technology transfer occurred because facts associated with the process travelled well. In this case, the wellness of travel is considered along multiple perspectives: extent, integrity, institutional support, consequences, and so on. We demonstrate that by focussing on how well facts travel one can gain a better appreciation of the process of technology transfer itself. Taken in conjunction with other chapters in this volume, our remarks resonate with some general observations regarding facts and their travel (i.e., integrity, vehicles, packaging, etc.).

We show that facts about technologies travel further and in advance of technologies (in their physical or material sense). In many instances, facts about TNPFP travelled independently of the fertigation technologies per se and independently also of the scientists who were promoting them. Nevertheless, this self-evident point has important implications in terms of thinking about how technologies travel as opposed to how facts about technologies travel. This is not to claim that all facts about technologies travel independently. But if they have an independent trajectory, then it is necessary to consider the differences in transmission mechanisms, the vehicles and the 'good companions.'

Further, if a technology transfer project operates in multiple, overlapping spaces, then the process is likely to be complex, non-linear and dynamic. It may involve the travel of many different types of facts (in our study, we grouped them as technological, technical, experiential and institutional), involve the interaction of various spaces and groups across these spaces and the travel of facts between these spaces and groups in multiple directions.

Our study reveals that technologies and facts about technologies could, and do, travel in bundles or packages.[32] Technical bits (i.e., scientific claims and procedures) are usually bundled with technological bits (i.e., material objects), both of which are, in turn, bundled with social and institutional bits (i.e., access to finance, supervision and support of those more knowledgeable, access to markets, etc.). The scientists believed that the main fertigation technologies would be adopted readily if they were bundled along with other technologies (crop spacing, pest control, marketing, etc.). But crucially, they remained sensitive to the environment in which they were received and applied. The institutional structure they used (farmer associations, close interaction of field scientists with individual farmers, extensive financial assistance, etc.) helped not only the technical and technological bits to travel, but also helped experiential and institutional facts to travel across various spaces. Thus, apart from the demonstration effects of 'this technology works' (a technical fact), there needed to be a transition to 'this technology works for me' (an experiential and institutional bundle of facts).

Our study further reveals the importance of trust for the travel of facts: trust in the technology being promoted (i.e., in its reliability), but also in the promoters of the technology (i.e., delivery on promises). These aspects of trust are different from the trust imputed to authority or to people in positions of authority, or to those in possession of specialist knowledge and expertise.[33] The trust we refer to is one that is gained through constructive dialogue and interaction. Thus, when the scientists delivered on their promises (delivery of finance and knowledge, close supervision of fertigation use, being accessible to solve problems, etc.) and the farmers experienced this fact, it enhanced the reputations of the scientists as well as the farmers' trust in them and improved the travel of technical facts. Trust in this sense is a social and institutional dimension that ensured wellness of travel.

Our study also highlights another institutional dimension: the importance of cooperative behaviour amongst farmers and its significance for the travel of both facts about technologies and technologies. This aspect manifests itself in institutions such as the farmer associations, which, as

---

[32] The way that facts travel in bundles can be seen, in the much more limited domain of scientific reports, in Merz (this volume).

[33] For example, the trust placed by the general population in the expertise of Harvey in confirming the age of Parr (Haycock, this volume) is an example of this trust. We use trust in a different sense, that is, mutual trust.

we have seen, were crucial for the success of the TNPFP. In this instance, as may be the case with other kinds of technologies, cooperative behaviour was crucial for the technology as well as facts about the technology to travel across and beyond various spaces. The significance of this aspect is to think about how some technologies may tend to 'work' in a collective sense, even though they may be 'applied' in an individualistic sense. This 'working' of technologies encompasses the idea of technological bundles, wherein the cooperative behaviour is an element of this bundle. Cooperation increased the depth and breadth of travel for TNPFP facts (technical, experiential, etc.) and along with it the transfer and adoption of modern agriculture.

Acknowledgements

This work was supported by the ESRC/Leverhulme Trust project, "The Nature of Evidence: How Well Do 'Facts' Travel?" (grant F/07004/Z) at the Department of Economic History, LSE. We would like to thank members of the Facts research group, especially Mary Morgan, for their comments and advice. We would also like to extend our gratitude to Dr. Vadivel and his colleagues at Tamil Nadu Agricultural University, without whose generous help and guidance this project could not have been undertaken, and to all the farmers and others who agreed to be interviewed.

Bibliography

Bandiera, O. and Rasul, I. 2006. Social Networks and Technology Adoption in Northern Mozambique. *Economic Journal*, 116(514), pp. 869–902.

Binswanger, H. P., Khandker, S. R., and Rosenzweig, M. R. 1993. How Infrastructure and Financial Institutions Affect Agricultural Output and Investment in India. *Journal of Development Economics*, 41(2), pp. 337–66.

Birner, R. and Anderson, J. R. 2007. How to Make Agricultural Extension Demand-Driven. International Food Policy Research Institute Discussion Paper, 00729.

Bonny, B. P., Prasad, R. M., Narayan, S. S., and Varughese, M. 2005. Participatory Learning, Experimentation, Action and Dissemination (PLEAD) – A Model for Farmer Participatory Technology Evolution in Agriculture. *Outlook on Agriculture*, 34(2), pp. 111–5.

Byres, T. 1983. *The Green Revolution in India*. Milton Keynes: The Open University Press.

Chambers, R., Pacey, A., and Thrupp, L. A. (eds). 1989. *Farmer First: Farmer Innovation and Agricultural Research*. Sussex: Institute of Development Studies.

Chambers, R. and Wickremanayake, B. W. E. 1977. Agricultural Extension: Myth, Reality and Challenge. In: Farmer, pp. 155–67.

Collins, H. and Evans, R. 2007. *Rethinking Expertise*. Chicago: University of Chicago Press.

Epstein, S. R. 2005. Transferring Technical Knowledge and Innovation in Europe, c.1200–c.1800. Department of Economic History, London School of Economics: Working Papers on the Nature of Evidence: How Well Do 'Facts' Travel?, 01/05.

Farmer, B. H. (ed.). 1977. *Green Revolution? Technology and Change in Rice-growing Areas of Tamil Nadu and Sri Lanka. Boulder.* Colorado: Westview Press.

Feder, G. and O'Mara, G. T. 1981. Farm Size and the Diffusion of Green Revolution Technology. *Economic Development and Cultural Change*, 30(1), pp. 59–76.

Feder, G., Just, R. E., and Zilberman, D. 1985. Adoption of Agricultural Innovations in Developing Countries: A Survey. *Economic Development and Cultural Change*, 33(2), pp. 255–98.

Foster, A. D. and Rosenzweig, M. R. 1995. Learning by Doing and Learning from Others: Human Capital and Technical Change in Agriculture. *Journal of Political Economy*, 103(6), pp. 1176–209.

Geroski, P. A. 2000. Models of technology diffusion. *Research Policy*, 29(4–5), pp. 603–25.

Griliches, Z. 1957. Hybrid Corn: An Exploration in the Economics of Technological Change. *Econometrica*, 25(4), pp. 501–22.

Harriss, B. 1977. Rural Electrification and the Diffusion of Electric Water-Lifting Technology in North Arcot District, India. In: *Farmer*, pp. 182–203.

Harriss, J. 1977. The limitations of HYV technology in North Arcot District: the view from a village. In: *Farmer*, pp. 124–42.

Howlett, P. and Velkar, A. 2008. Agri-Technologies and Travelling Facts: Case Study of Extension Education in Tamil Nadu, India. Department of Economic History, London School of Economics: Working Papers on the Nature of Evidence: How Well Do 'Facts' Travel?, 35/08.

Latour, B. 1999. *Pandora's Hope: Essays on the Reality of Science Studies.* Boston, MA: Harvard University Press.

Munshi, K. 2004. Social Learning in a Heterogeneous Population: Technology Diffusion in the Indian Green Revolution. *Journal of Development Economics*, 73(1), pp. 185–213.

Oudshoorn, N. and Pinch, T. 2005. *How Users Matter: The Co-construction of Users and Technologies.* Cambridge, MA: MIT Press.

Pretty, J. N. 1995. Participatory Learning for Sustainable Agriculture. *World Development*, 23(8), pp. 1247–63.

Rogers, E. M. 1962. *Diffusion of Innovation.* New York: Free Press.

Röling, N. and Engel, P. 1990. The Development of the Concept of Agricultural Knowledge Information Systems (AKIS): Implications for Extension. In: W. M. Rivera and D. J. Gustafson (eds.), *Agricultural Extension: Worldwide Institutional Evolution and Forces for Change.* Amsterdam: Elsevier.

Röling, N. and Wagemakers, A. (eds.). 1998 *Facilitating Sustainable Agriculture: Participatory Learning and Adaptive Management in Times of Environmental Uncertainty.* Cambridge: Cambridge University Press.

Ryan, B. and Goss, N. C. 1943. The Diffusion of Hybrid Seed Corn in Two Iowa Communities. *Rural Sociology*, 8(1), pp. 15–24.

Vadivel, E. (ed.). 2006. *Tamil Nadu Precision Farming Project: Expertise Shared and Experience Gained.* Coimbatore, India: Tamil Nadu Agricultural University.

Valeriani, S. 2006. The Roofs of Wren and Jones: A Seventeenth-Century Migration of Technical Knowledge from Italy to England. Department of Economic History, London School of Economics: Working Papers on the Nature of Evidence: How Well Do 'Facts' Travel?, 14/06.

Warner, K. D. 2008. Agroecology as Participatory Science: Emerging Alternatives to Technology Transfer Extension Practice. *Science, Technology, and Human Values*, 33(6), pp. 754–76.

Wise, M. N. (ed.). 1995. *The Values of Precision*. Princeton, NJ: Princeton University Press.

ELEVEN

ARCHAEOLOGICAL FACTS IN TRANSIT: THE "EMINENT MOUNDS" OF CENTRAL NORTH AMERICA

ALISON WYLIE

1. Introduction

Archaeological facts have a perplexing character.[1] They are often seen as tangible, less likely to "lie" and more likely to bear impartial witness to actual actions, events, and conditions of life than do, for example, the memories reported by witnesses or participants. At the same time, however, they are notoriously enigmatic and incomplete; they are sometimes described by critical archaeologists as inherently multivocal and malleable (Habu, Fawcett, and Matsunaga 2008). The anxiety that haunts archaeological interpretation, surfacing at regular intervals in sharply skeptical internal critique,[2] is that the tangible, surviving facts of the record so radically underdetermine any interesting claims archaeologists might want to make that archaeologically based "facts of the past" are inescapably entangled with fictional narratives of contemporary sense-making. And yet, these same internal critics make effective use of the recalcitrance of archaeological facts (of the record) to unsettle entrenched convictions that have given presumptive facts of the past purchase, that have allowed them to travel unchallenged.

This jointly solid and uncertain character of archaeological facts is the source of epistemic hopes and anxieties that are by no means unique to

---

[1] In commonsense parlance, and in much archaeological discussion, "archaeological facts" are the physical traces, artifacts, and features that constitute a material record of the cultural past. I use the term in this sense here, but will want to complicate this understanding of archaeological facts in what follows.

[2] The details of these recurrent debates are discussed in "How New Is the New Archaeology?" in Wylie (2002).

archaeology and that have everything to do with the ways in which archaeological facts travel. I consider here a set of cases, drawn from longstanding traditions of archaeological investigation of the earthen mound sites of the central river systems in North America, that illustrate strategies by which contemporary archaeologists appraise the integrity of archaeological facts in terms of what can usefully be described as their trajectories of travel. In the process I disentangle several different senses of "fact" that figure in these appraisals.

To anticipate: In what follows I rely on distinctions between facts of the record and mediating facts, and between two types of historical facts: facts of the past and narrated facts.[3] For purposes of this discussion, facts of the record consist of the surviving material traces on the basis of which archaeologists build reconstructive and interpretive claims about the cultural past that produced them (of which they are record). These inferential moves are mediated by facts that originate in fields ranging from nuclear physics to ethnography, ethnobotany to geology; that is to say, they depend on facts that travel into archaeology from collateral fields where the types of material that make up the traces of interest to archaeologists are a primary focus of inquiry, quite independent of their archaeological significance. As noted, I also distinguish between facts of the past – the actions, events, and conditions that actually happened in the past – and narrated facts about the past that are intended, in various ways, to capture, convey, interpret, and explain facts of the past. I draw this distinction in the provisional way recommended by Trouillot (1995, pp. 8, 26), not because I believe we have any independent, epistemically secure access to historical reality, but because this is a distinction on which we inevitably rely in the course of making and evaluating historical claims. As Trouillot puts it, there is no prospect for eliminating the systematic ambiguities inherent in the way we use the term "history" to refer both to events in the past and to the narratives by which we understand the past in the present. The line between history and fiction depends on a distinction between narrated facts about the past and facts of the past; although this distinction is undermined in innumerable ways, conceptually and in practice, it bears important epistemic weight.[4] Indeed, it animates the practice of archaeology insofar as it characteristically demands more of inquiry than fictionalization. A constructivism that

---

[3] This distinction tracks, in some respects, those drawn by Valeriani in discussion of the Italian debate about the epistemic status of history versus archaeology (this volume), and in her treatment of the traveling facts that constitute architectural history (2006, 2008).

[4] Trouillot's attention to the instability of this distinction resonates with Adam's discussion of the slipperiness of the distinction between fact and fiction (this volume).

systematically collapses this discussion undercuts the "cognitive purpose" of fields like history and archaeology; it "cannot give a full account of the production of any single narrative" (Trouillot 1995, pp. 11, 13).

I return to these distinctions in the conclusion. In what follows I will argue that it is the interplay between these types of facts that archaeologists exploit in making nuanced judgments about the credibility of claims about the past, and that these judgments depend fundamentally on appraising what can usefully be described as the trajectories of travel of "archaeological facts."

2. The Vagaries of Travel: One Hundred Sixty Years of Archaeological Research on "Eminent Mounds"

The earthen mound sites of the Mississippi, Tennessee, Illinois, and Ohio River valleys are among the most intensively studied archaeological sites in North America; they have been mapped, described, excavated, interpreted, and speculated about since the mid-nineteenth century.[5] These sites are typically attributed to two distinct cultural traditions. The earlier Hopewell sites consist of earthworks and settlements ranging from 200 BC to AD 400 (Middle Woodland), associated with horticulture based on indigenous domesticates and with assemblages of artifacts characterized by a distinctive design tradition that incorporates material traded from as far away as the Rocky Mountains and the Appalachians, the Gulf Coast, and the Great Lakes. The later Mississippian sites date to AD 950–1550 and are characterized by elaborate ceremonial complexes that include earthworks and extensive palisades as well as mounds, a related design tradition – the Southern Ceremonial Complex – and well-established practices of maize agriculture. These, then, are the received facts about the past that define this archaeological subject of inquiry.

As monumental as these sites are, the archaeological record of the "mound builders" has proven to be highly vulnerable to destruction. Even by 1848, when Squier and Davis published *Ancient Monuments of the Mississippi Valley*, the mounds and earthworks characteristic of these sites were rapidly being destroyed. Indeed, the motivation for this Smithsonian-sponsored survey was concern that, as the "tide of emigration" brought Euro-American travelers and settlers into these central river valleys, their rich "antiquarian" resources were rapidly being looted and plowed under (1998 [1848], pp. xxxi–xxxiv). There is a palpable sense of urgency in Squier

---

[5] I have discussed this research tradition in connection with "agnatology" (Wylie 2008).

and Davis's observation that the "sites selected for settlements, towns, and
cities, by the invading Europeans, are often those which were the especial
favourites of the mound-builders, and the seats of their heaviest population"; unless their material legacy could be documented immediately, all
record of these cultures would be lost (1998 [1848], pp. 6–7). This pattern
of destruction has continued apace: The vast majority of earthworks and
mounds documented by Squier and Davis and their nineteenth-century
successors[6] have been destroyed to make way for construction, or more
slowly dispersed by successively deeper and more destructive plowing as
agriculture was increasingly mechanized and industrialized. With no legal
framework for protecting archaeological sites on private land in the United
States and the trade in antiquities growing exponentially, even the most
aggressive campaigns to "save the past for the future" have proven to be
distressingly ineffectual.

There are, in addition, the vagaries of rapidly proliferating and evolving traditions of archaeological research to reckon with; in the course of
the last one hundred sixty years professional and avocational archaeologists
of various stripes have excavated and recorded the "facts" of these sites in
widely varying, often inconsistent ways. This variability is as much a function of shifting goals – changing interests in and competing understandings of the mound builders – as of pressures to professionalize embodied
in evolving standards of field practice and analysis. From the time these
sites were first reported, European travelers, traders, and settlers recorded
profound ambivalence about them: The mounds stood as a reproach to any
presumption that the rich lands along the interior waterways were unculti-
vated and unpeopled until the advent of Euro-American settlement. Despite
Jefferson's pioneering arguments (Jefferson 1787, pp. 97–100; Thomas 2000,
pp. 29–35), the working assumption, through the nineteenth and into the
twentieth centuries, was that none of the indigenous peoples living in the
region at the time of contact were capable of such monumental construction. Some "mysterious race," now vanished, must have achieved a level of
social complexity and sophistication that proved unsustainable or that was
destroyed by incursions of more primitive peoples (Squier and Davis 1998
[1948], p. 7); the "mound-builders" were linked to populations from locales
as diverse as Irish monasteries and the fictional lost Atlantis. For those
who accepted that they were an indigenous North American population,

---

[6] See Burns (2008) for an account of the formation of networks of agents who documented
and excavated sites in this region on behalf of the Peabody Museum and the Smithsonian
Institution.

the dominant problem was to locate the mound builders in a hierarchy of social, cultural forms that were presumed to lie along a linear trajectory of cultural evolution. The result was a selective practice of excavation and recording of these sites that focused on the highly visible, the monumental, and the exotic, and was structured by the question of who could possibly have built the mounds: Facts of ancestry figured prominently; the industry in measuring skulls and calculating evolutionary affiliation got under way in earnest, and evidence of their artistic accomplishment was routinely juxtaposed with supposed facts of "cannibalism," a penchant for elaborate ritual, and barbaric mortuary practices. The foundational assumptions of nineteenth-century anthropology, and the collecting interests of emerging research and educational institutions, structured the recovery and description of the archaeological facts.

In the 1930s, large-scale archaeological projects supported by the work projects administration (WPA) generated vast quantities of archaeological data, but despite a more open-ended research agenda and a commitment to build robust chronological and spatial schemes – a necessary step toward establishing key narrative facts about the cultural past – the quality of work was highly variable, even by the standards of the day. Some of the most detailed and systematic stratigraphic excavation and recording of mound sites ever undertaken was supported by the WPA in this period. But some projects were little more than artifact-collecting expeditions and, as Depression-era work projects, the emphasis was on fieldwork; often little provision was made for analysis and curation of the material recovered, much less public reporting. Sometimes nothing at all was published, even for sites that became the anchors for regional cultural histories. A case in point is Marksville, which was identified as a type site for the Middle Woodland in the Lower Mississippi River valley, significantly extending the range of Hopewell traditions into this region from the sites in Ohio with which they had chiefly been identified (McGimsey et al. 2005, pp. 1, 4). In other cases, only superficial summaries appeared, as at Shiloh Indian Mounds, a regionally significant Mississippian site on the Tennessee River, where a four-page report was the only publication produced by excavations that had opened up thousands of square feet (Welch, Anderson, and Cornelison 2003; Welch 2006, p. 26). Often even the most substantial publications were highly selective; many of the features reported in field notes went unmentioned in published reports, and when they were described, it was in the most general terms, without stratigraphic profiles or sufficiently detailed locational coordinates to allow even the reliable reidentification on the ground of the excavation

units in which they were exposed, much less an assessment of their chronological association with other mapped or excavated features (Welch
2006, pp. 30, 35–40). This pattern of expansive excavation and selective
recording and publication continued, on a smaller scale and with a focus
on typology and chronology, through the 1950s.

The "archaeological facts" that comprise the legacy of these midcentury
excavations – the assemblages of artifacts recovered and the associated field
notes, profiles, maps, and feature and artifact drawings – have suffered a
similarly patchy history of curation. Contemporary archaeologists routinely
describe the difficulties they encounter working with surviving collections
from the WPA-era excavations. Sometimes the problem is fragmentary
documentation that provides artifact assemblages little provenience (Welch
2006, pp. 23–4). Often, no records survive that could give artifacts secure
context even when, as in the case of Marksville, they have been widely used
to define distinctive cultural types and periods, setting the terms by which
facts of the record travel within archaeological contexts. In the case of the
enormous collections generated by the excavations at Marksville in the
1930s, McGimsey and his collaborators note that the original documentation "would have been of great value as the ceramic characteristics and culture history of the Marksville period were being defined" but that many of
the original field records had been lost (2005, p. 3). Describing the archaeological record of Mississippian sites in the Etowah Valley, Georgia, King
reports that, not only had all the documents been lost that might link artifact collections to specific excavation contexts, but "a substantial percentage
of the artifacts collected [by WPA excavation teams] were discarded after
the original analysis was performed"; all that remain are type collections
"composed of unique sherds and representative examples of more common
types" (King 2003, p. 36). Another all-too-common problem that King
encountered at Etowah is that even these surviving collections have been
dispersed; "working with the Etowah data is made more complex by the
fact that four different institutions sponsored excavations at the site, so collections are housed in six locations…[each of which] has its own history,
organizational system, and procedures for accessing collections" (King,
pp. 33–6, 50–2).

Ironically, although facts of the record have proven distressingly vulnerable to dispersal and attrition, a number of interpretative claims-cum-
facts about the past (narrative facts) have demonstrated remarkable staying
power; they "haunt our current understanding" (Cunningham, Goldstein,
and Gaff 2002, p. 1; see also Goldstein and Gaff 2003), setting "interpretive
frameworks…that persist in popular and even in scholarly reviews" (Muller

2003, p. 1; Muller 2002).[7] Even though the mound builder debates of the nineteenth century were resolved, at least in professional contexts, when excavation revealed burial populations whose morphology was well within the frame of that typical of contemporary Native Americans, the fascination with burials (the presumption that all mounds are mortuary sites), with the ceremonial and the savage (especially evidence of warfare and cannibalism), and with questions about how the mound builders fit into grand evolutionary schemes – whether they were civilized, or on a trajectory to civilization, or an example of arrested evolutionary development – dominated archaeological thinking well into the twentieth century and persists in museum presentations and the public imagination. The legacy of this interpretive tradition is an entrenched practice of selectively collecting and emphasizing archaeological facts (of the record and of the past) that fit comfortably with dominant narratives about precontact indigenous cultures as a history of culturally distant and vanished, alternately noble and savage, "others." In short, contemporary archaeologists working on these sites wrestle both with failures to travel – as the primary (in situ) archaeological record is destroyed and the secondary, recorded facts are lost or dispersed – and with the travel-hardy persistence of a canonical set of interpretive facts about the past that have long dominated archaeological thinking.

In the last fifty years, archaeologists have developed more sharply focused and technically sophisticated projects designed to refine the regional chronologies and culture-historical schemes that structure Hopewell and Mississippian research, and to develop a more fine-grained understanding of the internal structure and histories of particular mound sites. These are not just internal puzzles generated by antecedent research; they are foundational questions that must be resolved before archaeologists can assess claims about the relationships between specific sites and features or address broader questions about shifting interaction spheres on a regional scale or the internal organizational structure and power dynamics of particular site-based communities or, most provocatively, the meaning of the distinctive symbolic repertoire of precontact Hopewell and Mississippian cultures. The difficulty is that postwar land development has taken a substantial toll on what remained of the mound sites that were excavated in the 1930s and 1940s. Increasingly, the only surviving mound sites are protected state or national parks, subject to regulations that strictly limit any destructive

---

[7] The staying power of facts, once set in motion, is a theme that connects this discussion of archaeological facts to a number of other contributions to the Facts project; I note some specific points of resonance in what follows.

investigation of intact deposits. Archaeologists are under growing pressure
to find ways of enlisting old data – existing records and collections – to
answer new questions. To this end they must work against a tendency to
dismiss surviving collections and records as too fragmentary, partial, and
enigmatic to be useful for contemporary purposes, at the same time as they
explore creative ways of retrieving useable facts that survive, often unrecog-
nized, in maps and photographs, field notes, and collections that are all too
often all that remain of sites that have long since been destroyed.

3. Critical Histories of Travel

I am particularly interested in two broad strategies by which archaeologists
extract new facts from old that are exemplified by the work presented in a
symposium sponsored by the Society for American Archaeology: "Emblems
of American Archaeology's Past: Eminent Mound Sites of the Eastern
Woodlands Revisited" (Schroeder 2003, and discussion in Wylie 2008).
In this context a dozen archaeologists currently working on Hopewell and
Mississippian sites took stock of the trajectory of research through the 100
to 150 years they have been investigated, with the aim of assessing the potential for making effective use of the complicated legacy bequeathed them by
antecedent generations of archaeologists. In the process they illustrate what
I will refer to as strategies of secondary retrieval.[8] The facts (of record) that
archaeologists find lodged in existing archives and collections are rendered
useable for contemporary purposes by "repositioning" them in relation to
one another and to new facts of record, sometimes in a quite literal sense,
but also, crucially, in a more metaphorical Foucauldian sense by which they
are situated in the context of the research traditions that produced them.
In connection with this last, the practice of Eminent Mounds archaeologists shows how detailed histories of the travel of these collections, records,
and interpretations – itself an empirically grounded undertaking – can
play a critical role in the process of secondary retrieval, not only bringing discarded archaeological facts of the record to light, but also grounding
the adjudication of their epistemic integrity as a basis for framing factual

---

[8] The term "secondary retrieval" comes from Trouillot's discussion of the third of four
moments in the production of history: the generation of textual traces; the compilation of
these traces as an archive; the retrieval of these traces from the archive and the configura-
tion of them as facts to be built into historical narratives; the construction of narratives that
have retrospective significance (Trouillot 1995, p. 8, 26). The archaeological counterpart to
the creation and recurrent exploration of the "archive" is what interests me here; see Wylie
(2008) for more detailed discussion of Trouillot's account.

claims about the past (narrative facts). The effect of these strategies is to put archaeological facts back into circulation, to send them off on new travels. Here are two examples drawn from the projects reported in the "Eminent Mounds" session that foreground these strategies and throw into relief key conditions that have an impact on how well archaeological facts travel.

3.1 Reassessing Attributions of Mortuary Function and Interpretations of Funerary Traditions

One area where the influence of nineteenth-century interests and assumptions is especially clear is in the preoccupation with mortuary remains. Given an intense interest in skeletal morphology as the key to determining the identity and affiliation of the mound builders, early investigators paid particular attention to mounds that were burial sites, and to evidence of what were taken to be especially exotic funerary practices. Their assumptions, and their records and interpretations, have had a profound impact on Eminent Mound archaeology, structuring patterns of (selective) recovery, description, and analysis of facts of the record that have set in motion a number of presumptive facts about the past that have proven to be resolute travelers. Certainly, there are many spectacular mortuary sites, and some of them yield just the kind of mass burials, dispersed and fragmentary remains, and evidence of violent death that are the stuff of mound builder legend. Famous examples are Mound 72 at the Mississippian site of Cahokia (outside East St. Louis), where 272 burials were excavated in the 1960s and 1970s, or Aztalan, a culturally related Mississippian village in Wisconsin at which similarly complex internment practices have been the basis for attributions of cannibalism that have proven hard to dislodge (Cunningham, Goldstein, and Gaff 2003, p. 2). It should be noted, however, that these are often not the most prominent features on mound sites; archaeologists report great variability in the function of mounds, ranging from refuse dumps, platforms on which various kinds of structures were built (some of which seem to have been the locus of ceremonial activities), and elements of astronomical alignments, as well as cemeteries and crematoria.

Although this understanding of the complexity of mound sites is now well established, it is still a matter of conventional wisdom – a staple of popular accounts and of museum presentations – that all mounds are mortuary sites. Puzzled by the persistence of these well-traveled interpretive facts, an archaeologist working at Fort Ancient in Ohio traced the origins of these claims about this site to the reports of excavators in the 1890s and 1930s (Connolly 2003, pp. 3–4; Connolly and Lepper 2004, pp. 85–113). Connolly discovered that, far from presenting robust evidence of mortuary remains, in one case, the original excavator described a puzzling *lack* of skeletal material, and in another he speculated about the possibility that bone fragments, long since disappeared from collections, might be human.[9] This quite straight-forward example illustrates how consequential it can be to trace circulating narrative facts back to the facts of record that are their purported ground and warrant: This is a matter of undertaking a secondary retrieval of facts of the record and of holding interpretive facts accountable to them.

At the famous Mississippian sites of Aztalan and Cahokia, recent excavations bear witness to the intensive use of (some) mounds as funerary sites that would seem to reinforce dominant interpretative narratives of cannibalism and "deviant" ritual (Balter 2005, p. 613), a legacy of nineteenth-century fascination with the Mississippian "other" as exotic and barbaric. Rather than holding conventional interpretations accountable to newly recovered or neglected facts of the record, a crucial strategy here has been to reassess the background assumptions that inform conventional interpretations of these funerary remains. Goldstein (who has worked at both Cahokia and Aztalan) argues that attributions of cannibalism or human sacrifice are only plausible if archaeological interpretation is informed by a narrowly ethnocentric set of assumptions about mortuary practice (2001, 2006). When the facts that mediate these interpretations are scrutinized and supplemented by insights drawn from broader ethnohistoric sources than has been typical, it becomes clear that the collective burial of disarticulated and dispersed skeletal material is the archaeological signature for a variety of mortuary traditions that involve elaborate preparation of the dead and secondary burial, but not necessarily cannibalism or human sacrifice. Indeed, some of these traditions are to be found in the heartland of western European tradition. Consider, for example, the mortuary practices typical for royal and aristocratic members of European dynasties (Babenberg and Habsburg). As described by Weiss-Krejci (2005), these involve all kinds of body processing, including evisceration, defleshing, treatment with salts and dyes, separate burial for disarticulated body parts, as well as temporary storage or exhumation, relocation, and dispersal in a series of secondary burials – practices that produce just the kinds of mortuary signatures taken to be evidence of the barbaric and exotic in North American mound sites.[10] In this case, it is the role of quite another kind of archaeological fact that provides critical

---

[9] This persistence of (narrative) archaeological facts in the absence of corroborating evidence, or even in the face of counterevidence, bears some similarities to the "imaginative dislocations" described by Wallis in connection with the Eyam plague narrative (2005).

[10] I thank Lynne Goldstein for bringing Weiss-Krejci's analysis to my attention.

leverage in reassessing received "facts about the past": mediating facts about how particular material signatures could have been produced and about the conditions under which one causal-cultural pathway would more likely be instantiated than another.[11]

3.2 Reassessing Site-Specific Culture Histories, Regional Interaction Spheres, and Evolutionary Trajectories

The secondary retrieval of archaeological facts, as undertaken by Connolly at Fort Ancient, typically involves not just searching out critical anchoring facts, but often the labor-intensive process of reconstructing how surviving material was recovered, how surviving fragments (of material and of data) relate to one another, and how they relate to what has not survived. As Welch describes the groundwork laid by a colleague for understanding the history of research and surviving records of excavations at Shiloh since the 1860s, it took decades to assemble scattered documents, and then weeks of work with collections held by the National Museum of Natural History to "piece together what is recorded and to discover what information is truly missing" (Welch 2006, pp. 23–4, 28). As tedious and painstaking as it is, this labor of secondary retrieval and quite literal repositioning of facts of the record can yield quite dramatic, destabilizing results.

In the case of the Mississippian site of Jonathan Creek in Kentucky, Schroeder (2005) has constructed integrated geographic information system (GIS) maps that incorporate all the locational data recorded by the generations of archaeologists who have surveyed or excavated a particular mound site, in the process cross-checking their accuracy across existing records (comparing photographs and maps of various eras) and against data derived from new fieldwork (e.g., testing for old trenches and geological markers that make it possible to tie features recorded on archival maps to coordinates on contemporary maps).[12] Schroeder demonstrates, through

---

[11] Contrast this strategy of critical analysis focused on mediating assumptions with Goldstein and Gaff's use of direct archaeological testing to assess common assumptions about Aztalan (2002). For a more detailed account of the strategies by which archaeologists deploy facts drawn both from archaeological subjects and from interpretive sources, see essays on analogical reasoning and "The Constitution of Archaeological Evidence" in Wylie (2002, pp. 136–53, 185–99).

[12] There are some intriguing similarities between the strategies by which archaeologists aggregate localized data points into structural and distributional facts (which they then use to establish or to challenge consequential narrative facts) and the bioinformatics practices described by Leonelli by which small (local) facts are normalized and recontextualized so that they can be assembled into large facts (this volume and Leonelli 2008).

analysis of this systematically coordinated and repositioned data, that it is impossible to sustain the WPA-era claim, formative for much subsequent work, that Jonathan Creek was occupied by two successive, ethnically distinct populations. The divergent architectural styles identified in the 1930s and 1940s by Webb, the original excavator, show complex patterns of overlap and juxtaposition that suggest simultaneous or seasonal occupation rather than a pattern of temporal alternation (Schroeder 2005, p. 65). In building this argument, Schroeder does more than just impugn factual claims about particular features of the site and its occupants. She offers a critical history of the theoretical presuppositions and methodological conventions that shaped the work of Webb at Jonathan Creek (and WPA-era archaeologists generally), showing how Webb's impressionistic archaeological field observations could authoritatively ground an expansive narrative about ethnic group migration and interaction. The inferential tracks on which his facts traveled were supplied by a conception of "archaeological cultures" according to which stylistic differences must mark the boundaries between static, culturally autonomous ethnic groups; it was assumed that stylistic variability within a site and across a region must be explained in terms of the migration of populations (Schroeder 2005, pp. 57–9). By contextualizing Webb's archaeological practice in this way, Schroeder calls into question a set of much broader, travel-hardy narrative facts: accepted facts about cultural difference that underpin the categories of description and analysis in terms of which Webb retrieved and documented what became the surviving archaeological facts (of the record) with which Schroeder now works.

Often, this conjoint process of secondary retrieval – the recovery, synthesis, and reanalysis of facts of the record, as well as the appraisal of the conditions of their initial retrieval – takes archaeologists back to the field. Where they lack chronological control or details of provenience they reopen the trenches excavated by earlier generations of archaeologists with the aim of locating surviving traces of recorded features in the walls and balks; sometimes this allows them to build a repertoire of stratigraphic profiles that make it possible to tie these features into a site-wide chronology, refining and substantially correcting histories of site occupation. In some cases the results have destabilized broader regional as well as local facts about the mound builders, with ramifying implications for the repertoire of nineteenth- and early twentieth-century narratives about prehistoric cultures that underpin archaeological conventions and dominate popular thinking about Mississippian and Hopewell sites.

For example, it is conventionally assumed that the major Hopewell and Mississippian sites must have been occupied continuously, showing sustained growth in size and density as they attained their status as regional centers and extended their influence into the hinterland, until they suffered precipitous collapse and were abandoned. The cultural markers of distinctive stylistic traditions that appeared across a region – commonalities evident in the structure and distribution of earthworks and various classes of material culture and, by inference, in ceremonial practice – are assumed to have diffused from dominant population centers to smaller sites through lines of regional influence or actual migration. These local and regional histories are, in turn, understood in terms of the conventions of a linear evolution from bands to tribes to chiefdoms to states. The lines of disagreement have long been drawn between those who are inclined to push the mound-building cultures of North America toward one or the other end of this continuum. On one hand, there are those who emphasize the internal complexity, the degree of social differentiation, hierarchy, and centralization of power associated with emergent mound centers, characterizing them as protostates on the model of state formation familiar from Mesopotamia and central Mexico. And on the other hand, critics of this line of thinking see these communities as inherently unstable chiefdoms that realized variable degrees of complexity; they emphasize the repetitive structure and relative autonomy of local polities that periodically coalesced into regional networks but did not develop the infrastructure – the social hierarchies and divisions of labor – presumed necessary to sustain a functioning state and its projects of monument building.[13]

When the complexities of refined site chronologies and occupational histories are taken into consideration, however, neither set of interpretive conventions fits these sites well. Internal site chronologies routinely show that even the most substantial mound sites were periodically abandoned, sometimes for as much as 100 years at a time in occupational histories of 450 years (Sullivan 2009). Even when mound sites were continuously occupied, they cycled through periods of expansion and contraction; often their periods of major fluorescence were not the culmination of a history of successively larger and more visible occupation (King 2003, pp. 60–4, 81–3, 140–3). At a regional level, although there is evidence of a distinctive Hopewell architectural grammar marked by standard units of measure (Connolly 1998, pp. 85–113), astronomical alignment in the internal structure of Mississippian sites (Kelly 1996), and widely distributed stylistic conventions (e.g., of the Southern Ceremonial Complex), it is increasingly

---

[13] This dynamic of debate is described in a number of contexts. See, for example, Milner and Schroeder 1999, pp. 96–9.

implausible that these commonalities can all be accounted for in terms of patterns of population movement and cultural diffusion. In some cases, sites identified as regional centers prove to have been abandoned during the very periods in which their influence was assumed to have been at its height (Sullivan 2009). Other sites that had been interpreted as outposts, subject to the influence of regional centers, show persistent and puzzling anomalies, which suggest that they were more likely manifestations of a locally derived tradition that assimilated some features of the regional culture; McGimsey describes the Hopewell aspects of Marksville, presumed to define the southern limits of Hopewell influence, as a thin "veneer" overlaid on a robust local tradition (2005, p. 11). Moreover, many local traditions prove to have been highly variable within the regions and periods of their influence. As in the case of Jonathan Creek, within-site stylistic diversity that had been interpreted as evidence of a sequence of culturally distinct occupations proves to have been contemporaneous, challenging any assumption that precontact cultures were sharply bounded, internally homogenous, static, and aligned with distinct populations.

The upshot is that as enigmatic as they are, the "facts of the record" originating in these intensively studied, much-speculated-about sites do prove to have a capacity to travel that exceeds, and disrupts, the conceptual foundations of the research traditions that set them in motion, as Valeriani argues (this volume). There is growing consensus that conventional assumptions about cultural evolution, succession, and interaction – the "restrictive and static cultural categories" derived from evolutionary schemas – must be fundamentally reassessed (Muller 1995, pp. 321–4, 335–6; Muller 1999, pp. 157–8; Muller 2003, p. 20;). Cultural complexity cannot be equated with stratification (Goldstein 2001), or assumed to mark a stage on the path toward stratification; the facts about mound builder cultures generated by the secondary retrieval, reanalysis, and repositioning of facts of the record – their patterns of cycling "through periods of formation, florescence, and fragmentation" – undermine the expectation that they were on track to become "truly stratified socio-political systems" (Milner and Schroeder 1999, pp. 96, 103). These precontact cultures do not fit any of the models of social and cultural formation projected by conventional evolutionary schemas. Some recommend a thorough overhaul of this framework. They direct attention to a range of ethnohistoric cases in which chiefly elites exercise political authority through diverse social mechanisms that do not necessarily give rise to or anticipate statelike structures; these, they suggest, offer resources for explaining how Mississippian and Hopewell societies could have produced monumental earthworks and mound sites without exaggerating their stability or the degree of vertical hierarchy (King 2003, pp. 140–3; Cobb and King 2005, pp. 167–92). This is a matter of repositioning facts about the mound builders in the context of new mediating facts about the range of possibilities by which communities can mobilize to take on ambitious projects (like building large-scale earthen monuments), coordinating collective effort, and engaging in highly complex cultural practices that extend across regions and over long periods.

4. Archaeological Facts and Their Travels: Three Questions

What, then, counts as a fact in archaeological contexts? And what ensures that some archaeological facts travel well, altogether too well in some cases, while others prove to be highly vulnerable to misrecognition and attrition? As I suggested at the outset, there are a number of different kinds of "facts" at issue here, each with distinctive trajectories of travel and capacities for success in traveling. I close by enumerating four of these, adding a fifth, and identifying two strategies by which the integrity of traveling facts is adjudicated in archaeological contexts.

Facts of the archaeological record, my point of departure, are most obviously conditioned in their travels (temporal, spatial, and disciplinary) by their own intrinsic physical characteristics, and by the conditions of their deposition and preservation.[14] They include the full range of artifacts and material traces, produced by both routine human behavior and intentional action, that make up the built environment that cultural actors produce and that constrains their action. In the case of the cultural past studied by Eminent Mounds archaeologists, these include, for example, the monumental earthworks themselves and an array of material traces that testify to their date and mode of construction; the uses to which they were put; the size of the communities that built them; their social relations and the subsistence practices that sustained them; and their motivating beliefs and intentions, as well as the local and regional histories in which they were enmeshed.

These surviving traces only become components of an archaeological "record," however, when they are retrieved, documented, and curated. Consequently, facts of the (archaeological) record include not only the primary surviving material but also facts about its composition, provenience, and associations generated by the process of recovery and analysis. For example, post-molds excavated by Webb at Jonathan Creek, as well

---

[14] See, for example, Schiffer's influential discussion of the cultural and environmental formation processes (1996).

as his original maps and notes, and also the spatial and temporal patterns Schroeder identified when she constructed a composite GIS-based map of these features, are all facts of the (archaeological) record. So, too, are the chemical signatures of the source contexts in which traded material originated (e.g., obsidian) and of the firing temperature and production techniques used to produce distinctive ceramic artifacts, the ratios of decayed (radioactive) C^{14} to (stable) C^{12} and C^{13} in organic material (the basis for calculating cutting or burning dates), the isotope values of bone marrow extracted from the skeletal remains of individuals, and differences in these values across populations. It follows that the travel fortunes of archaeological facts depend on the technical resources and dynamics of research traditions and the motivating ambitions of practitioners, collectors, and curators, as well as an immensely complex range of political-economic and institutional factors that, together, determine which traces will be retrieved, documented, and curated as facts of the (archaeological) record.

Mediating facts play a critical role in the trajectories and success of travel for archaeological facts. Facts of the record only have standing, as such, given elaborate conceptual and technical scaffolding.[15] These are facts about the properties of various constituents of the record, and about the conditions (causal processes, cultural practices) that could or likely did produce surviving material traces and that affect their preservation, transmission, and recovery. Put to work in archaeological contexts, these interpretive resources, either developed internally or drawn from collateral fields, make it possible to identify facts of the record as travelers, and to reconstruct the conditions of production, use, and deposition by which they have traveled from, and can be linked to, particular events and conditions in the past. Most are, by nature, facts that transgress disciplinary boundaries, so their capacities to travel also depend on a range of factors that include, for example, institutionally enabled transfers of technical skills and resources (e.g., post–World War II support for the development of radiocarbon dating); the accidents of cross-field interaction and individual interest (GIS); and, crucially, their comfortable fit with the conventional wisdom, professional or public, that underpins archaeological categories of description and analysis (the narrated facts of received ethnohistory).

Examples of mediating facts at work in the archaeology of Eminent Mounds are the geological facts that underpin stratigraphic analysis, making it possible to establish building and occupational sequences. The sourcing of artifacts, the reconstruction of how they were produced, and residue analysis that

---

[15] See Haycock's discussion of scaffolding, this volume.

suggests how they were used all depend on facts of physical chemistry and material science that have traveled from their home contexts into archaeology. Facts of astronomy provide the framework that enables the identification of systematic patterns of alignment between sites and of features within sites. Experimental archaeology generates intriguing facts about how much labor is required to produce an monumental earthwork or mound (much less than has typically been supposed), and ethnoarchaeology provides a fine-grained empirical understanding of how ceramics are produced, reused, discarded, and how distinctive stylistic features diffuse in communities that use what they make (rather than producing for a market). The ethnography of feasting practices and performative ritual suggests a range of models for understanding how mounds might have been used, while comparative ethnohistories of burial practices suggest diverse ways in which the mortuary deposits of mound builder fame could have been produced. Finally, the ethnography of "tribes" and "chiefdoms," and more recently of "house societies," as well as the historical sociology of state formation processes, are all instrumental in suggesting how social groups that undertake the collective projects of monument construction could be organized.

The goal, of course, is to establish *narrative facts about the past* ranging from highly localized facts tethered to particular material traces, through empirically grounded inferences about site histories and their occupants (their migrations and interactions), to the factual underpinnings of framework assumptions about cultural differences and cultural dynamics. The research traditions that make up Eminent Mounds archaeology have generated an enormous body of narrative facts. These include facts about the function of particular artifacts or features of the kind archaeologists have painstakingly reassessed in recent years (e.g., the presumption that the Fort Ancient mounds were cemeteries, that Jonathan Creek was protected by a palisade, and that disarticulated skeletal material at Cahokia was the product of human sacrifice) as well as the convention-disrupting facts about histories of site occupation, interaction, and internal diversity they have secured by means of secondary retrieval and the repositioning of localized facts (e.g., the cycling patterns documented at Hiwasee Island, the appreciation that Marksville was a Hopewell outpost, and that apparently distinct cultural groups coexisted at Jonathan Creek). And they extend to such framework-anchoring facts (many now disputed) as the conviction that Hopewell and Mississippian cultures are distinct, that they must have been chiefdoms or insipient states (given their complexity), and that cultural affinities across space reflect the migrations of distinct culture-bearing peoples.

Although these narrative facts are set in motion and authorized by facts of the record, they have their own distinctive circuits of transmission and reception, structured by lineages of disciplinary training and practice, and by a context-specific repertoire of narrative conventions.[16] This is a primary source of the epistemic anxiety that archaeologists express about their facts: that these narrative frames have a life of their own; that they determine what can be recognized as a fact of the record, what mediating facts will be brought into play, what survival and circulation patterns they will have; that facts about the past reduce to narrative convention. On this view there is no distinction between fact and fiction; historical and archaeological facts just are whatever we narrate them to be.

In practice, however, the patterns of interdependence among archaeological facts, and their capacity for travel, is a source of epistemic possibility. Archaeological facts, like Trouillot's historical facts, prove not to be "infinitely susceptible of invention" (1995, p. 21). To stabilize any claim about the past is an accomplishment that depends on a complex articulation of resources – material, technical, and conceptual. The Eminent Mounds cases illustrate a subset of the strategies of triangulation by which archaeologists use critical points of convergence between, and friction among, different types of archaeological facts of the record to assess their integrity as facts about the past (Wylie 2002, pp. 205–10). It is the intransigent materiality of facts of the record, and the contingent independence of the mediating facts that allow their interpretation as facts of the record, that animates the presumption that there is a difference between narrative facts about the past and *facts of the past*, the fourth type of fact at issue here. As tenuous a construct as they are, archaeological facts (of the record) routinely bear witness to a past that proves not to be as imagined, not to fit any of our familiar narrative templates. We require more than fictionalization, as Trouillot puts it, precisely because facts of the record have a capacity to challenge even deeply held foundational narrative facts. We appreciate this distinction most clearly – we sense that "the facts" (of the past) have been revealed – when facts of the record do not conform to established narrated facts. It is only given this possibility that it makes sense to insist that it matters whether narratives are "fact or fiction" and to impose "tests of credibility" on them (Trouillot 1995, pp. 11, 13).

---

[16] See, for example, Wallis's (2005) discussion of the role that narratives play in making facts travel and Merz's account (this volume) of how the records of experiments may be understood in narrative terms.

The strategies by which archaeologists exploit these epistemic possibilities are all, fundamentally, a matter of making the trajectories of travel themselves an object of critical scrutiny. On the first such strategy, that of *secondary retrieval*, archaeologists cross-check narrative facts against facts of the record that are presumed to anchor them, and they reposition facts of the record, often extracting facts that were not originally recorded or deployed in building narrative accounts of the past. On the second, *recontextualizing facts of the record*, archaeologists expand the repertoire of mediating facts embodied in background knowledge, techniques, and skills of analysis by which facts of the record are linked (interpretively) to facts of the past. Both strategies depend on building a critical historiography of archaeological facts that serves not just to deconstruct illusions of epistemic security, but also to reanimate and recalibrate these facts. In the cases considered here, a fifth type of fact plays a pivotal role: archaeological facts in a Foucauldian sense, *genealogical facts* about the complicated travels of all the kinds of material, interpretive, and narrative facts that constitute archaeological practice. By understanding these circuits and conditions of travel, archaeologists put new facts of the record into circulation, they hold both new and old facts about the past accountable to them, and they identify a range of questions that have not previously been asked of them. Taken together, these constitute tests of credibility that depend jointly on the capacity of facts of the past to travel with integrity, and on the capacity of archaeologists to discern where and how their travel may be obstructed.

Acknowledgments

I'm grateful for the support of the Leverhulme Trust (grant F/07004/Z), which made it possible to attend the FACTS projects workshops when this chapter was in formation, and has given me the opportunity to present it in a number of contexts while visiting Reading University (Department of Archaeology) as a Leverhulme Trust Visiting Professor (January through July 2010).

Bibliography

Balter, Michael. "'Deviant' Burials Reveal Death on the Fringe in Ancient Societies." *Science* 310 (2005):613.

Burns, J. Connor. "Networking Ohio Valley Archaeology in the 1880s: The Social Dynamics of Peabody and Smithsonian Centralization." *Histories of Anthropology Annual* 4 (2008):1–33.

Cobb, Charles R., and Adam King. "Re-Inventing Mississippian Tradition at Etowa, Georgia." *Journal of Archaeological Method and Theory* 12 (2005):167–92.

Connolly, Robert P. "Architectural Grammar Rules at the Fort Ancient Hilltop Enclosure." In *Ancient Earthen Enclosures of the Eastern Woodlands*, edited by Robert C. Mainfort and Lynne P. Sullivan, 85–113. Gainesville, FL: University of Florida Press, 1998.

"From Authority to Guide: The Archaeologist in Public Interpretation." In *Society for American Archaeology, 68th Annual Conference*. Milwaukee, WI, 2003.

Connolly, Robert P., and Bradley T. Lepper, eds. *The Fort Ancient Earthworks: Prehistoric Lifeways of the Hopewell Culture in Southwestern Ohio*. Columbus, OH: Ohio Historical Society, 2004.

Cunningham, Peter, Lynne Goldstein, and Donald H. Gaff. "Reexamining and Reinterpreting Aztalan: Making Old Data Useful by Integrating It with New Approaches." In *Society for American Archaeology, 68th Annual Conference*. Milwaukee, WI, 2003.

Goldstein, Lynne. "Ancient Southwest Mortuary Practices: Perspectives from Outside the Southwest." In *Ancient Burial Practices in the American Southwest: Archaeology, Physical Anthropology, and Native American Perspectives*, edited by Douglas R. Mitchell and Judy L. Burnson-Hadley, 249–53. Albuquerque, NM: University of New Mexico Press, 2001.

"Mortuary Analysis of Bioarchaeology." In *Bioarchaeology: The Contextual Analysis of Human Remains*, edited by Lane A. Beck and Jane E. Buikstra, 375–88. Burlington, MA: Elsevier, 2006.

Goldstein, Lynne, and Donald H. Gaff. "Recasting the Past: Examining Assumptions About Aztalan." *The Wisconsin Archaeologist* 83, no. 2 (2002):98–110.

Habu, Junko, Clare Fawcett, and John M. Matsunaga, eds. *Evaluating Multiple Narratives: Beyond Nationalist, Colonialist, Imperialist Archaeologies*. New York: Springer, 2008.

Jefferson, Thomas. *Notes on the State of Virginia*. (Edited with an introduction by William Peden.) Chapel Hill, NC: University of North Carolina Press, 1954 (1787).

Kelly, John E. "Redefining Cahokia: Principles and Elements of Community Organization." In *The Ancient Skies and Sky Watchers of Cahokia: Woodhenges, Eclipses, and Cahokian Cosmology*, edited by Melvin L. Fowler. Special Issue of *The Wisconsin Archaeologist* 77 (1996):97–119.

King, Adam. *Etowah: The Political History of a Chiefdom Capital*. Tuscaloosa, AB: University of Alabama Press, 2003.

Leonelli, Sabina. "Circulating Evidence Across Research Contexts: The Locality of Data and Claims in Model Organism Research." Department of Economic History, London School of Economics: Working Papers on the Nature of Evidence: How Well Do 'Facts' Travel? 25, 2008. http://www2.lse.ac.uk/economicHistory/pdf/FACTSPDF/2508Leonelli.pdf

McGimsey, Charles R., Katherine M. Roberts, E. Edwin Jackson, and Michael L.Hargrave. "Marksville Then and Now: 75 Years of Digging." *Louisiana Archaeological Society Bulletin* 26 (2005).

Milner, George R., and Sissel Schroeder. "Mississippian Sociopolitical Systems." In *Great Towns and Regional Polities in the Prehistoric American Southwest and Southeast*, edited by Jill E. Neitzel, 95–107. Albuquerque, NM: University of New Mexico Press, 1999.

Muller, Jon. "Regional Interaction in the Later Southeast." In *Native American Interactions: Multiscalar Analyses and Interpretations in the Eastern Woodlands*, edited by Michael S. Nassaney and Kenneth E. Sassaman, 317–40. Knoxville, TN: University of Tennessee Press, 1995.

"Southeastern Interaction and Integration." In *Great Towns and Regional Polities*, edited by Jill E. Neitzel, 143–58. Albuquerque, NM: University of New Mexico Press, 1999.

"The History of Archaeology in West Virginia." In *Histories of Southeastern Archaeology*, edited by Shannon Tushingham, Jane Hill and Charles H. McNutt. Tuscaloosa, AL: University of Alabama Press, 2002.

"Kincaid Mx1, Pp1." In *Society for American Archaeology, 68th Annual Conference*. Milwaukee, WI, 2003.

Schiffer, Michael B. *Formation Processes of the Archaeological Record*. Salt Lake City: University of Utah Press, 1996.

Schroeder, Sissel. "Emblems of American Archaeology's Past: Eminent Mounds Sites of the Eastern Woodlands Revisited." In *Symposium organized for the Society for American Archaeology, 68th Annual Conference*, Milwaukee, WI 2003.

"Reclaiming New Deal-Era Civic Archaeology: Exploring the Legacy of William S. Webb and the Jonathan Creek Site." *CRM, The Journal of Heritage Stewardship* 2 (2005):53–71.

Squier, Ephraim G., and Edwin H. Davis. *Ancient Monuments of the Mississippi Valley*. Washington, DC: Smithsonian Classics of Anthropology, 1998 [1848].

Sullivan, Lynne P. "Archaeological Time Constructs and the Construction of the Hiwassee Island Mound." In *TVA Archaeology: Seventy-Five Years of Prehistoric Site Research*, edited by Erin Pritchard and Todd Ahlman, 181–212. Knoxville: University of Tennessee Press, 2009.

Thomas, David Hurst. *Skull Wars: Kennewick Man, Archaeology, and the Battle for American Identity*. New York: Basic Books, 2000.

Trouillot, Michel-Rolph. *Silencing the Past: Power and the Production of History*. Boston: Beacon, 1995.

Valeriani, Simona. "The Roofs of Wren and Jones: A Seventeenth-Century Migration of Technical Knowledge from Italy to England." Department of Economic History, London School of Economics: Working Papers on the Nature of Evidence: How Well Do 'Facts' Travel?, 14, 2006. Available at: http://www2.lse.ac.uk/economicHistory/pdf/FACTSPDF/1406Valeriani.pdf

"Behind the Façade: Elias Holl and the Italian Influence on Building Techniques in Augsburg." *Architectura* 38 (2008):97–108.

Wallis, Patrick. "A Dreadful Heritage: Interpreting Epidemic Disease at Eyam, 1666–2000." Department of Economic History, London School of Economics: Working Papers on the Nature of Evidence: How Well Do 'Facts' Travel?, 2, 2005. Available at: http://www2.lse.ac.uk/economicHistory/pdf/FACTSPDF/FACTS2-Wallis.pdf

Weiss-Krejci, Estella. "Exarnation, Evisceration, and Exhumation in Medieval and Post-Medieval Europe." In *Interacting with the Dead: Perspectives on Mortuary Archaeology for the New Millennium*, edited by Gordon F. M. Rakita, Jane E. Buikstra and Lane A. Beck, 155–72. Gainesville, FL: University Press of Florida, 2005.

Welch, John R. *Archaeology at Shiloh Indian Mounds, 1899–1999.* Tuscaloosa, AL: University of Alabama Press, 2006.

Welch, Paul D., David G. Anderson, and John E.Cornelison. "A Century of Archaeology at Shiloh Indian Mounds." In *Society for American Archaeology, 68th Annual Meeting.* Milwaukee, WI, 2003.

Wylie, Alison. "*Thinking from Things: Essays in the Philosophy of Archaeology.* Berkeley, CA: University of California Press, 2002.

"Agnotology in/of Archaeology." In *Agnotology: The Making and Unmaking of Ignorance,* edited by Robert N. Proctor and Londa Schiebinger, 183–205. Stanford: Stanford University Press, 2008.

PART FOUR

COMPANIONSHIP AND CHARACTER

TWELVE

PACKAGING SMALL FACTS FOR RE-USE: DATABASES IN MODEL ORGANISM BIOLOGY

SABINA LEONELLI

1. Introduction

Model organisms, such as fruit flies, mice and zebra fish, are the undisputed protagonists of twenty-first-century biology. Their prominent position as experimental systems has been further enhanced by the recent sequencing of their genomes, which opened up new opportunities for cross-species comparisons and inferences (the so-called 'post-genomic era'[1]). Such comparative research requires that facts about model organisms be able to travel across a multitude of research contexts. Indeed, the very idea of focusing on a limited set of organisms stems from the desire to bring together as many facts about these organisms as possible, in the hope to increase the scientific understanding of their biology and thus use them as representatives for the study of other species. Moreover, the high costs associated with the production of facts make their use beyond their context of production into an economic, as well as a scientific, priority.

Fulfilling this goal is complicated by the diversity of disciplinary approaches, methods, assumptions and techniques characterising biological research. Each research group tends to develop its own epistemic culture, encompassing specific skills, beliefs, interests and preferred materials.[2] Further, biologists tend to adapt their methods and interests to the features of their organism of choice, thus amplifying the existing diversity among

---

[1] For information on genomics, see Dupré and Barnes (2008); on the epistemology of model organism research, see Ankeny (2007).
[2] The de facto pluralism characterising biology has been widely discussed in the social and philosophical studies of science (e.g., Mitchell 2003, Knorr Cetina 1999 and Longino 2002).

research communities.[3] This pluralism in approaches makes it difficult to make facts travel to contexts other than the one in which they have been produced, as researchers do not share a common terminology, conceptual apparatus, tacit knowledge or set of instruments. The global nature of biological research makes travel even harder: Not only do facts need to cross disciplinary and cultural boundaries, but they also need to travel great distances, becoming accessible to biologists regardless of their geographic location.

Most of the current work in the field of bioinformatics is devoted precisely to resolving the tension between the local nature of facts about organisms and the need for them to circulate across widely different research contexts and locations.[4] Bioinformaticians, and particularly database curators, use digital technology to package facts for travel. Their work is defined by the need to serve a wide variety of database users across the globe, each looking for data fitting their own interests and methods. To a curator, successful travel is marked by the re-use of facts within new research contexts. However, making facts available online does not automatically involve making them usable: Whether facts are adopted across contexts is the result of packaging strategies developed by curators through years of specialised training and dialogue with users.

This chapter examines these packaging strategies to address a key question in contemporary science: What counts as successful re-use of facts? This question lies at the heart of the study of travelling facts, which needs to discuss not only the conditions for the journey of facts from one realm to another, but also the conditions for their acceptance or rejection upon arrival in a new context. This is also the question that curators have to answer when packaging facts for travel. As I shall illustrate, 'good packaging' consists of developing labels that facilitate the retrieval and adoption of facts by prospective users. More specifically, facts need to be de-contextualised from their original locus of production, while at the same time retaining 'travelling companions' to facilitate their re-contextualisation into new

---

[3] It is common practice to name communities in experimental biology on the basis of the organism that they study (as in 'the worm community,' denoting the ensemble of biologists using *Caenorhabditis elegans*, and 'the Arabidopsis community,' using the plant *Arabidopsis thaliana*).

[4] Mansnerus, this volume, offers an account of circulation of knowledge in the context of modelling, which differentiates between the effects of such circulation at the receiving end, while Schneider, this volume, in discussing architectural style pays more attention to the ways in which receivers deal with facts.

research settings.[5] Balancing these two requirements against each other is not easy, nor, given the ever-changing nature of the facts and practices involved, are there universal and enduring ways to compromise between de-contextualisation and re-contextualisation. It is, therefore, curators' responsibility to constantly update their work to reflect the nature and potential destinations of travelling facts. The result is a dynamic process, whose functioning depends on the degree to which curators manage to capture the changing wishes and constraints of practicing biologists.

2. Small and Big Facts About Organisms

As a starting point for my discussion, I differentiate between two types of facts typically found in model organism biology. The first is what I call a *big fact*. This is the type of fact that attracts the attention of most scholars of science, since it constitutes what is generally seen as the end result of scientific research: knowledge about the world that help us to interact with it. Big facts in biology usually consist of a more or less general description of a biological entity, one of its components or one of its functions (as in 'gene X regulates the development of trait Y'). They are expressed propositionally and travel mostly through publication in academic journals, though some of them reach vast non-academic audiences.[6] In this chapter I focus on the travel of the second type of facts, which I call *small facts*. These are the physical traces left by an experimental apparatus, such as images, numbers, dots on a slide and material objects (such as stains on an embryo resulting from in situ hybridisation). Small facts almost never travel beyond scientific circles and are not expressed in propositional form.

Especially in genomics, increasing quantities of small facts are produced in digital formats (i.e., extensible markup language [XML] files capturing DNA sequences) or reformatted so as to be exchangeable through the Internet (e.g., stained embryos are photographed and circulated in the form of images) – indeed, these are the ones I shall be focusing on, as they are the ones that are most easily incorporated into digital databases.[7] Whether they are digital

---

[5] The problem users face in choosing a relevant information set parallels the problem of defining when comparisons are meaningful, or effective, to make facts travel, discussed by both Burkhardt and Ramsden, both this volume.

[6] Other chapters in this volume focus specifically on what happens to big facts that travel beyond the scientific community. See, for instance, the contributions by Adams, Ramsden and Oreskes.

[7] Communicating facts effectively is a considerable challenge in any science; see Merz, this volume, for how facts are communicated in images.

or not, small facts remain essentially material objects: They are the physical result of an interaction between a researcher working with specific instruments and a biological sample like a tissue, a cell or a whole organism. Their physicality, which determines the ways in which they can be used within the context of experimental research, is a major factor setting them apart from big facts: Small facts constitute the material grounds on which knowledge claims (big facts) about biological entities are extracted and validated. Another big difference between big and small facts lies in their ability to travel *solo*. Each big fact has a distinct individuality and can travel alone as well as with other big facts – as a headline in a newspaper or as a line in a textbook. Small facts tend instead to travel in groups (e.g., 'datasets'). One small fact does not usually have much evidential weight. To become significant in a research context, small facts find strength in numbers: The more small facts are grouped together, the stronger their identity and evidential value.[8]

My definition of small facts encompasses anything that biologists might refer to as data. It is similar to the definition of data as 'marks' provided by philosopher Ian Hacking,[9] insofar as it is a procedural definition: It characterises as a small fact anything that has been produced by an instrument under laboratory or field conditions, while differentiating these traces from the propositional statements (descriptions, explanations, hypotheses) made when trying to interpret their significance. To define small facts solely through the procedures through which they are produced might seem too broad because it encompasses a huge variety of objects, as well as too restrictive, because it does not take into account the various degrees of preparation underlying the production of different types of small facts.[10] For the purposes of this paper, however, I gloss over the significant differences among types of experimental results and focus instead on the common status that they enjoy among practicing scientists: that is, the status of 'raw data' used to validate or discredit hypotheses.[11]

---

[8] This can be contrasted with a somewhat different pattern presented by Ankeny, this volume, whereby medical cases act as a vehicle to get particular facts to cohere together to make generic facts.

[9] 'Un-interpreted inscriptions, graphs recording variation over time, photographs, tables, displays' (Hacking 1992, p. 48).

[10] For instance, Hans-Jörg Rheinberger (unpublished) provides an illuminating examination of the relation between traces and data in the case of sequencing, where he distinguishes between the sequence gel produced by radioactive tracing and the polished version of those marks (the chain of letters widely known to represent nucleotide sequences) regarded as the official and 'transportable' result of the experiment.

[11] As I discuss later, there are many cases in biology where small facts produced in an experiment remain unused. This does not affect my discussion: Regardless of whether they end

As widely documented within the history, philosophy and social studies of science, the production of small facts is highly regimented and includes various types of interventions before, during and after any experiment. Small facts could not be produced without recourse to both tacit and articulated knowledge. Further, the conditions for the production of small facts are carefully engineered on the basis of specific expectations, interests, hypotheses, experimental settings and instruments. The production of small facts aims at the validation of big facts: Strictly speaking, there are no such things as raw data. At the same time, however, small facts exhibit a biological significance that transcends their role as evidence for a specific experimental hypothesis, and justifies their treatment as raw data by researchers. This is because the experimental context in which they are produced does not wholly determine their evidential value. As intuited by Pierre Duhem (1974 [1914]) a century ago, small facts are the result of experimenters' interactions with real entities. No matter how tightly controlled an experiment is or how well known the entities already are to scientists, what small facts end up revealing about those entities is not wholly predictable, nor can it be entirely captured by any single big fact. In Duhemian terms, the evidential value of small facts is underdetermined: Small facts exhibit more or less significance depending on the context in which they are used.[12]

Model organism biology well exemplifies the underdetermined evidential value of small facts. In that context, the same set of small facts can often be used as evidence for a variety of big facts. Even more strikingly, small facts about organisms are not always created to serve as evidence for a *specific* big fact.[13] Often they are created because biologists have acquired new instruments enabling them to obtain information about entities of interest ('high-throughput technologies,' thus named because of their ability to produce vast datasets in a short time). In these cases, it is not obvious how

---

up being used or not, small facts are always produced in the hope that they may serve as evidence for one or more claims.

[12] As in the case of the seminal work by Bogen and Woodward (1988) on the relation between data and phenomena, philosophers have tended to forget some of Duhem's lessons and focus solely on the evidential value held by small facts within their context of production. It is assumed that small facts are always created to function as evidence for a given big fact (an hypothesis or claim in need of testing) and that their significance is tied to the context in which they were originally produced, as small facts can only be interpreted by researchers who are familiar with every detail of the setting in which they were created. I critique these views in Leonelli (2009a).

[13] On data-driven research versus hypothesis-driven research, see Kell & Oliver (2003), Krohs & Callebaut (2007) and Rheinberger (unpublished).

that information should be interpreted once it is produced. An example of this type of data-driven research is the shotgun technique used to sequence genomes, which produces billions of data points awaiting analysis and eventual interpretation in the form of big facts.

In order to capture the underdetermined evidential value of small facts, I wish to focus on the procedures (rather than the theoretical lens) through which those facts are produced. In the eyes of researchers, these procedures are the most important characteristic of small facts. They define the type of intervention through which small facts are obtained, the type of entity on which such intervention is carried out and the physical appearance of small facts, which in turn determines the modalities through which they can be transported to new places and used as evidence for new claims. The procedures through which small facts are produced make them unique as a source of information. And indeed, as I show in Section 4, it is information about these procedures that ends up functioning as their 'travelling companions.'

3. Packaging in Bioinformatics

Let us now examine the strategies used by database curators to make small facts travel. To place curators' work in context, I should note that traditional scientific institutions, including funding bodies and the peer review system, have hitherto favoured big facts as the preferred outcome of scientific research. Accordingly, the commonly accepted measure of a scientist's worth is her ability to discover and publish new big facts. Further, publications in scientific journals are not good vehicles for the travel of small facts: Small facts are only included as proof that the big fact of interest has been empirically tested. This may seem logical, as small facts are only valuable insofar as they are used as evidence for big facts. However, this system does not take the underdetermined evidential value of small facts into account. Given the typically short length of scientific papers, most data obtained in any single experiment are not selected for publication and are discarded without any opportunity to be of use. Further, the small facts that are actually published are classified as evidence for a single big fact. This means that interest in that big fact becomes the only means to find and retrieve those small facts – a situation not exactly conducive to their adoption by different research contexts. Thanks to this publication system, most small facts either are thrown away or are untraceable to anyone who has no direct interest in the project that used them first. As a result, the evidential value of small facts is not maximised.

Biologists are well aware that this communication regime makes small facts unusable to anyone other than their producers and their closest peers.

This is why databases are gaining attention as a system devoted exclusively to the disclosure of small facts, which complements vehicles for the travel of big facts (such as journals).[14] Making small facts travel requires apposite infrastructure: vehicles that physically store small facts and transport them outside of the context in which they have been produced. Given the sheer size and diversity of datasets to be circulated, these vehicles must be capable of storing and organising large amounts of small facts. They need to be accessible to users with disparate expertises and interests, which requires a user-friendly interface and the possibility to choose among several types of searches. Further, they should enable users to quickly scan through the available small facts for any given area and perform comparisons among datasets to find possible correlations.

Online databases have the potential to meet all of these requirements. They are available through the Internet, which minimises the constraints imposed on the size and types of small facts travelling through them and eliminates the efforts and time involved in making them physically accessible around the world. Many databases, especially the ones developed through public funding, can be accessed free of charge or other restrictions. They can and often do provide differential access: Thanks to the flexibility of digital interfaces, users can choose parameters for their queries depending on their interests and expertise. Moreover, their computational capabilities mean that databases can incorporate tools for automated data analysis, which helps users check for correlations and patterns across datasets – an indispensible help when needing to restrict one's search from billions of small facts to a manageable sample.

When looking at databases as vehicles for small facts, it is easy to take the metaphor of 'packaging' seriously. This is because the process of packaging small facts for dissemination bears remarkable similarities to the process of packaging items to be dispatched through the mail. The tractability of travelling items is crucial in both cases: Standard shapes and dimensions help the packaging and circulation of the mail just as they help the packaging and circulation of small facts. Indeed, curators are involved in wider efforts within the biological community to standardise formats for different types of small facts.[15] Further, small facts are objects whose ability to travel depends on material aids and infrastructure designed for this purpose,

---

[14] For the idea of publications and databases as pertaining to two separate communication regimes, see Hilgartner (1995).

[15] The progressive standardisation of the format of small facts has greatly helped curators' efforts, as standardisation simplifies the process of grouping different types of small facts within the same databases. For a detailed study of the issues involved in the standardisation of data formats, see Rogers and Cambrosio (2007) on the case of microarray data.

as well as interventions by people other than their senders and receivers. Human activities and material environments are equally important to the travel of small facts. Post offices, trucks, drivers, letter carriers and mail sorters play a similar role to databases and their curators. There would be no travel without the digital platform provided by databases and the work put in by curators to design and use them as a vehicle for small facts. And just as the mail is a service designed to satisfy senders and receivers, the need for small facts to travel is generated by the laboratory cultures in which data are produced and re-used in another setting.

4. Two Types of Labels

There are also important differences between packaging an object for express delivery and packaging a small fact for dissemination, and it is these differences that this chapter aims to explore. In both cases, whether travel is successful depends on whether what is packaged arrives at its destination without being damaged or lost; which destination this will be depends on the way in which objects are labelled. However, in the case of small facts, labels *should not determine* the destinations to which the facts will travel. There is no doubt that the labels chosen by curators have a strong influence on the direction that small facts will take. This is unavoidable, since the function of labels is precisely to make small facts retrievable by potential users. Without labels, facts would not travel at all. Yet for successful re-use to take place, the journey that small facts ultimately undertake should be determined as much by their users as it is by their curators. Curators cannot possibly predict all of the ways in which small facts might be used. This would involve familiarity with countless research programmes around the world – as well as a degree of scientific understanding and predictive ability that transcends the abilities of one individual or group. Therefore, the best way to explore and maximise the evidential value of small facts is to enable as many researchers as possible to use small facts in their own way and within their own research context.

Given these premises, labelling becomes the most challenging component of the packaging process. Curators are required to create labels that, while making small facts retrievable by database users, do not prevent users from making their own selection of small facts. These labels need to indicate the information content of small facts without adding indications – such as a mailing address – about where the facts could be delivered. Giving small facts the flexibility to travel wherever they might be needed constitutes a crucial characteristic of their packaging, which makes it much more sophisticated than the packaging of objects for travel to an already well-defined destination.[16] Choosing appropriate labels to classify small facts is crucial to their successful re-use. Within this section, I intend to demonstrate just how difficult a task this is for curators.

4.1 Relevance Labels

Enhancing the facts' usability involves making them visible and accessible to as many researchers as possible. Curators thus label small facts in a way that make them attractive to users in new contexts: that is, according to their relevance to investigating biological entities. This labelling system, known as 'bio-ontologies', consists of a network of terms, each of which denotes a biological entity or process. Small facts are associated with one or more of these terms, depending on whether they are judged to be potentially relevant to future research on the entities to which the terms refer. For instance, gene *VLN1* has been found to interact selectively with an actin filament known as F-actin (Huang et al. 2005). This is an interesting finding, given the crucial role played by the actin protein in several cellular processes, including motility and signalling. Still, the actual functions of *VLN1* are still unknown: Apart from its interaction with F-actin, there are no big facts yet to associate with the small facts about *VLN1*. Database curators tracked the available small facts about *VLN1* and they classified them under the following terms: 'actin filament binding', 'actin filament bundle formation', 'negative regulation of actin filament depolymerisation' and 'actin cytoskeleton.' Thanks to this classification, users interested in investigating these processes will be able to retrieve the small facts about *VLN1* and use them to advance their understanding.

Depending on which entities they aim to capture, many bio-ontologies are in use in contemporary bioinformatics.[17] One of the most popular

---

[16] In this respect, the focus of my analysis differs from the analysis of circulation offered by historians Kapil Raj (2007) and Mary Terrall (2008). They focus on the mediation strategies used by early modern travellers to exchange facts between Europe and the East and on cases where wished-for objects are delivered to a scientific destination (a naturalist's collection). Although this approach highlights a relocation of objects across space and time, it seems to capture the activity of collecting rather than the activity of circulating evidence, as it does not emphasise the possibility that travelling objects be used in unexpected ways depending on the contexts through which they travel.

[17] See Baclawski and Niu (2006). I restrict the present discussion to the bio-ontologies listed in the Open Biomedical Ontologies consortium (Smith et al. 2007).

ones, from which I took the previous example, is gene ontology, which encompasses three types of biological objects: cellular processes, molecular functions and cellular components (Ashburner et al. 2000). Since their introduction in the late 1990s, bio-ontologies have come to play a prominent role in databases of all types, ranging from genetic databases used in basic model organism research to medical databases used in clinical practice (Augen 2005, p. 64). One of the main reasons for this success is the way in which bio-ontology terms are chosen and used as labels for the classification of small facts.

Curators select these labels according to two main criteria. The first criterion is their intelligibility to practicing biologists, who need to use those labels as keywords in their data searches. In a bio-ontology, each biological entity or process currently under investigation is associated with one (and only one) term. This term is clearly defined so that researchers working in different areas can all understand what it is supposed to denote.[18] Often, however, different groups use different terms to formulate big facts about the same entity. This makes it difficult to agree on one term that could be used and understood by everyone interested in that entity – and in the small facts relevant to its study. Curators resolved this problem by creating a list of synonyms for their chosen label. This means that researchers can search for small facts associated with a given entity both by using the official label employed by the bio-ontology and one of the listed synonyms for that label: Either way, they will be able to retrieve the small facts associated with the entity of interest.

The second criterion for the selection of labels is their association with datasets. The idea is to use only terms that can be associated with existing datasets: Any other term, whether or not it is intelligible to bio-ontology users, does not need to be included, as it does not help to classify small facts. Curators create an association between a dataset and a term when they have grounds for assuming that the dataset provides information about the entity denoted by that term. This happens mainly through consultation of data repositories, where small facts are categorised as resulting from the experimental manipulation of the entity denoted by the term (e.g., sequence data: small facts about the molecular composition of specific stretches of DNA) and of publications using data as evidence to establish a

---

[18] For example, nucleus is defined as 'a membrane-bounded organelle of eukaryotic cells in which chromosomes are housed and replicated. In most cells, the nucleus contains all of the cell's chromosomes except the organellar chromosomes, and is the site of RNA synthesis and processing. In some species, or in specialised cell types, RNA metabolism or DNA replication may be absent' (Gene Ontology website, February 2008).

big fact about the entity denoted by the term (as for *VLN1* data; I discuss this case further in the next section).

Thanks to bio-ontologies, researchers can check which small facts might be relevant to the object of their research. The focus on objects rather than methods or specific traditions makes it easier for researchers to bridge the epistemic cultures in which small facts are originally produced. In this way, researchers with widely different backgrounds (in terms of methods and instruments used, discipline or even theoretical perspective) can access the same pool of small facts and assess their relevance to their research. This enormously increases the chance that database users spot small facts produced in other fields that are relevant for their own research purposes. Therefore, it becomes more likely that the same small facts are used as evidence towards the validation of various big facts about the same entity. Thus, labels such as bio-ontologies constitute a promising first step towards the packaging of small facts for successful re-use. They are not, however, sufficient for this purpose.

As I discussed in the previous section, the evidential value of small facts is underdetermined. This, of course, does not mean that any small fact can be used as evidence for any big fact. On the one hand, the successful re-use of small facts depends on the information that researchers manage to extract from them. For instance, consider the famous case of the DNA photographs taken by Rosalind Franklin in 1952 and reviewed by James Watson without Franklin's permission. From a quick glance at photograph 51, Watson was able to see evidence for his ideas on DNA structure, while Franklin, who did not share those ideas and was a more careful experimenter, did not interpret the image in the same way – a divergence that arguably led to Watson and Crick being credited with the discovery of the double helix in 1953. Franklin was not simply 'wrong': She used her own interpretation of the images as a guide to excellent work on viruses (Maddox 2002). This episode illustrates that there is no single 'right interpretation' of small facts. Interpretation depends on a user's background and interests, which again highlights the need for curators to package facts in ways that enable the emergence of local differences in interpretation.

4.2 Reliability Labels

On the other hand, the emergence of such differences in interpretation, and thus the successful re-use of small facts, depends on the users' awareness of the experimental procedures through which the small facts were originally produced. As I stressed earlier, these procedures define several important characteristics of small facts. Their format, the actual organism used in the experiment, the instrument(s) with which they were obtained, the laboratory conditions at the time of production: All these elements are crucial in determining the quality, and thus the reliability, of small facts. This means that in order to re-use data found through a database, users need to be able to check, if they so wish, the conditions under which small facts have been obtained.

This is why curators devised a second type of label to classify information about the provenance of small facts. These labels are referred to as 'evidence codes,' and they provide essential information about the procedures through which small facts are produced. They include categories for data derived from experimental research, as in IMP (inferred from mutant phenotype), IGI (inferred from genetic interaction) or IPI (inferred from physical interaction); data derived from computational analysis, as in IEA (inferred from electronic annotation) or ISS (inferred from sequence similarity) and even information derived from informal communication with authors (TAS – traceable author statement) and intervention by curators (IC – inferred by curator). Evidence codes are associated with each set of small facts that shares the same provenance. Once users have found facts they are interested in, they can click on the related evidence code and start to uncover the procedures through which the facts have been produced.

Without this second type of labels, and the information retrieved through them, small facts could hardly be re-used. First, researchers who are interested in their evidential value would not necessarily be convinced of their reliability. The reliability of small facts is a function of who produced them, for which reasons and in which setting. Without access to this information, there is no justification for users to trust the small facts displayed in a database. Second, without knowing where data come from, users would not know how to align those facts with the evidence they already have. Adding a new set of data to an existing research project means having some means of comparing the new facts with the facts already produced, especially in case of non-overlapping or even conflicting information. Drawing such a comparison means, in turn, being able to evaluate the similarities and differences between the procedures through which the two sets of facts have been produced. Knowing that both sets have been obtained from the same type of organism (for instance, fruit flies) would enhance a user's willingness to treat all facts at his disposal as compatible. Finding that one set of facts comes from experimental research, while the other is predicted through simulation, will instead warn the user that the two sets might not have the same evidential value.

5. Good Packaging: De-Contextualisation for Re-Contextualisation

What makes databases into good packages for small facts is the opportunity afforded to their users to evaluate both the *relevance* and the *reliability* of the facts in question. The two labelling systems enable users to disentangle the activity of searching and comparing data from the activity of assessing the reliability and significance of data. Thanks to bio-ontologies, researchers accessing a database can find out which existing datasets are potentially relevant to the study of the entities and processes in which they are interested. Once they have restricted their search in this way, they can use evidence codes to examine information about data production. This second type of label enables them to assess the reliability of the data that they located through bio-ontologies, and eventually discard data that are found wanting, according to the users' epistemic criteria.

Remarkably, the consultation of evidence codes does not necessarily reduce the existing gap (if any) between the epistemic cultures of the producers and the users of the facts. Users get access to as accurate a report as possible about the conditions under which small facts were originally obtained. This does not mean that they need to know and think precisely what the producers know and think about those small facts. Rather, the consultation of evidence codes enables users to recognise disagreements with producers concerning suitable experimental conditions, to reflect on the significance of such disagreements and to form their own opinions on the procedures used to obtain small facts. Any judgements on the reliability of data necessarily depends on the user's viewpoint, interests and expertise – which is why curators abstain as much as possible from assessing the quality of small facts and choose a labelling system allowing each user to form her own opinion.

On the basis of these insights, I argue that packaging small facts for successful re-use involves two complementary moves. The first move, for which database curators are entirely responsible, involves the *de-contextualisation* of small facts from their context of origin.[19] The labelling of facts through bio-ontologies ensures that facts are at least temporarily decoupled from the local features of their production, which enables users to evaluate their potential relevance to their research purposes without having to deal with a chaotic sea of information. When choosing and applying bio-ontology

---

[19] Schell, this volume, offers a very different example of how de-contextualisation helps facts travel.

terms, database curators operate in ways similar to librarians when classifying books, or archivists when classifying documents: Small facts are labelled so that users coming to the database can use those classificatory categories to search for a content-relevant item and borrow it for their own purposes.

Identifying which facts to borrow from a database is a crucial first step for researchers interested in using them. Yet it does not help them to decide how to use the facts once they have borrowed them. In other words, while helping to de-contextualise small facts for circulation, bio-ontology labels do not help to re-contextualise small facts for use in a new research setting. This *re-contextualisation* is the second move required for the successful re-use of data, and it is achieved with the help of the second type of labels: the evidence codes. In selecting this second type of label, the analogy between curators and librarians falls through, as libraries do not generally need to provide information about the circumstances in which a book or document was obtained in the first place. Indeed, the classification of information about the provenance of data is a rather different process from the classification of their potential biological relevance. It is a genealogical exercise in which curators investigate and reconstruct the sources and history of the small facts that they annotate. Small facts are material objects that need good travelling companions in order to be adopted and re-used across contexts. Evidence codes give access to the qualifications that endow small facts with what Mary Morgan, in her introduction to this volume, calls 'character.'

Let us explore this idea in more detail. De-contextualisation through bio-ontologies is a way for small facts to lose the personality attributed to them in their original research context: The whole point of de-contextualisation is to make small facts extremely adaptable, which can only be achieved by stripping them of as many qualifications as possible, leaving them free to travel as objects in search of a new interpretation. By contrast, re-contextualisation through evidence codes enables users to evaluate the character of small facts by assessing their provenance. This second step is necessary to qualify the value of small facts as evidence, and thus building an interpretation of their biological significance in a new research setting. In this sense, the process of re-contextualisation is reminiscent of work conducted by curators in a very different setting: museum collections, whose visitors can best form an opinion about the cultural significance of the objects in display when they are given information about the history of those objects and their creators.[20]

---

[20] The ways in which small facts acquire character through evidence codes can be usefully compared to other cases of travelling objects in this volume, such as Wylie's 'archaeological facts' and Valeriani's building structures.

By enabling users to access de-contextualised small facts, databases provide the differential access needed to make small facts travel across contexts. By providing evidence codes, databases facilitate the re-contextualisation of small facts, while at the same time making it possible for them to shift character and significance, depending on their new location. This modality of re-use is particularly important in model organism biology, where the same small facts might acquire entirely different interpretations when examined by biologists working on different species and/or dissimilar research cultures. Through their vision of re-contextualisation, curators are attempting to enable biologists to pick up new small facts without necessarily having much in common in terms of their goals and expertise. For instance, researchers investigating the regulatory functions of specific genes are using databases to check what data are available on their gene of interest, how those data were produced and on which species. This enables them to compare what is known about the behaviour of the gene across species, without having to become a specialist on each type of organism and experimental procedure involved.

6. The Role of Curators

While illustrating how databases make small facts travel, I already hinted at ways in which databases are challenging existing social structures and conventions governing the dissemination of scientific results.[21] They are giving visibility and usefulness to small facts that, contrary to the big facts, used to be discarded by their producers or jealously kept in their laboratories for further research. In most scientific contexts, small facts used to be subject to public scrutiny only when providing crucial evidence for a big fact: They would not be made public before the big fact was published in a journal, and the small facts that did not serve an immediate purpose as evidence in that case would be discarded. Many factors have acted as an incentive to make small facts travel across contexts. Among them are the testimony of the few scientific communities that did exchange small facts at the pre-publication stage, a collaborative strategy that proved to be extremely successful, especially in the case of model organism research,[22] and the emphasis by funding bodies on the value of small facts as public goods,

---

[21] Hine (2006) and Leonelli (2009b) discuss many of the challenges posed by databases to scientific social orders.

[22] One example is the vast community of researchers working on the plant *Arabidopsis thaliana* (Leonelli 2007).

especially following the controversy on the importance of preserving open access to the results of the Human Genome Project (Sulston and Ferry 2002). Even in the presence of these factors, the opportunity of free exchange provided by databases challenges the competitive ethos prevalent within biological research, thus making it more and more difficult for researchers to avoid donating their data.

An even more important challenge posed by databases concerns the role of curators as a new type of expert within biology, whose relations to existing experts are not yet clearly established. As I illustrated, curators are aware that their work of de-contextualisation is essentially at the service of the activity of re-contextualisation by the users. De-contextualisation is the means through which small facts are made fit to travel, and re-contextualisation is the ultimate aim of travel. Ideally, therefore, users should be able to re-contextualise small facts in ways that depend solely on their own backgrounds and interests. In practice, however, curators also need to develop a range of skills enabling them to choose labels (both bio-ontologies and evidence codes) that mirror as closely as possible the developments and expectations of the potential users of small facts.

Curators achieve a high level of fit between their work and the work of database users through a variety of interventions. For instance, take the activity of *extracting* small facts from publications and repositories. To do this, curators are forced to single out publications that they consider reliable, updated and representative for specific datasets. When gathering available data on a specific gene (such as the unknown flowering object gene [UFO] in *Arabidopsis thaliana*), curators need to choose one or two publications that best represent data relevant to a given gene product for the purposes of classification. They cannot compile data from each relevant publication, as it would be too time-consuming: Even just a keyword search on PubMed on 'UFO Arabidopsis' results in thirty-five journal articles, only one or two of which will be used as reference for an annotation. Thus, curators choose what they see as the most up-to-date and accurate publications on a specific gene product, which as a consequence become 'representative' publications for that entity.

Further, once curators settle on a specific publication, they have to assess which small facts therein contained should be extracted and/or how the interpretation given within the paper matches the terms and definitions already contained in the bio-ontology. Does the content of the paper warrant the classification of given data under a new bio-ontology term? Or can the contents of the publication be associated with one or more existing terms? These choices are impossible to regulate through fixed and objective standards. Indeed, bioinformaticians have been trying to automate the process of extraction for several years without success. The very reasons why the process of extraction requires manual curation are the reasons why it cannot be divorced from subjective judgement: All the choices involved are informed by a curator's expertise and his or her ability to bridge the original context of publication and the context of bio-ontology classification.

Performing curation tasks such as extraction presupposes skills honed through specific training and years of experience. Curators are veritable 'packaging experts,' and their combination of skills is crucial to producing both bio-ontologies and evidence codes.[23] Their expertise includes, on the one hand, some familiarity with various fields of biological research. This gives them the cross-disciplinary understanding necessary to recognise and respect the diversity characterising epistemic cultures (and thus, terminologies and methods) within experimental research. On the other hand, curators need to have some experience 'at the bench.' This enhances their awareness of what users need to find through evidence codes (e.g., protocols and search parameters). Curators working on Gene Ontology, for example, are biologists by training and motivation: Their decision to extend their expertise towards computer science and bioinformatics was primarily due to their interest in improving data analysis tools for model organism research as a whole. The curators' hands-on knowledge of experimental work is reflected in the development of bio-ontologies and enhances their intelligibility to experimenters. At the same time, only through a more generalist expertise can curators assess which terms to use, how to define them and how to relate them with each other. Curation is no job for a specialist with a narrow experimental focus; nor is it a job for a computer scientist with no clue of how research at the bench is conducted.

7. User Perspectives

By taking upon themselves the task of choosing the appropriate package for small facts, curators make important decisions on what counts as relevant small facts for any specific research project. Most users are happy to trust them with this role, as they do not want to spare time and energy from their research to deal with choices about packaging. For this same reason, however, users are reluctant to invest effort in understanding the choices made by curators. Users want an efficient service, thanks to which

---

[23] The expertise of the curators is critical here, as indeed, the expertise of the scientists in Howlett and Velkar's account of technology transfer, this volume.

they access a database, type a keyword, get the relevant data and go back to their research. By so doing, users often do not understand the extent to which the packaging affects the travel of small facts and the ways in which they will be re-used.

Curators are well aware that their interventions influence where and how small facts will travel. They are willing to recognise that it is their professional duty to serve the user community as best as they can, and they feel both responsible and accountable for their packaging choices – indeed, they are actively seeking scientific recognition for their service as packaging experts (Howe et al. 2008). They are also aware that it is impossible to conform to the expectations and practices of rapidly changing fields without being in constant dialogue with the relevant user communities. This is also because, aside from one-to-one dialogue and website statistics on which parts of a database are most popular with users, there is currently no reliable way for curators to systematically evaluate how users are using information in the database. Many researchers are not yet used to citing databases in their final publications – they would rather cite the papers written by the original producers of the data, even if they would have not been able to find those papers and associated data without consulting a database. Curators thus cannot assess which research projects have made successful use of their resources, unless researchers report their achievements to them directly.

Yet, many attempts to elicit feedback fail because of users' disinterest in packaging practices and their inability to understand their complex functioning. The gulf between the activities and expertise of curators and users tends to create a problematic system of division of labour. On the one hand, curators invite users to critically assess their work and complain about what they might perceive as 'bad choices.' On the other hand, users perceive curators' work as a service whose efficiency should be tested and guaranteed by service providers rather than the users. They thus tend to trust curators unconditionally or, in the absence of trust, simply refuse to use the service.

The tensions between database users and curators are exemplified by a recent attempt to package and re-use small facts about leaves. AGRON-OMICS is a European project sponsored by the Sixth Framework programme bringing together plant scientists from a variety of laboratories and disciplines, including molecular, cellular and developmental biology. Its goal is to secure an integrated understanding of leaf development by gathering and analysing data extracted from the model organism *Arabidopsis thaliana*. A crucial component of this project is precisely the search for efficient tools to circulate small facts among members of the group and to the research community at large. The question of labelling was uppermost in the minds of the group co-ordinators from the outset in 2006. What categories could be used to circulate data gathered by researchers so steeped in their own local terminologies and practices?

The first meeting of the project, a two-day workshop titled 'Ontologies, Standards and Best Practice', was devoted to tackling this question.[24] Participants included the main scientific contributors to AGRON-OMICS and the curators of the databases that were most likely to be of use, such as Geneinvestigator, the Arabidopsis Reactome, the Gene Ontology and the Plant Ontology. Curators did most of the talking, both through presentations explaining what their tools could do and through hands-on workshops teaching researchers to use them. Most questions raised concerned systems for tracking the relevance and reliability of small facts; users and curators certainly agreed on the importance of keeping the focus on these two factors. Overall, the workshop was successful in alerting researchers to the importance of finding good packages to make their facts travel. Remarkably, however, this lesson came with an increased awareness of the difficulties plaguing these efforts, and particularly of the problems associated with labelling small facts for re-use.

Many of the scientists attending displayed mistrust for the work of curators, which they saw as far removed from actual biological research. The very need to de-contextualise facts was seen as potentially problematic, despite evidence for the necessity of this process to make facts travel. There were complaints that curators, in their tight collaboration with computer scientists, tended to favour a polished labelling system over one that would actually help experimenters; it was also remarked that the synonyms system devised by curators to accommodate terminological pluralism only works if curators are aware of all existing synonyms for a given label. Further, some researchers were dazzled by the multitude of tools available for labelling (well over twenty were mentioned at the meeting, most of which researchers were not yet acquainted with). Although some labels, such as Gene Ontology, are fairly well established across a number of databases, there are many cases of databases developing their own labelling systems without regard for the ones already in place. This leads to a proliferation of labels, which is confusing to most users, who feel they are wasting time in learning to use all those systems and in assessing each label's merits relative to others. Although some scientists appreciated the idea of being able to choose

---

[24] The workshop, which I attended, took place at the Department of Plant Systems Biology, VIB-UGent (Gent, 21–23 May 2007).

among different labelling tools, this was often associated with an interest in developing those tools themselves.

Dialogue between users and curators over these difficulties resulted in both sides increasing their understanding of labelling processes. Curators walked away with a better idea of the needs and expectations of AGRON-OMICS researchers. Users, however, retained a degree of scepticism in curators' work. Indeed, precisely as they were learning to appreciate the scope and implications of curators' work, AGRON-OMICS scientists saw the importance of selecting appropriate labels for their facts, as well as the power that this brings over the eventual re-use of those same facts. Therefore, they resolved to take over some of that work to ensure that the labels used to package facts were perfectly suited to their research needs. One of the action points agreed upon at the end of the meeting was the creation of two new bio-ontologies: one for *Arabidopsis* phenotypes and one for *Arabidopsis* genotypes. The main rationale for this effort was the perceived absence of suitable labels dealing with these biological entities. Also, developing their own labels would ensure that scientists take over the packaging – and thus the modalities for future re-use – of small facts of particular importance to their project.

8. Regulating the Packaging Process

The successful re-use of small facts requires a highly dynamic system of labels. Curators have the crucial function of mediating between the needs of local research contexts and the need to devise standards that can be used by all.

In other words, curators are not simply responsible for making small facts travel: They are responsible for making small facts travel *well*, which involves communicating with users to make sure that facts are indeed being re-used.

In the case of AGRON-OMICS, the potential tensions between curators and users were resolved by making these two figures overlap. It is not clear, however, if this is a good solution. In the absence of a generalist curator aiming to serve the whole biological community, the labels used for packaging might end up serving the needs of the AGRON-OMICS group over and above the needs of other scientists, thus hampering the successful re-use of those same data in other quarters. Further, as I already mentioned, not all scientists are willing to invest time and effort towards the creation of good packages for small facts. Part of AGRON-OMICS funding is explicitly directed at the study and testing of packaging tools for small facts, which means that they can employ people to work on bioinformatics and they have resources for developing and maintaining communication with curators at the international level (thus preventing the danger of narrowing their vision to their own project). The same is not true of other projects, especially smaller projects with more specific goals.

A more general solution could be to enforce some mechanisms of communication between curators and users so that curators receive frequent feedback from the widest range of users, thus ensuring that their packaging strategies are indeed serving the needs of users as they evolve through time. In other words, packaging – and particularly the de-contextualisation processes for which curators are responsible – is in need of external regulation. An example of such regulatory mechanisms is the requirement to submit data to databases in appropriate formats when publishing a paper. This has recently been implemented by *Plant Physiology*, a major journal in plant science, in collaboration with The Arabidopsis Information Resource (TAIR), the main database for *Arabidopsis* research. Researchers wishing to submit a paper to *Plant Physiology* are required to submit to TAIR all the data created during their project. This forces them to become acquainted with the labelling system adopted by TAIR (which includes both evidence codes and the Gene Ontology). The experience might encourage direct involvement by experimenters in the development and use of bio-ontologies.[25]

Yet another effective regulatory measure is the introduction of institutions that are responsible for implementing the packaging and setting standards for it. That the rise of regulatory institutions would support the development and maintenance of 'good packaging practice' will not come as a surprise to the readers of this volume, as the importance of such structures is emphasised by many other analyses of how facts are packaged for travel (e.g., the case of technology transfer in Northern India examined by Howlett and Velkar). In the case of bioinformatics, many prominent packaging efforts have been centralised in few loci, such as the European Bioinformatic Institute and the Gene Ontology Consortium in Hinxton, Cambridge.[26] This institutionalisation of packaging prevents the proliferation of labels and therefore enhances their power to cross contexts. It also helps to train packaging experts, who can teach users how to deal with labels and vehicles; and it enables the creation of feedback mechanisms through which users can provide constructive critiques to curators, thus bettering

---

[25] See Ort and Grennan (2008).
[26] Leonelli (2009b) discusses the institutional history and status of 'labelling centres' such as the Gene Ontology Consortium.

their packaging strategies and the resulting re-contextualisation processes. A downside of institutionalisation is the centralisation of power on what counts as good packaging. This involves a potential loss of diversity in packaging strategies, as it gives particularly prestigious and well-funded groups the opportunity to shape the choice of labels according to their own preferences and interests. Another problem is that existing centres, despite the support they receive from funding bodies and user communities, are struggling to cope with the immense amounts of small facts to be curated.[27]

And yet, despite the efforts of so many highly qualified minds, it is still not clear whether this packaging system will end up working as desired. This is not because the system is not well designed and maintained, but rather because we do not yet know whether the tensions between curators and users will be resolved in a way that is satisfactory to both.

9. Conclusions

My analysis has focused on the strategies devised by scientists to cope with the need to access and re-use the billions of small facts produced by contemporary research, and particularly high-throughput technologies. The case of bioinformatics illustrates the complexity of making facts travel even within the supposedly narrow boundaries of the scientific world. Travel across research contexts involves crossing large distances, both in geographic and in epistemic terms. This requires a lot of effort, especially in the case of small facts, which are produced in extremely large numbers, travel only in groups and with the help of specific travelling companions (e.g., labels) and require purpose-made vehicles to move around.

What is most interesting for our purposes is precisely the purpose-made character of this new apparatus vis-à-vis its efficiency as a packaging tool. Given the care and thoughtfulness in creating and perfecting packaging strategies using the latest technologies, the case of database curation constitutes an ideal case of travelling facts. This is a case where a whole system of labels, communication strategies and digital vehicles has been explicitly created to make sure that facts travel well.

Yet the value of the packaging process is ultimately dependent on the efficiency with which curators and users communicate about their

[27] Recourse to 'crowdsourcing' or 'wikification', that is, user involvement in annotating databases, has been hailed as a solution to these problems (Leonelli 2009b). Yet, given the high level of specialised expertise required to curate small facts, it is difficult to know whether it would work. The degree to which good packaging requires centralisation remains an open question.

respective needs and interests. As I illustrated, scientists have recently had to acknowledge that a whole apparatus of innovative technologies, expertises and institutions was needed to make these facts travel: Without curators, apposite labels and databases, it would be difficult to enact the processes of de-contextualisation and re-contextualisation needed to make small facts travel.

Acknowledgments

This work was supported by the ESRC/Leverhulme Trust project, 'The Nature of Evidence: How Well Do "Facts" Travel?' (grant F/07004/Z) at the Department of Economic History, LSE. Special thanks go to the scientists who took time to be interviewed and provide feedback in the course of this research. I am particularly grateful to the co-ordinators of AGRON-OMICS, Fabio Fiorani and Pierre Hilson; Klaus Mayer and colleagues at the Munich Information Centre for Protein Sequences; Sean May and colleagues at the Nottingham Arabidopsis Stock Centre; Michael Ashburner, Midori Harris and colleagues at the Gene Ontology; Sue Rhee and colleagues at The Arabidopsis Information Resource; and participants to the BBSRC work-shop 'Data Sharing in the Biomedical Sciences' (Edinburgh, June 2008).

Bibliography

Ankeny, R. (2007) Wormy logic: Model organisms as case-based reasoning. In Creager, A. N. H., Lunbeck, E. and Wise, N. eds. *Science Without Laws: Model Systems, Cases, Exemplary Narratives*. Chapel Hill, NC: Duke University Press, pp. 46–58.

Ashburner, M. et al. (2000) Gene Ontology: tool for the unification of biology. *Nature Reviews: Genetics*, 25, pp. 25–9.

Augen, J. (2005) *Bioinformatics in the Post-genomic Era. Genome, Transcriptome, Proteome and Information-Based Medicine*. Indianapolis IN: Addison-Wesley.

Baclawski, K. and Niu, T. (2006) *Ontologies for Bioinformatics*. Cambridge, MA: The MIT Press.

Bogen, J. and Woodward, J. (1988) Saving the phenomena. *The Philosophical Review* 97, 3, pp. 303–52.

Duhem, P. (1974 [1914]) *The Aim and Structure of Physical Theory*. New York: Atheneum.

Dupré, J. and Barnes, S. B. (2008) *Genomes and What to Make of Them*. Chicago: Chicago University Press.

Hacking, I. (1992) The self-vindication of the laboratory sciences. In Pickering, A. ed. *Science as Practice and Culture*. Chicago: Chicago University Press, pp. 29–64.

Hilgartner, S. (1995) Biomolecular databases: New communication regimes for biology? *Science Communication* 17, pp. 240–63.

Hine, C. (2006) Databases as scientific instruments and their role in the ordering of scientific work. *Social Studies of Science* 36, 2, pp. 269–98.

Howe, D., et al. (2008) The future of biocuration. *Nature* 455, pp. 47–50.

Huang et al. (2005) Arabidopsis VILLIN1 generates actin filament cables that are resistant to depolymerization. *Plant Cell* 17, pp. 486–501.

Kell, D. B. and Oliver, S. G. (2003) Here is the evidence, now where is the hypothesis? The complementary roles of inductive and hypothesis-driven science in the post-genomic era. *Bioessays* 26, pp. 99–105.

Knorr Cetina, K. (1999) *Epistemic Cultures*. Cambridge, MA: Harvard University Press.

Krohs, U. and Callebaut, W. (2007) Data without models merging with models without data. In Boogerd, F. C., Bruggeman, F. J., Hofmeyr, H. S. and Westerhoff, H. V. eds. *Systems Biology: Philosophical Foundations*. Amterdam and Oxford: Elsevier, pp. 181–213.

Leonelli, S. (2007) Arabidopsis, the botanical Drosophila: From mouse-cress to model organism. *Endeavour* 31, 1, pp. 34–8.

(2009a) On the locality of data and claims about phenomena. *Philosophy of Science* 76, 5, pp. 737–49.

(2009b) Centralising labels to distribute data: The regulatory role of genomic consortia. In Atkinson, P., Glasner, P. and Lock, M. eds. *The Handbook for Genetics and Society: Mapping the New Genomic Era*. London: Routledge, pp. 469–85.

Longino, H. (2002) *The Fate of Knowledge*. Princeton, NJ: Princeton University Press.

Maddox, B. (2002) *Rosalind Franklin: The Dark Lady of DNA*. London: Harper Collins.

Mitchell, S. (2003) *Biological Complexity and Integrative Pluralism*. Cambridge: Cambridge University Press.

Ort, D. R. and Grennan, A. K. (2008) Plant physiology and TAIR partnership. *Plant Physiology* 146, pp. 1022–3.

Raj, K. (2007) *Relocating Modern Science: Circulation and the Construction of Knowledge in South Asia and Europe, 1650–1900*. Houndmills and New York: Palgrave Macmillan.

Rheinberger, H. (unpublished) *From Traces to Data, From Data to Facts*.

Rogers, S. and Cambrosio, A. (2007) Making a new technology work: The standardisation and regulation of microarrays. *Yale Journal of Biology and Medicine* 80, pp. 165–78.

Smith, B. et al. (2007) The OBO Foundry: coordinated evolution of ontologies to support biomedical data integration. *Nature Biotechnology* 25, 11, pp. 1251–5.

Sulston, J. and Ferry, G. (2002) *The Common Thread: A Story of Science, Politics, Ethics, and the Human Genome*. Washington, DC: Joseph Henry Press.

Terrall, M. (2008) *Following Insects Around: Tools and Techniques of Natural History in Réaumur's World*. Talk delivered at the Sixth Joint Meeting of the BSHS, CSHPS, and HSS. 5 July 2008, Keble College, Oxford.

Online Resources

AGRON-OMICS website: http://www.agron-omics.eu/

Gene Ontology website: http://www.geneontology.org/

THIRTEEN

DESIGNED FOR TRAVEL: COMMUNICATING FACTS THROUGH IMAGES

MARTINA MERZ

1. Introduction

Visual images can be effective devices for communicating facts.[1] Yet this does not imply that whenever images propagate the facts automatically come along – nor do facts that travel in images always travel well. The relation of *images, facts and their travels* is more complex. The complex relationship will be explored in this text for the case of microscopy images in the field of nanotechnology and their travels both through scientific publications and popular media.

Nanotechnology researchers produce images using probe microscopy, such as scanning tunnelling microscopy (STM) and atomic force microscopy (AFM), and electron microscopy.[2] Unlike optical microscopy, which resolves structures in the range of millimetres and fractions thereof, these types of microscopy operate at the level of atoms and attain atomic resolution. Scientists use the instruments to image and analyse atomic and molecular structures. But importantly, probe microscopes also allow researchers to produce and manipulate such nanoscale structures. Through the exploitation of quantum mechanical effects, these instruments are employed to produce objects (e.g., materials) with novel properties. This potential and practice is considered a defining and characteristic constituent of

---

[1] In accordance with constructivist science studies this article takes as a fact what is established as a fact through material and discursive practice within an epistemic community.

[2] A note on terminology: The notion "image" refers to visual images only and not to other kinds of images such as metaphors. "Nanotechnology" is employed in this text as a synonym for both nanotechnology and nanoscience. This choice is motivated, first, by a preference to increase readability and, second, based on the understanding that the distinction between the two is often used in contingent ways in the concerned communities.

nanotechnology (cf. Baird et al. 2004; Mody 2004; Daston and Galison 2007, chap. 7; Hennig 2010).

The lab-produced images of atomic or molecular structures are among the most important outcome of nanotechnology practice. A small selection of these lab images, suitably edited, has found their way into scientific publications through which the researchers communicate their findings to their peers. An even smaller selection of the images, edited in other ways, has been diffused through alternative channels (news media, web sites, etc.) to the public.[3] Images that originate in scientific laboratories carry facts. Scientists package facts of different kinds in the form of images and visual displays to transfer them from their context of production – the scientific laboratory – to other contexts.[4] How these packages are designed for travel and how users unpack them later on is the focus of this article.

Within and across the scientific field images do not travel easily on their own. To travel well, they require good company: labels and instructions for use, an accompanying explanatory or contextual text.[5] But above all, they are rarely to be found without the companionship of related images or other visual representations. The travelling companions are not just there for the ride, but are essential epistemic elements in the way that the scientific culture of nanotechnology produces and communicates facts. In contrast, the diffusion of images from science into other spheres follows its own rules and guidelines. Images may be stripped of their companions, become iconized or re-contextualized in novel ways. Thus, whether facts travel well by unaccompanied images is judged according to distinct standards of evaluation in different communities.

In the following, I will first feature the case of an emblematic image from nanotechnology that has been diffused widely within the public realm (Section 2). I will next turn to microscopy images in research articles, to their packaging in composite visual displays and to the role images play in the transferral of facts from an article's authors to its scientific readers more generally (Section 3). How a composite visual display is unpacked by a fellow scientist is analysed to illustrate the fact–image travel dynamics (Section 4). The text concludes with a discussion of the sense in which facts travel well by way of images (Section 5).

---

[3] Besides images that originate in the scientific laboratory, a wide range of other images has become associated with nanotechnology in popular media (Lösch 2006; Milburn 2008; Nerlich 2008; Landau et al. 2009); such images will not feature in this article.

[4] The case of Calhoun's images associated with his rat experiments, discussed by Ramsden in this volume, offers an example of this.

[5] See Leonelli (this volume) for a related discussion on the issue of labels and packaging.

2. The IBM Logo: Facts, Images, Icon

One of the images most closely associated with nanotechnology in the public imagination is the IBM logo. It can be downloaded from a variety of Internet sites, as shown in Figure 13.1 – that is, with specific contrasts, shapes and colours – and it frequently appears in print media (Baird and Shew 2004; Hennig 2010). This image will be introduced first from the perspective of its viewers before turning to the question of the image's scientific origins.

2.1 Dissemination in the Public Realm

What is the image about? To assess the response of viewers to this question, I confronted twenty people of different professional and educational backgrounds with a colour print of the image that contained neither a legend nor any other additional information. It turned out that a majority of the people had not previously seen the image. These first-time viewers identified the image merely with the word "IBM," the company's logo. This answer came in variations, for example, "the brand IBM," "IBM: the computer producer (logo)" or "publicity for IBM." Other respondents associated the image with the process of its production, for example, "writing produced

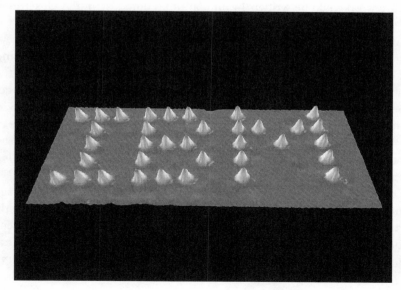

Figure 13.1. IBM logo composed of individual atoms. Image originally produced by IBM.

by 'nano'-technology," "representation by a nano-microscope" and "a small joke from the IBM research lab." This modest assessment suggests that the viewers' reading of the IBM image decisively depends on whether they have been previously exposed to it and on the availability of background information. Assuming that viewers have such background knowledge, what is the image about? This text argues that an important reason why the IBM image has come to symbolize nanotechnology is that crucial facts about the power of nanotechnology are associated with the image – but for the facts to actually reach the viewer (i.e., for the facts to travel well, see next), the image needs to be accompanied by additional information. This information may come in the form of a legend or of accompanying text in another format. Consider an example from the web site of the British Science Museum.[6] The Museum's website guides online visitors through the antenna "Nanotechnology: small science, big deal" to the rubric "See for yourself," from there to the "Exhibition sneak preview," where the IBM logo (as shown in Figure 13.1) is exhibited under the header "The smallest world" with the legend "Each blue blob is a xenon atom arranged using a microscope." In addition, the image is accompanied by the text:

Each blue blob on this image is a xenon atom. Scientists working for IBM used a scanning tunnelling microscope to move the atoms around and write their company logo. Each atom is one tenth of a nanometre wide, so their entire word could be written 14 million times onto a stamp. A closer look at this material leads to the proposition that more than one kind of fact is involved. The combination of image, legend and supplementary text conveys three types of factual statements. The building blocks of the letters I, B and M consist of individual atoms – in this case, xenon atoms – the scale of the entire composition being in the range of nanometres. This constitutes a fact about the imaged phenomenon (*phenomenal fact*). The atoms were moved into place by a STM – this represents a fact about the employed course of action to produce the phenomenon (*procedural fact*). An STM has been used both to move atoms *and* to visualize the result of this manipulation – this is a fact about the apparatus and technology used (*technological fact*).[7]

Once the reader is aware that the blobs represent single atoms, the image unfolds its suggestive power: Because the pattern (I, B, M) is so manifestly artificial, impossible to imagine as a product of anything but an intentional act, the procedural fact is inscribed in it as much as the fact that the technological capacity exists to perform the task. The accompanying

---

[6] www.sciencemuseum.org.uk
[7] Howlett and Velkar (this volume) also use the term "technological fact" in this way.

text only adds the details: that an STM was used, that xenon atoms were placed on a nickel surface. The replies of the second set of respondents (noted earlier) point at this association of the IBM logo with procedural and technological facts.

Calling the image "a small joke from the IBM research lab," as another respondent earlier did, identifies it with the corporation's scientific and technological project. But more than merely the trace of a joke, the image is a forceful reminder of and renders homage to the company behind the project. The instrument used to produce the sample, the STM, had been invented by IBM researchers G. Binnig and H. Rohrer, who received the Nobel Prize for this invention in 1986, whilst other researchers of the corporation, D. M. Eigler and E. K. Schweizer, had produced the nanoscale IBM logo in their lab (Binnig et al. 1982; Hennig 2004, 2006).

To reiterate, although the IBM logo image is widely diffused, it requires an accompanying text to ensure that the procedural and technological facts travel with it and are well received. In the terminology of this volume: Without supporting material, these facts do not travel well; they remain concealed in the image and may go unnoticed by the viewer. The viewer instead may take the image to be an expression of other facts, such as that of the power of IBM. However, once the message about the underlying facts has been received, the image alone will suffice for viewers to recall the encapsulated procedural and technological facts.[8] The careful crafting of the image according to established "macroscopic viewing conventions"[9] (Hennig 2004, p. 15) helps to render the image accessible and recognizable by a wider public. As a result, the IBM logo image has today become an element of nanotechnology's iconography. Yet, as an icon, the image no longer only stands for procedural and technological ability; it has also come to symbolize nanotechnology's expected potential and the scientists' power over nature.

2.2 Scientific Communication

Scientists acknowledge that the single IBM logo image is a carrier of procedural and technological facts. When I asked a physicist what the image represented to him, he asserted that "it shows the capacity of the researchers to control the position of atoms that they can place without mistake." But how

---

[8] Contrast this with the case of the silhouettes of raptors that are put on windowpanes to keep birds from flying into windows, a measure that has no scientific backing (Burkhardt, this volume).

[9] My translation.

was this image first introduced into the scientific community? It was published in 1990 by IBM researchers Eigler and Schweizer in a three-page letter with the title "Positioning single atoms with a scanning tunnelling microscope" in the journal *Nature*. With its claim and demonstration that the STM can be used to position individual atoms on a surface with atomic precision, the article raised considerable interest in the scientific community.[10]

When comparing how the public IBM logo image and the visual displays in the *Nature* article talk to their respective audiences, a number of differences come to the fore. First, it is not surprising that the scientific article contains a wealth of detailed textual information about the experimental process that supports the central claim, since, as a general rule, images in scientific articles are always embedded in other types of material.

Second, the scientific article does not exhibit a single (isolated) image of xenon atoms but instead presents a composition of six adjacent images that come in two columns of three images each (Figure 13.2).

The six images exhibit a temporal "sequence of STM images taken during the construction of a patterned array of xenon atoms on a nickel (110) surface" (legend, Eigler and Schweizer 1990, p. 525), which shows successive stages of the construction process. The composite visual display reinforces the impression of procedure and process: It visually documents and demonstrates the fact that the IBM-pattern can be produced, step by step. In the body of the text, the figure is introduced as "a sequence of images taken during our first construction of a patterned array of atoms, and demonstrates our ability to position atoms with atomic precision" (ibid.). This first figure is accompanied by two other figures. The second consists of a schematic rendering of how the microscopy tip attracts an atom and moves it across the surface.

The third is composed of another block of six images but uses an alternative form of representation to show "various stages in the construction of a linear chain of xenon atoms on the nickel (119) surface" (legend, ibid., p. 526). All three figures make a factual statement about *procedure* by explicitly exhibiting the ability of the researchers to position atoms – in contrast, the public IBM logo image conveys the procedural fact in a more implicit manner. Although not easy to decipher from the images, the fact that the atomic structure had been both produced and visualized by an STM (*technological fact*) was mentioned right at the beginning of the legend. After all, this is what made the publication so noteworthy. Concerning the explicit rendering

---

[10] In the words of the logo's scientific creators: "This capacity has allowed us to fabricate rudimentary structures of our design, atom by atom" (Eigler and Schweizer 1990, p. 524).

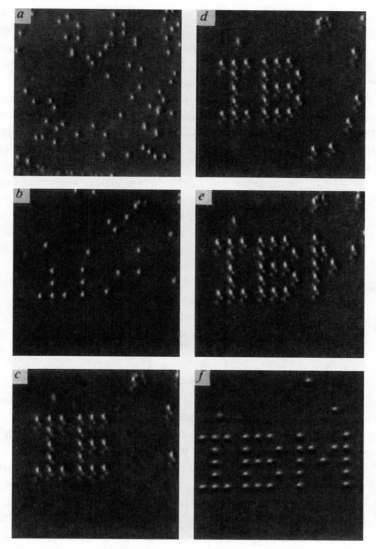

Figure 13.2. "A sequence of STM images taken during the construction of a patterned array of xenon atoms on a nickel (110) surface. Grey scale is assigned according to the slope of the surface. The atomic structure of the nickel surface is not resolved. The <1 10> direction runs vertically. **a**, The surface after xenon dosing. **b–f**, Various stages during the construction. Each letter is 50 Å from top to bottom." (Legend and figure as in Eigler and Schweizer 1990, p. 525).

From: Reprinted by permission from Macmillan Publishers, Ltd.: Nature (Eigler, D. M. and E. K. Schweizer (1990), "Positioning Single Atoms with a Scanning Tunnelling Microscope," *Nature* 344(5 April):524–6), copyright 1990, http://www.nature.com/nature.

of the temporal dynamics, one may wonder whether the *Nature* article is an exceptional case due to its declared aim to establish the success of a novel procedure. It will be argued later that procedural facts, which are characteristic outcomes of nanotechnology research, are typically represented in the form of microscopy images embedded in composite visual displays.

A third difference concerns the visual characteristics of the images. The public version of the IBM logo image (Figure 13.1) turns out to be not simply the last image in the sequence of six (Figure 13.2) shown in the scientific article. Instead, it consists of a careful reconstruction and redesign that shares primarily the abstracted IBM pattern of the individual atoms with its scientific counterpart(s). In the *Nature* article, the images come in black and white, they are a little blurry and the contrast between background and signal isn't optimal. Close inspection reveals that in three of the sequence's six images the atoms appear double, due to the STM tip having been "dirty" (Hennig 2010). In contrast, the public image of the single-atom IBM logo seems polished, shiny and colourful. The representation of the atoms as illuminated blobs with shadows is the result of an adaptation to macroscopic viewing conventions (ibid.). This distinction of image design according to scientific conventions and according to the preferences of public media points to a more general trend, caught tellingly in the opposition of a "rhetoric of rough" for the case of science and a "semiotics of smooth" as illustrated by the public IBM logo image (Curtis 2007). Of course, design conventions of probe microscopy images (and other kinds of images) are today heavily debated in the scientific community. Also whilst researchers follow the trend towards more sophisticated and colourful renderings, they insist, at the same time, that images deemed good according to scientific standards may (need to) look "dirty" to less experienced people.

The assessment of the STM imagery as it appears in the *Nature* article suggests tentative conclusions, which may serve as hypotheses for the further investigation of facts that travel in images within a scientific community. First, STM images emphasize procedural capacity (procedural fact), while other types of facts are communicated less explicitly. Second, an image rarely comes alone; it is typically accompanied by other images and visual representations. The question then arises how these features relate to how well facts travel in scientific publications.

3. Images in Scientific Articles

As a central medium of scientific communication, the research article in the sciences cannot be imagined without the presence of images, diagrams, tables, graphs and other types of visual representations. The science studies literature[11] has addressed the scientists' production, transformation and diffusion of visual representations with an emphasis on image multiplicity, aptly caught by Bruno Latour's pithy phrase:

> An isolated scientific image is meaningless, it proves nothing, says nothing, shows nothing, has no referent. (Latour 2002, p. 34)

Image multiplicity has been discussed with respect to the *production* of images in the scientific laboratory (Amann and Knorr Cetina 1990; Lynch 1985) and the *diffusion* of visual representations from the laboratory to the public (Latour 1990). In both cases, the studies drew attention predominantly to "serial" relations of images (Lynch and Woolgar 1990b, p. 6) – that is, the directed transformations of visual representations that render the underlying phenomena of investigation progressively "more visible, stable, and measurable" (ibid.) or, to put it in Latour's terms, the "cascade of ever simplified inscriptions" (Latour 1990, p. 40). In contrast, "transversal" (i.e., non-sequential) relations between visual representations – as I will call them – have received little attention (Alač 2004; Bastide 1990; Lynch 1990; Myers 1990).

In what follows, the transversal relations of images and other visual displays *within* a scientific text will be the centre of attention. This will require the exploration of the mutual contextualization of the figurative elements within an article, considering not only the relations of these elements among each other but also the figurative and the textual elements. For this purpose, it seems fruitful to conceive of a visual display as "an autonomous surface that is nonetheless contained within a text" (Lynch 1990, p. 155). From this perspective, instead of reducing visual representations by default to the role of merely illustrating the text, the relation of visual display and text is open to negotiation. Based on the hypothesis that something interesting is happening to the underlying facts when figurative elements are assembled into composite visual displays, this section will first look at the visual fingerprint of displays and articles. It then recapitulates three types of facts that travel in images, and finally assesses scientists in their roles as readers and writers to learn more about the relation of facts and images.

---

[11] The influential collection of articles, *Representation in Scientific Practice*, edited by Lynch and Woolgar (1990a), put the analysis of scientific representational practices on the agenda of science studies. For a comprehensive introduction to the analysis of the social practice of scientific imaging and visualization, see Burri and Dumit (2008).

3.1 Composite Visual Displays

When leafing through the pages of journals in the field of nanotechnology, it is eye-catching that, as in the case of the aforementioned *Nature* article, the visual displays – that is, what is subsumed and bracketed under the label "Figure" in an article – are predominantly composed of *several* images, curves and schemas together. In the following, the term "composite visual displays" refers to such compositions that join and gather several images, curves and/or schemas within a common frame (visual display), complemented by a joint legend and under the header of a specific figure number.[12]

To get a better grip on the typical form of an article's composition of textual and visual elements, two hundred articles have been assessed numerically. The articles selected for analysis were taken from two journals: the first is the journal *Nanotechnology*, which is dedicated to covering research in nanoscale science and technology from a multidisciplinary perspective; the second is the journal *Advanced Materials*, one of the top international materials science journals, read by materials scientists, chemists, physicists and engineers of various orientations and the nanotechnology community.[13] From each journal one hundred articles were considered, twenty-five each from 1992, 1993, 2007 and 2008.

Visual material has a high status in both journals, as measured by the space it occupies in an article: The relation of visuals (including legends) and running text is roughly one to three. Indeed, a microscopy image rarely appears on its own (this happens in only 5 per cent of articles in 2007/2008); in about 40 per cent of the articles, several microscopy images are exhibited in direct juxtaposition (47 per cent in 2007/08); in about 30 per cent of the articles microscopy images are combined with curves (48 per cent in 2007/2008); in about 15 per cent they are combined with schemas (21 per cent in 2007/2008). Thus, in many cases, composite visual displays (which contain an average of about four elements in 1992/1993 and seven elements in 2007/2008) have a more complex internal referential structure than the sequence of the IBM article.

---

[12] The journal *Advanced Materials* calls such structures "multi-panel images," but this terminology is not followed here to avoid confusion. The term "image" is reserved for a single visual display, typically in contrast to other types of visual displays such as schemas or curves.

[13] The journal *Nanotechnology* was founded in 1990 and currently has an ISI Impact Factor of 3.3 (in 2007). The journal *Advanced Materials* has an ISI Impact Factor of just above 8 (in 2007) and celebrated its twentieth anniversary in August 2008.

This analysis confirms what other sources (e.g., poster presentations, further journals) suggest. First, images are abundant: Today, only one in eight articles contains no microscopy image at all.[14] Second, an image is rarely presented on its own in scientific publications. This text argues that *the companionship of other visual representations ensures that the facts that are embedded in images travel well across the scientific community*. The contention is based on the understanding that an image acquires meaning in the context of and when juxtaposed with other images, schemas and curves.

3.2 Embedding Facts in Images

As the discussion of the IBM logo image suggests, factual statements of different kinds are embedded in and can be uncovered from individual microscopy images as well as from an ensemble of images. This potential of images to carry various kinds of facts resembles that of material objects, which can "store and communicate" facts as diverse as material, technical, user-related facts and so on. (Valeriani, this volume). The factual multivalence of scientific images is related, one may assume, to their "semiotic openness" in combination with "their being regarded as the simultaneous voice of technoscientific authority and as expressions of nature" (Burri and Dumit 2008, p. 305). This characteristic might account for the willingness of readers to assign fact status to images, while the specific kind of factual statement is co-determined by the image's context. The factual statements are about (at least) three kinds of entities:

- *Phenomenon*: Microscopy images "reveal" (in the scientists' terminology) that the underlying phenomenon or object of investigation has a certain shape. In the case of probe and electron microscopy, this is typically a factual statement about the atomic or molecular structure of the sample of interest.
- *Procedure/process*: The prime factual statement that is communicated by way of an image (or visual display) does not necessarily concern the rendered phenomenon, in particular with respect to its atomic structure. Instead, the image can highlight more specifically that the exhibited features are the result of a certain experimental procedure or process. The image of the IBM logo, for example, conveys the fact that individual atoms can be deliberately positioned on a surface in

---

[14] The reasons for the abundance of visual displays in scientific articles cannot be discussed in detail here, but one may assume that the progress of image reproduction technologies and the reduced cost of image production play an important role.

a selected pattern by following a certain procedure, detailed in the accompanying text. The fact that this exercise was conducted with a specific kind of atom and a specific choice of surface is of lesser importance, although it is, of course, relevant to fellow scientists who might want to replicate the experiment.

- *Technology*: In other cases, the communicated factual statement primarily relates to the technology used to produce what the image reveals. It is a statement about technological capacity and might. This concerns not only the specific form of microscopy used but also other kinds of visualization technology, such as image analysis software, whose importance for the production of the published images should not be underestimated. Factual statements about technology are specifically important in periods in which a technology becomes newly established. In these cases, the focus shifts from the portrayed (phenomenon) to the portraying instrument: The images may reveal the capacity of the underlying technology, while the imaged phenomena assume an instrumental role. Probe microscopy (such as STM, AFM) has been established over the last two decades (Mody 2004). As a consequence, probe microscopy images are today no longer associated with the underlying technological facts to the same degree as they used to be.

Thus, microscopy images can carry various kinds of factual statements. These may be statements about the visualized phenomenon, about the procedure and process followed to bring about the phenomenon or about the underlying technology. An image *can* be but does not need to be associated with one specific factual statement – there is no one-to-one correspondence between image and fact. An image may also carry various kinds of statements to different degrees. An example is once again the IBM logo image, which reveals process and technology alike. This is a typical feature of probe microscopy, as it is used both to visualize and to produce the imaged atomic structure, which leads to a tight coupling of instrument and production procedure.

Typically, the image itself does not determine the underlying factual statement. Rather, the kind of factual statement is assigned through the interaction with the image environment, that is, with the other visual displays and the texts around it. In a way then, it is an emergent property. As a result, facts are not only embedded in (and emerge from) individual images but also in composite visual displays.

3.3 Visual Narratives: "By Looking at the Figures
I Should Get the Story"

A research article attracts the attention of peers not only because of its scientific quality and innovative character. Visibility depends decisively also on the reputation of the scientific journal in which the article appears and on how the results are "packaged." How scientists select and present images and visual displays in their publication thus plays an important role in producing visibility for their results. Scientists are confronted with this issue from two complementary perspectives: as authors and as readers. They have expertise both in packaging facts for travel in the form of images and visual displays, and in extracting them from the article with its textual and visual material as readers, these two forms of expertise feeding on each other.

How then do scientists acquaint themselves with the scientific literature in their field of expertise?[15] In a first step, they identify publications of potential interest according to an article's title, authors and abstract, and then download the articles for closer scrutiny. In a next step, the readers turn to the images and their legends:[16]

> I read the abstract, I look at the pictures and the legend – that's what I do first. And then it depends, whether I quickly scan through the article, how important it is. (Interview)

The reader's move from images and legends towards the surrounding main text has a correspondence in how scientists write an article. The first step consists of putting the figures together, the next "to write the article around the figures" (Interview). Authors add the legends first and then, step by step, the remainder of the text. The preposition "around" suggests that the visual displays are conceived of as the article's centre and core. The centrality of the figurative characterizes the article as it determines both the reader's and the author's focus of attention.

But how should this visual centre be considered? As the assessment in Section 3.1 suggests, the scientists group images and visual representations in composite visual displays, resulting in only a few figures per article. The reason scientists give for not exhibiting individual images is that this

---

[15] The following is based on qualitative interviews with senior scientists in the field of nanoscale science.

[16] In his seminal work on the genre of the experimental article, Bazerman (1988, chap. 8) also discusses how physicists read physics literature. The insignificance of visual displays in this case – Bazerman mentions physicists "perhaps scanning figures" (ibid., p. 243) only in passing – reminds us that visual displays were not of central importance to all physics specialties at that time.

"makes the story very hard to read" (Interview). Instead, "putting information together" (Interview) in the form of composite visual displays is seen as a way to package related facts into a "local story," which ensures that the facts become optimally accessible to the readers (cf. Section 4).

What do the researchers mean when they say that a visual display tells a story? One may suppose that the scientists' stories, like narratives, "create a sense of *why* things happen" (Hayles 1999, p. 10). Such stories emphasize, among other things, temporal sequence and causality (ibid.; cf. also Bruner 1990). Visual displays and their individual components provide accounts of the underlying processes (temporal sequence) and they allow the viewers to construct an account of why the represented events occur (causality). Stories embedded in visual displays involve two kinds of causes: On the one hand, the scientists' motivations and procedures are the cause of a certain experimental course of action, which an article's readers attempt to decipher from the visual displays; on the other hand, the outcome of experiments is interpreted as having certain natural causes, which refers back to phenomenal facts and their interpretations. These two kinds of causes are entangled in the stories that scientists uncover from visual displays.

Consider once more Figure 13.2, the composite visual display that shows how the IBM logo image came into being. The story embedded in the figure simultaneously emphasizes temporal sequence and causality: the sequence of STM images presenting the deliberate and successful act of the researchers to produce the I-B-M pattern of individual atoms step by step. Such stories are a way to communicate different kinds of facts. Although the story associated with the IBM figure highlights a procedural fact, a story may also evolve around facts about phenomena and/or technology. In order to allow readers to construct a story, a visual display has to exhibit a certain complexity and, typically, a composite nature. In many cases, this involves (in contrast to Figure 13.2) a figure that contains different forms of visual representation, such as a combination of images and schemas. It should be noted, though, that stories that materialize around visual displays typically do not have the densely layered texture that is characteristic of fictional narratives.[17] Perhaps surprisingly, this feature does not seem to be necessary to create a sense of why things are happening in an experimental setting.

Clustering visual representations in the form of composite visual displays means that the spatial relationship between the visual representation and its associated referring text is broken up – after all, each figure corresponds

---

[17] On the role of narrative in helping facts to travel, see Adams (this volume).

to a precise passage in the text that refers to it (Myers 1990, p. 249). The distance between image and referring text is yet another indication for the high degree of autonomy of the narrative that is created by visual representations and their legends within a composite visual display. Despite the autonomy that engenders "local stories" (as the interviewee put it), there is a sharp awareness of the correlation between the different visual displays in one article and the requirement that they sum up the main content and results of the paper. This requirement is summarized by the advice a senior scientist gives to his student:

> By looking at the figures I should get the story. If the figures are not telling me the story, you are missing one or two. Or you didn't choose them properly. (Interview)

The quote implies that the visual displays *together* should be comprehensive and represent all central moves and outcomes of the article. From this perspective, the main text can be interpreted as an extended legend to the figures, supplemented by information about the motivation of the work and how it relates to similar work.

4. Communicating Facts in Composite Visual Displays

In what follows, a specific composite visual display will be examined more closely from two contrasting perspectives: first, the authors' perspective as reconstructed from the article in which the figure appears (see Section 4.1) and, second, the perspective of a reader (see Section 4.2). I had asked senior scientist Barbara[18] – a physicist by training who has worked in nanoscale science for many years – to select articles that she planned to read. One of them is the article that will be featured here. The publication is authored by a group of scientists from Beijing and recently appeared in the prestigious journal *Physical Review Letters* under the title "Probing Superexchange Interaction in Molecular Magnets by Spin-Flip Spectroscopy and Microscopy" (Chen et al. 2008).

4.1 Zoom in on a Composite Visual Display

The article's main running text is combined with four composite visual displays and one table. Of the four pages, about one and a half are covered with figures and their legends. The two columns of page two are each

---

[18] This is a pseudonym.

half-filled with a colour figure[19]: FIG. 1 (corresponding to Figure 13.3) presents a combination of microscopy images and schematic representations; FIG. 2 exhibits four graphs arranged in the form of a square. On page three, FIG. 3 joins three graphs and two schematic renderings. On the last page, FIG. 4 combines two schemas, three measurement curves with a joint heading, a superposition of an image and a schema and, finally, a visual rendering of a calculated structure. Therefore, the article is a good example of the present trend towards more complex composite visual displays.

The four figures seem to segment the paper and align the fourteen paragraphs, which are not subdivided into sections. After the first two paragraphs, which introduce the reader to the topic, provide a motivation and present an outline of the article, two to three subsequent paragraphs each are associated consecutively with each of the four figures: They introduce the molecular structure (FIG. 1), the results of scanning tunnelling spectroscopy measurements (FIG. 2), the spin-flip spectra (FIG. 3) and the superexchange mechanism (FIG. 4). The article closes with an outlook and acknowledgements, followed by the list of references.

Let us take a closer look at the first figure (see Figure 13.3). It is the only one dominated by microscopy images and it is also the one that attracts reader Barbara's attention (cf. Section 4.2). The figure is visually structured by two types of elements: the three annotated STM images (a, c, d) on the one hand, the two schemas in the upper-right corner (b) and at the figure's bottom (e) on the other hand. The three images seem to form a unit, an impression that is reinforced by the same colours being used; the two schemas appear to constitute a frame to this image block. Upon consideration of the composition and the content of the visual representations, there does not seem to be a clear entry point into the figure. The detailed representations of (c) and (d) are unlikely to constitute a starting point, nor is the complex rendering of schema (e) with its two inserts, but both (a) and (b) look like probable candidates – and it will be shown that they are.

Insight into the authors' perspective on the composite can be gained by considering the alphabetic order of its visual representations, which suggests a reading order, and by unravelling the figure's legend and the corresponding running text.

Image (a): FIG. 1(a) presents an overview of the multilayered molecular structure of interest. The legend introduces it as an STM image of

---

[19] To avoid confusion, the figure numbers are labelled in capitals when the numbering of Chen et al. (2008) is concerned and in lowercase when this article's numbering is referred to.

"self-assembled multilayers of CoPc molecules"[20] that sit on a lead surface. The statement emphasizes both the imaging technology and that the structure is lab-produced. Indeed, the procedure is described in the running text for readers interested in the details: "Cobalt phthalocyanine (CoPc) molecules were then thermally sublimed onto the Pb islands at room temperature to form a self-assembled monolayer with square lattice pattern. Subsequent sublimation of CoPc was performed at sample temperature of ~ 120K to form the ordered multilayer structures, as shown in Fig. 1(a)" (p. 1). The running text clearly labels image (a) as associated with the *procedural fact* by referring to it at this precise spot. The image also gives rise to the appreciation that it had been produced by STM (*technological fact*), the details of which (voltage and current) are presented in the legend.

Schema (b): The schema depicts the molecular structure of one such Cobalt phthalocyanine molecule and thus stands in for a *phenomenal fact*. It presents a full figure within the composite figure, as it has its own legend – explanations of the colour codes of the chemical elements, which make up the molecule – and also comes with a scale (5 Å), which makes it stand on its own. Accordingly, the running text adds little further information.

Images (c) and (d): The next two "zoom-in STM images" are presented as a package (albeit with individual numbers), as they come with the same size, scale and orientation. They show details of the "stacking geometry of CoPc in the multilayer structures" with a focus on the first and second layers (c) and the second and third layers (d). As zoom-ins, they are directly associated with STM image (a) of which they present a more detailed and fine-grained picture, which is further detailed in the running text, but they are also intimately connected with schema (e). They show how a layer is positioned with respect to the next one and how the molecules are arranged (*phenomenal fact*).

Schema (e): The schema provides a model of how the different layers sit on top of each other, indicating the orientations and displacements of the molecules in differing layers. Inserts, linked up by arrows with two layers each, show in more detail how molecules that sit on top of each other are twisted against each other. The schema summarizes the relevant features of the stacking geometry, drawing heavily on information from the STM images. As such, it constitutes an "image of synthesis" (Allamel-Raffin 2006). What about the factual status of this schematic representation? The corresponding caption leaves no doubt: It represents the structure (*phenomenal fact*) associated with a determined "stacking geometry" and a quantified spacing between molecular levels.

---

[20] This and the following quotations refer to Chen et al. (2008) unless otherwise indicated.

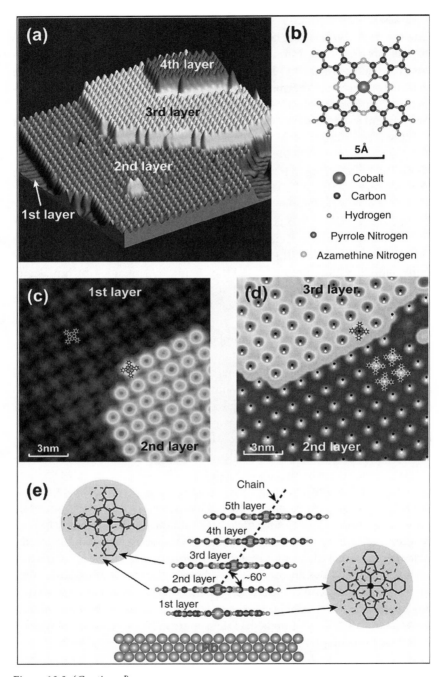

Figure 13.3. (*Continued*)

Figure 13.3. "FIG. 1 (originally in colour). CoPc multilayers on Pb. (a) STM image (V = 0.9 V, I = 0.03 nA) of the self-assembled multilayers of CoPc molecules on Pb(111) film (26 ML thick). (b) Molecular structure of CoPc. (c) STM image (V = 0.6 V, I = 0.1 A) showing the relative stacking of the 1st and 2nd CoPc layers. (d) STM image showing the relative stacking of the 2nd and 3rd CoPc layers. The white and black dots indicate the centers of the molecules on the 2nd and 3rd layers, respectively. (e) Stacking geometry of CoPc molecules. The molecular layers are spaced 3.5 ± 0.1 Å apart. The inserts show the orientation and displacement between molecules in adjacent layers."

The construction logic underlying the composite visual display suggests to move from Figure 13.3(a), which specifies the *technology* used and the *procedure* followed to generate the multilayers of CoPc molecules, to Figure 13.3(b), (c) and (d), which provide information about the *phenomena*, that is, the stacking geometry of these multilayer structures, and finally to Figure 13.3(e), which synthesizes the phenomenal statements. Although facts about procedure and technology are conveyed by the figure and its legend, the emphasis of this composite is on the phenomena – at least, this is the impression gained by reconstructing the authors' perspective. In the next section, a reader approaches the same composite figure in an attempt to make sense of it.

4.2 A Reader's Perspective

Only readers can provide the ultimate proof that facts are communicated through images in scientific articles. Readers can show how successful they are in unpacking visual displays to access the underlying facts. For this reason, it seemed fruitful to observe how scientists read a scholarly publication, how they attempt to extract facts from the imagery and how they reason in this process. To make reading observable, it, above all, has to be made audible.[21] Hence, I had asked senior scientist Barbara to go through

---

[21] For a discussion of how thinking and other allegedly immaterial practices can be observed cf. Knorr Cetina and Merz (1997).

the articles she had previously selected for closer inspection audibly, letting me observe how she made sense of the articles and their visual displays. Her initial monologue, followed by a few questions of mine requesting further specification, was recorded and transcribed, and presents the empirical material of this section.

Barbara had selected the article by Chen et al. (2008) on the basis of its title and abstract. She takes the article, runs over the pages and flips it open at page two, turning to FIG. 1 (Figure 13.3 here) and FIG. 2 that exhibits four graphs. During the seven minutes that we spend on the article, her eyes and fingers move back and forth between the individual components of the first figure and between the two adjacent figures. The movements would have created an interesting zigzag pattern if recorded visually.[22] With some limitation, this zigzagging movement becomes traceable also when looking at the transcript.

She starts to muse about what the figures show by addressing the article's first figure in the following way:

> First, what is important, what is all this about, which molecule on which surface? Here (FIG. 1b) I see the molecule, then I read the caption, on lead, so it's on lead 1–1–1, then I look at the corresponding image (FIG. 1a), and then I already see, okay, in this case we are dealing with multilayer growth, so second, third, fourth layer. This is STM and it is turned. All right, and next: here is again schematically how it looks like (FIG. 1e). What do I look at next? Then I see the resolution is good (FIG. 1c–d), here you see single molecules, now these are phthalocyanines. (Interview transcript 01.09.22)

The scientist does not pause at any one image or schema but creates a narrative that smoothly moves from one element to the next. Considering the statements in turn allows one, once again, to associate them with the different types of facts.

Schema (b) and legend: Barbara's opening question "what is all this about?" inquires into the specific sample that was investigated. An answer is provided by a combination of the schematic representation, which shows the chemical structure and composition of the concerned molecule, and the legend, which informs her that these molecules sit on a lead surface. Although the composite visual display is visually dominated by three microscopy images, Barbara's first gaze is directed at a schema – which in this instance provides a first insight into the structure of the phenomena under investigation *(phenomenal fact)*.

---

[22] Ideally, the observation of scientists making sense of visual displays should be video-recorded.

Image (a): She then turns her attention from the schema towards the large microscopy image (a), which shows that the molecules do not cover the lead surface uniformly but come in layers, four layers being clearly visible, as each layer has a different colour and is annotated correspondingly. This image, on the one hand, provides further insight into the phenomena under investigation *(phenomenal fact)*. On the other hand, Barbara interprets it as hinting at the underlying manufacturing process *(procedural fact)*, which she addresses by the notion "multilayer growth" – the layers do not sit naturally on the lead surface: they have to "grow," which requires a dedicated technical procedure. Barbara also acknowledges that an STM was used to visualize the structure *(technological fact)*.

Schema (e): The next visual representation in focus is once again a schema, the second in the display, which Barbara scans only quickly with the words "here is again schematically how it looks like" *(phenomenal fact)*.

Images (c) and (d): Barbara then turns to the two smaller images, but instead of scrutinizing the details of the molecules' stacking configurations, she is merely interested in the images' resolution, which she identifies as good because single molecules are visible. This fact about the imaging arrangement *(technological fact)* provides trust in the experiment as a whole. Technological facts are not only associated with the capacity of technologies but also with the skills and precision of the scientists that handle them, as this case illustrates. The statement with which Barbara closes her consideration of the entire figure "now these are phthalocyanines" seems to draw all components of the composite together. With this statement she also comes full circle with her initial question "what is all this about, which molecule on which surface?"

Of course, this is not the entire story. The article contains another three figures, and Barbara's account on the article's visual displays is more extended, while she does show most interest in FIG. 1. The following considerations will use the aforementioned for illustration, but will take other cases into account as well. This will allow me to draw out some specifics of how facts are communicated through images and visual displays.

4.3 Observations

An image contains indications about *different kinds of facts*: A microscopy image may provide factual information about phenomena, procedure and technology alike. Whether one of these types of facts will emerge as a dominant feature will depend on the image's context as much as on the reader's specific expertise and interest. For example, the legend may suggest the type of fact to be communicated by an image. Alternatively, an entire article may be explicitly associated with one type of fact. An example is the *Nature* article that displays the IBM logo images to communicate the procedural fact of this type of structure's fabrication.

The explored cases show that a specific type of fact is rarely communicated in isolation by an image. Factual statements about procedure and technology are closely associated (e.g., Figure 13.2), an image that exhibits a phenomenal fact may also communicate a procedural fact (e.g., Figure 13.3) and so on. It seems that scientists have a preference for communicating loose fact bundles of this kind. This may explain their penchant for composite visual displays, which create an environment in which *facts about procedure, phenomena and technology circulate and are tied together.* How Barbara traces the individual components of the figure to make sense out of it is a telling illustration: phenomenal fact (b and legend) → phenomenal fact and procedural fact (a) → technological fact (a) → phenomenal fact (b) → technological fact (c, d) → summary of entire figure.

When comparing the (reconstructed) authors' and readers' accounts of the composite visual display Figure 13.3, one first notes, importantly, that there is no indication that (major) facts are lost on the way between author and reader. However, authors and readers tie visual representations and facts together in a different manner – and it can be assumed that there is a wide variety of ways to do so on both the authors' and the readers' side. One reason for this heterogeneity is that both address the visual displays on the basis of their respective background knowledge and purposes (Bazerman 1988, chap. 8). Consider the *reading order*. A composite visual display comes with a reading order suggested by its authors. In the discussed case it moves through the visual representations of Figure 13.3 sequentially, from (a) to (e), communicating first procedure and technology, then phenomenal facts with increasing detail and breadth, ending with an image of synthesis (cf. Section 4.1). Reader Barbara does not follow this order but selects her own, based on her personal interest, experience and knowledge: She zooms in on the schematic representation (Figure 13.3) of the specific molecule first because she is curious about the particular "magnetic atom in such an organic molecule" and what one might be able to do with it. The multilayer structure of such molecules (Figure 13.3) is of interest to her in the next step. The composite visual display allows readers to be flexible in the reading order they choose. The spatial composition of the display – typically visual representations are not aligned but rather grouped loosely within a rectangular frame – invites readers to choose their own passageway through the display to make sense of it.

Composite visual displays provide *flexible fact retrieval mechanisms* not only because they allow viewers to determine their own reading order. They are characterized also by *multiple cross-referencing* of their individual components, each acquiring meaning and fact status in view of the others (Bastide 1990; Alač 2004;), which helps the facts to circulate within the composite. Cross-referencing employs a variety of reference forms: for example, comparison with an alternative (sample, instrumental representation, visual representation, etc.), zoom (i.e., change of scale), abstraction, concretization (e.g., image vs. schema) and so on. The discussed case of FIG. 1, for example, contains referencing relationships of zoom (a versus c and d), of abstraction (a, c and d vs. e) and of comparison (c vs. d). Such transversal relations and mutual contextualisations between individual visual representations in a composite display assist viewers in constructing a story by associating the circulating facts.

A composite presents an assembly of different kinds of facts and cross-reference associations between the different visual representations. The reader fits these elements together to create a sense of what happens in the underlying experiment and why things happen. Also the *legend* assists this endeavour, acting as supplement and intermediary between the components of a complex visual display. It fills in what remains open in the visual material. For example, in the case at hand, it indicates that the substrate consists of lead, on top of which the molecule layers are deposited, as the images show. It also provides the parameters of the STM, and it offers assistance with interpreting the white and black dots in the microscopy images. The legend also provides a link between the display and the main running text. As such, it enables the facts to be tied to their qualifiers and to contextual information about their origin, their experimental underpinnings and their expected value.

The question of how well facts travel in images and composite visual displays *within* a scientific community is thus less one of facts getting lost or being misinterpreted.[23] It is rather a question of whether flexible fact retrieval mechanisms such as composite visual displays exist that enable scientists at the receiving end to construe a rich story of phenomenal, procedural and technological facts, full of dynamics and hints at the underlying experimental practice.

---

[23] The travelling of facts within an expert community is the concern of several chapters in this volume; see, for example, the contributions by Leonelli, Valeriani and Whatmore and Landström.

5. Conclusions

This text has attempted to show how microscopy images in the nanometre range can be effective devices to communicate facts. Yet, it has also argued that facts do not automatically travel when images are diffused.[24] Above all, this is the case when images travel from science to the public. The case of the IBM logo image illustrated that the image may lose its (procedural and technological) facts when being diffused to *non-expert communities* who had not been exposed to the image previously. Thus, the facts did not travel well in the sense that viewers could not easily unravel them. To travel well, these facts need to be packaged more thoroughly with legends and explanations that accompany the image. To travel well, they rely on a certain degree of scientific understanding on the viewers' side. This seems to be a typical feature of communicating facts by images. The semiotic openness of images (cf. Section 3.2) implies, on the one hand, that they are ruled by requisites (e.g., prior knowledge and experience) while they are, on the other hand, highly suggestive and draw their force from association with visual conventions. This is why the facts embedded in the IBM logo image are easy to recall once the main message has been received but remain buried otherwise.

The communication of facts through images follows other rules within the *scientific community*, mainly because the viewer's preconditions are different. Members of a scientific community share visual conventions, expertise concerning how to produce, edit and interpret images, and so on. They are skilled both in writing and reading scientific publications, which implies that they can swap perspectives. These are skills that distinguish scientists from public viewers. When comparing how facts are communicated by images to the public and within a scientific community, the most striking difference is that a single image is rarely used to transmit facts. It is as if scientists need more than one image to be convinced. In addition, a microscopy image typically allows readers to uncover more than one kind of fact. These two features combined enable a complex visual transmission of facts in research articles. Microscopy images are typically embedded in composite visual displays, the strength of which is that they embed factual statements that not only refer to the phenomena but also to the followed procedure and the adopted technology. One reason why facts travel well in such arrangements is that they allow scientists to construct narratives about the underlying experiment: They tell a story (or allow readers to tell

---

[24] Schneider (this volume) makes the same point in the context of architecture.

a story) about procedures, technologies and the structure and behaviour of phenomena all at once. This introduces a temporal order into the interpretation of visuals: The spatial relations within visual displays can thus be transformed into temporal ones. This move inverts the process that Rheinberger (2006: p. 352) describes as typical for laboratory practice: to convert the spatiotemporal arrangement of a lab into a two-dimensional frame through a variety of notation and inscription practices. Although the temporal order uncovered from visual displays is a deeply reconstructed one, which does not provide a faithful account of an experiment's history, it still provides scientists with a useful guideline for how to unravel facts from visual displays.

Acknowledgements

The research underlying this paper was funded by the Swiss National Science Foundation as part of the project "Epistemic Practice, Social Organization, and Scientific Culture: Configurations of Nanoscale Research in Switzerland." It has also benefited from generous funding provided by the Leverhulme Trust/ESRC grant "The Nature of Evidence: How Well Do 'Facts' Travel?" in the context of two extended research stays with the "Travelling Facts" team at the LSE in London. I thank the Facts team, the participants of the book workshop at LSE and, especially, the book's two editors for their constructive criticism. I am grateful to the nanotechnology researchers for introducing me into the intricacies and routine tasks of their research.

Bibliography

Alač, Morana (2004), "Negotiating Pictures of Numbers," *Social Epistemology* 18 (2–3):199–214.

Allamel-Raffin, Catherine (2006), "La complexité des images scientifiques: Ce que la sémiotique de l'image nous apprend sur l'objectivité scientifique," *Communication & Languages* 149:97–111.

Amann, Klaus and KarinKnorr Cetina (1990), "The Fixation of (Visual) Evidence" in Michael Lynch and Steve Woolgar (eds.), *Representation in Scientific Practice.* Cambridge MA: MIT Press, 85–121.

Baird, Davis and Ashley Shew (2004), "Probing the History of Scanning Tunneling Microscopy" in Davis Baird, Alfred Nordmann and Joachim Schummer (eds.), *Discovering the Nanoscale.* Amsterdam: IOS Press, 145–56.

Baird, Davis, Alfred Nordmann, and Joachim Schummer (eds.) (2004), *Discovering the Nanoscale.* Amsterdam: IOS Press.

Bastide, Françoise (1990), "The Iconography of Scientific Texts: Principles of Analysis" (translated by Greg Myers) in Michael Lynch and Steve Woolgar (eds.), *Representation in Scientific Practice.* Cambridge MA: MIT Press, 187–229.

Bazerman, Charles (1988), *Shaping Written Knowledge: The Genre and Activity of the Experimental Article in Science*. Madison WI: University of Wisconsin Press.

Binnig, Gerd et al. (1982), "Surface Studies by Scanning Tunneling Microscopy," *Physical Review Letters* 49(1):57–60.

Bruner, Jerome (1990), *Acts of Meaning*. Cambridge MA: Harvard University Press.

Burri, Regula Valérie and Joseph Dumit (2008), "Social Studies of Scientific Imaging and Visualization" in Edward J. Hackett et al. (eds.), *The Handbook of Science and Technology Studies*. Cambridge MA: MIT Press, 297–317.

Chen, Xi et al. (2008), "Probing Superexchange Interaction in Molecular Magnets by Spin-Flip Spectroscopy and Microscopy," *Physical Review Letters* 101(19):197208 (4 pages).

Curtis, Scott (2007), "The Rhetoric of the (Moving) Nano Image." Talk presented at the Workshop *Images of the Nanoscale: From Creation to Consumption*. NanoCenter, University of South-Carolina, Columbia SC. October 2007.

Daston, Lorraine and Peter Galison (2007), *Objectivity*. New York: Zone Books.

Eigler, D. M. and E. K. Schweizer (1990), "Positioning Single Atoms with a Scanning Tunnelling Microscope," *Nature* 344(5 April):524–6.

Hayles, N. Katherine (1999), "Simulating narratives: what virtual creatures can teach us," *Critical Inquiry* 26(1):1–26.

Hennig, Jochen (2004), "Vom Experiment zur Utopie: Bilder in der Nanotechnologie," *Bildwelten des Wissens. Kunsthistorisches Jahrbuch für Bildkritik* (Berlin: Akademie Verlag) 2(2):9–18.

    (2006), "Changes in the Design of Scanning Tunneling Microscopic Images from 1980 to 1990" in Joachim Schummer and Davis Baird (eds.), *Nanotechnology Challenges: Implications for Philosophy, Ethics and Society*. Singapore: World Scientific Publishing, 143–63.

    (2010), *Bildpraxis: Visuelle Strategien in der frühen Nanotechnologie*. Bielefeld: transcript, forthcoming.

Knorr Cetina, Karin and Martina Merz (1997), "Floundering or Frolicking – How does Ethnography Fare in Theoretical Physics? (And what Sort of Ethnography?)," *Social Studies of Science* 27(1):123–31.

Landau, Jamie et al. (2009), "Visualizing Nanotechnology: The Impact of Visual Images on Lay American Audience Associations with Nanotechnology," *Public Understanding of Science* 18(3):325–37.

Latour, Bruno (1990), "Drawing Things Together" in Michael Lynch and Steve Woolgar (eds.), *Representation in Scientific Practice*. Cambridge MA: MIT Press, 19–68.

    (2002), "What Is Iconoclash? or Is There a World Beyond the Image Wars?" in Bruno Latour and Peter Weibel (eds.), *Iconoclash: Beyond the Image Wars in Science, Religion and Art*. Cambridge MA: MIT Press, 16–38.

Lösch, Andreas (2006), "Anticipating the Futures of Nanotechnology: Visionary Images as Means of Communication," *Technology Analysis & Strategic Management* (Special Issue on the Sociology of Expectations in Science and Technology) 18(3–4):393–409.

Lynch, Michael (1985), "Discipline and the Material Form of Images: An Analysis of Scientific Visibility," *Social Studies of Science* 15:37–66.

(1990), "The Externalized Retina: Selection and Mathematization in the Visual Documentation of Objects in the Life Sciences" in Michael Lynch and Steve Woolgar (eds.), *Representation in Scientific Practice*. Cambridge MA: MIT Press, 153–86.

Lynch, Michael and Steve Woolgar (eds.) (1990a), *Representation in Scientific Practice*. Cambridge MA: MIT Press.

Lynch, Michael and Steve Woolgar (1990b), "Introduction: Sociological Orientations to Representational Practice in Science" in Michael Lynch and Steve Woolgar (eds.), *Representation in Scientific Practice*. Cambridge MA: MIT Press, 1–18.

Milburn, Colin (2008), *Nanovision: Engineering the Future*. Durham & London: Duke University Press.

Mody, Cyrus C. M. (2004), "How Probe Microscopists Became Nanotechnologists" in David Baird, Alfred Nordmann and Joachim Schummer (eds.), *Discovering the Nanoscale*. Amsterdam: IOS Press, 119–33.

Myers, Greg (1990), "Every Picture Tells a Story: Illustrations in E. O. Wilson's Sociobiology" in Michael Lynch and Steve Woolgar (eds.), *Representation in Scientific Practice*. Cambridge MA: MIT Press, 231–65.

Nerlich, Brigitte (2008), "Powered by Imagination: Nanobots at the Science Photo Library," *Science as Culture* 17(3):269–92.

Rheinberger, Hans-Jörg (2006), *Epistemologie des Konkreten*. Frankfurt/Main: Suhrkamp.

FOURTEEN

USING MODELS TO KEEP US HEALTHY: THE PRODUCTIVE JOURNEYS OF FACTS ACROSS PUBLIC HEALTH RESEARCH NETWORKS

ERIKA MANSNERUS

1. Introduction: Travellers' Tales

We are, perhaps, so familiar with narratives about knowledge construction that we may have forgotten to observe what actually happens to facts after their construction. The traditional study of the making of facts has mainly paid attention to activities that take place behind closed doors at their construction sites, often laboratories.[1] But how do such facts then travel across the scientific world? And when they 'arrive' somewhere – how do they accommodate themselves into different environments? Do they change their identities – slightly or considerably – or do they stay stubbornly as they are? By increasing our understanding of these matters, we can learn more about the nature and progress of model-produced knowledge, and about facts' 'wider life.'

This chapter addresses the core question: How do the characteristics and functions of travelling facts equip them to contribute to building new models? Can seeing how they are utilised in novel surroundings allow us a fresh perspective on their conventionally understood nature – as 'hard' facts validating scientific findings? Might they be plastic to a degree – like precious metals –valuable, but malleable enough to be 'worked'? And if so, what does that say about facts themselves?

This chapter analyses how model-produced facts about bacterial infections (in this case, *Haemophilus influenzae* type b *Hib* and *Streptococcus pneumoniae Pnc*) and preventive public health measures against the severe

[1] Cf. for example, Latour and Woolgar (1979/1986); Knorr Cetina (1981).

diseases they cause were adopted into a set of infectious disease models built in different research communities to examine population-level transmission dynamics, vaccination effects and herd immunity. To consider wider questions about facts and their travels, it complements this analysis by developing two perspectives on travelling facts:

- First, in considering how model-produced facts travel beyond their production sites and become model-adopted facts, used as evidence in other domains, it examines how they are *disseminated, communicated* and *cross-fertilised* across modelling communities, and also how their travels help form research communities.
- Second, it analyses how the different receiving communities that adopt travelling facts, *identify* and *characterise* them, and elaborates on how they use them in different *functional roles* to shape their own knowledge practices. Do they have *stubborn* characteristics, retaining their identity, or can they 'change colour' to accommodate themselves easily to new environments? Do they become *enriched* with added information, or *slimmed* down to their essentials? And in terms of their functionality in their new domains, how can they aid in opening new research areas, mediating different approaches and carrying information to facilitate novel applications?

By 'facts' I refer to knowledge claims that are generally accepted within a community, and that – once documented – can reliably be used by and acted on when they move beyond their first 'production site' into novel contexts.[2] It should be emphasised that a fact is not understood through its propositional character, nor is it given a truth-value. Yet this definition should not be read in relativistic terms. On the contrary, facts carry a degree of integrity when they travel, and their factual claims remain consistent despite any changes they undergo as they travel across scientific communities, reflecting the stability that comes from their continuous process of use and reuse.[3] In a way, when they travel, facts can be thought of as being 'in the making,' but retaining a degree of integrity that defends them from relativistic interpretations. And 'facts' should be distinguished from the set of

---

[2] Becker (2007,12) presents a community approach to facts: '(…] *facts are only facts when they are accepted as such by the people to whom those facts are relevant,'* However, this approach leaves aside the importance of usability and applicability of facts in both producing and receiving communities.

[3] As Mary Morgan writes in the introductory essay facts have a 'virtue of steadfastness or sturdiness … but also of a certain degree of useful mutability around the edges' (Morgan, this volume).

computational techniques and algorithms usually understood as templates.[4] The detailed analysis of our case revealed that techniques and methods did indeed travel alongside the facts as their 'partners.' But this chapter focuses particularly on *how far* facts maintain their stable core integrity when they move to receiving domains, and how they function there.

Facts' characteristics and functionality can be seen as two sides of a coin: Together, they tell us how facts become identified and used in new domains.[5] Considering how actors (e.g., modellers, epidemiologists) identify and characterise them when using them as evidence for new research is one side of the answer to what happens on their post-production travels. But facts may also play different functional roles in their new domains, or function in different ways. In this novel functioning, they can help in opening new research questions by supplying forgotten or otherwise ignored knowledge, by mediating between different approaches, solutions or tasks or by storing knowledge.[6] (In this respect, we may link them with research on how models function in scientific work.[7])

This is not to imply that facts have *agency* – that they are capable of challenging evidence, as if functioning of their own volition. Here, facts are observed as part of the social context that either produces or applies them – normally a community of researchers (modellers, epidemiologists, etc.) – who recognise facts, adopt them from scientific publications and apply them in their own research. Agency, of course, remains with the actors who characterise and use them.[8]

The structure of this chapter is as follows. Section 2 discusses how facts 'graduate' from being *model-produced* to *model-adopted* facts, and illustrates how they 'carry' knowledge into new scientific domains via three different abstract types of journey: *dissemination*, *communication* and *cross-fertilisation*. Section 3 employs these understandings to examine the case of how facts produced in a parent model of infectious disease transmission (the

---

[4] According to Humphreys (2004), a template is a rather universal set of equations, methods or techniques, adaptable to different fields.

[5] The characterisation of facts and of their functional roles emphasises the dynamic nature of knowledge. It acknowledges the heterogeneity and plurality of knowledge through the various uses facts are associated with in their new domains. This differs from a representationalist account presented by Jovchelovitch, who sees heterogeneity and plurality of knowledge as being enabled by different modalities of representations (2007, 3).

[6] They may even embed traces of the materiality incorporated in their production and translation (Cf. Latour's notion of inscription devices in Latour & Woolgar 1979/1986).

[7] Morgan and Morrison (1999).

[8] I have analysed the ways in which epidemiologists *act with* facts when renewing UK vaccination strategies (Mansnerus, 2009b).

case of Hib) journey to other transmission models (studying both *Hib* and Pnc) and how their travels facilitate the formation of research networks. Section 4 discusses how facts are identified in new contexts and analyses the 'two sides of the coin': how they are *characterised* and what *functional roles* they play in these new adoptive domains. Section 5 concludes by reflecting, in the light of the main findings, on the nature of factual knowledge and the question of how facts relate to evidence.

2. Models, Travelling Facts and Research Networks

How do facts spread outwards across communities from where they were initially produced? This section discusses model-produced and model-adopted facts, identifies *dissemination, communication* and *cross-fertilisation* as three different types of journeys made by travelling facts –and elaborates on how such travelling supports the formation of research communities.

2.1 Model-Adopted and Model-Produced Facts

Our current understanding of models emphasises their capabilities to function in scientific work, mediate processes and facilitate practices.[9] Yet, models built by interdisciplinary research teams are not only primary research objects, but also act as facilitators for integrating knowledge from different fields of study. Such integrative modelling practices produce models that are specifically *tailored* – built, used and applied for explicit purposes, in particular, for answering specific research questions in interdisciplinary communities.[10] Although these studies have enriched our understanding of the increasing importance of modelling in science and have given us vivid accounts of their heterogeneity as research objects, the question of generalisable evidence produced by modelling received little attention. Models in their local contexts have tended to be represented as the primary interest of analysis. Our focus here is not on models themselves, but on the facts produced in the models and disseminated via them to different domains, which we will study at a later point in this chapter by analysing the characteristics and functional roles of facts. Our focus is thus on model-adopted facts, which form the initial 'model assumptions' brought into models. And,

---

[9] We address models as being representations of a phenomenon under scrutiny or as tools and instruments for investigation. Along with analysing the functions of models as objects of research, we have a growing interest in understanding the practices that shape them. Cf. Morgan and Morrison (1999).

[10] See Mattila (2006b).

as an output, we gain model-produced facts as a similarly heterogeneous set of claims derived from the modelling process. (Both sets of model-related facts should be understood broadly to include estimates, parameter values, simulated results or commonly held knowledge in the field.)

2.2 The Three Different Journeys of Facts

Considering more closely the travels of model-produced facts from their original construction site to become model-adopted facts received into a new model or community, we can conceptualise three different types of journey – each of different 'reaches' – as illustrated in Figure 14.1.

The basic form of travelling is simple *dissemination*, and can be defined as the one-way distribution of model-produced facts, where their findings, information and templates are scattered as seeds to fall on fertile or stony ground. (In the research context, this process occurs via journals, at conferences and anecdotally through professional and social contacts – the very stuff of which networks are made.)

Figure 14.1 illustrates these simple journeys as dotted lines, where some of these 'seeds' (those that make contact with M1, M2 and M3) can be seen to have 'taken root' and become model-adopted facts, taken on board for

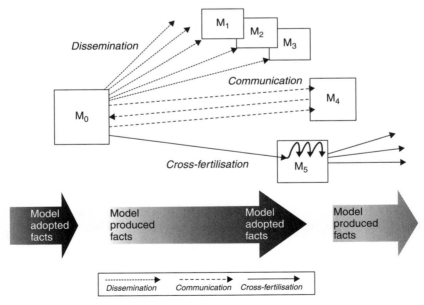

Figure 14.1. How facts journey between models.
From: © E. Mansnerus & J. Morgan.

their potential usefulness. There, another type of journey sees them as *communicating* facts, with their adoptive modellers checking back with their originating model to ascertain parameter values or transmission patterns so the travelling facts can make a reliable contribution in their new setting. (This is illustrated in Figure 14.1 by the dashed lines entering within the boundary of Model 4, and their iterative returning and revising 'travels' as the model is being built represented by their repeated 'to-and-fro' journeys).

At the most productive type of journey, model-adopted facts can be seen as *cross-fertilising*, interacting with, modifying and being modified by their new model (see M5). This results in them yielding 'observable fruits' – beneficial outcomes for both their new model and, as a 'new generation' model-produced fact, to continue a reproductive lifecycle in a new journey towards further identities in associated (or even distant) settings.

This concept – of facts travelling on different journeys to differing effects – can tell us about model-based knowledge in general, and also give a particular perspective on our main question: What happens to facts after their construction?[11] Their 'productive' journeys can also provide a practice insight about their *characteristics* and *functional roles*, helping explain why and how certain facts are adopted into new domains.

2.3 Travelling Facts Helping to Network Researchers and Create Communities

Figure 14.1 has illustrated the different types of ties that can be involved when facts move from being model-produced to model-adopted. But the clustering of models in Figure 14.1 suggests another outcome of facts travelling – that they seem to facilitate the formation of research networks. (With reference to the case discussed in the next section, one interviewee recounted how a referee process had first pointed him to the facts published in a Helsinki paper on Hib models, and this encouraged him to meet the modellers at a conference and share ideas.[12]) The primary adoption of a fact furthers contact between different modelling groups, and facilitates and enhances personal relations between researchers. Their travels (with their computational templates partners) help both to form communities and to shape their practices – in fact, networks seem to be built upon the epistemic

---

[11] Leonelli (this volume) discusses the travelling of experimental knowledge across information databases.

[12] Interview on epidemiological modelling and knowledge transfer, conducted at Health Protection Agency (HPA), Centre for Infections, Colindale, June 2006.

primacy of travelled facts. We shall see later how these communities represent a social network that can influence how facts are identified, characterised and put to work. But first, we will examine the case of a specific set of models related by travelling facts to make concrete the abstract notions of types of ties and of outputs, travels and fruitfulness discussed here.

3. The Goodnight Kiss and its Research Network

This study focuses on facts established in a set of infectious disease models built in research communities (mainly in Finland and the UK) to examine the population-level transmission dynamics, vaccination effects and herd immunity of Hib and Pnc bacteria. The Helsinki project was a multidisciplinary modelling project running from 1994–2003 at the National Public Health Institute in collaboration with the University of Helsinki and the Technical University of Helsinki.[13] This section examines a particular set of models generated from a 'parent' model – the Helsinki 'Goodnight Kiss' model (GNKM) – to identify the characteristics of facts and trace their functional roles.[14] We pay special attention to the factual claims circulated between the various models, so the main body of the research materials are publications.[15] The case was analysed in two phases. First, in February–April 2007, all publications reporting models on either bacteria (Hib or Pnc) published by the Helsinki modellers before 2004 (n = 8) were searched for cross-citations in other papers via the ISI Web of Knowledge.[16] Forty-two citations were found, resulting in fifty publications to be studied

---

[13] From 1 January 2009 The National Public Health Institute became The Institute for Health and Welfare.

[14] The study is also informed by previous research on interdisciplinary modelling practices at the National Public Health Institute, Helsinki, Finland during 2001–2004 and by the author's active participation in a course on infectious disease modelling organised by the London School for Hygiene and Tropical Medicine and Health Protection Agency, UK 2007.

[15] The first author listed in the publications documenting the parent 'Helsinki model' is Auranen. In the paper, the models are referenced with the publication, which may give an illusion that, since all facts seem to move between publications from the same author, 'nothing travels.' However, the Helsinki models form a set of ten models, published between 1996 and 2004, based on research conducted in a six-member interdisciplinary research group.

[16] The Helsinki modellers were researchers (both junior and senior) who started the modelling research in an INFEMAT project (1994) and had disciplinary backgrounds in mathematics/statistics, computer science and epidemiology. The INFEMAT project formed an interdisciplinary research group between the University of Helsinki, the National Public Health Institute and Helsinki University of Technology. Although the analysis in this chapter uses only the scientific publications from the project, it is informed by interviews and a long-term ethnography conducted as a part of my PhD study (2001–2003). Auranen and Leino, researchers at the National Public Health Institute, were key informants and collaborators during my study.

altogether. Second, the search results were studied to find what kind of cross-referencing was involved, and three different types were identified.[17] The first involved generally only acknowledging the existence of the group,[18] while the second focused mainly on computational techniques or methods used[19] or initially developed in the papers. In the third category (the focus of this chapter), the referencing was to factual claims that are treated as firmly based assumptions, such as existing knowledge about phenomena, model-based estimates and parameter values and model-produced facts. I chose three published models – Auranen et al. (1996), Leino et al. (2000) and Auranen et al. (2000), which were cross-referenced twenty-one times – to analyse in detail for this chapter and on which to base my consideration of the travels of facts.[20] Auranen et al. (1996) and Auranen et al. (2000) provide detailed examples of how travelling facts helped established research networks, while Leino et al. (2000) is a source for analysing what happens to these facts on their journeys. The analysis carries a dual focus – both on the communities of practitioners and on the facts themselves: The chapter therefore discusses both the different forms of fact transmission in relation to modelling practices, and how facts are adopted and what kind of functional roles they play in new utilizing communities. This section studies where and how facts moved across the various research domains, and then looks more closely at how they are characterised and function in their new, adoptive domains.

3.1 Infectious Disease Transmission: Hib and Pnc

Understanding the travels of factual claims through a complex network of infectious disease transmission models necessitates some familiarity with the epidemiological phenomena and the research questions involved, and of their public health significance.

The 'villains' in this story are Hib and Pnc. Hib bacteria (also known as Pfeiffer's bacillus after its discoverer Robert Pfeiffer, who first isolated the

---

[17] Howlett (2008) studied how facts travel between two disciplinary communities, anthropology and economics, by analysing citations in the disciplinary journals. He developed the notions of a listening tree and a talking tree to describe the processes of exchange.

[18] For example, Jackson et al. (2005)

[19] For example, Cooper et al. (2004).

[20] The choice is based on the observation that these articles were referenced either by other studies in the Helsinki group or by the wider Helsinki research community (those working for the same departments) and by various foreign studies on different topics (ranging from infectious disease studies to smoking).

germ in 1892[21]) colonises the nasopharynx and is transmitted in droplets of saliva when we sneeze or cough – or when we kiss our loved ones goodnight. Hib can cause severe (often life-threatening) disease among small children (and adults), including bacterial meningitis, septicemia, otitis media and arthritis. Vaccine development began in the 1930s after Pittman isolated the different strains of the bacteria (Pittman, 1931, 1933), but the first line of vaccines (the polysaccharides) were not introduced until the 1970s. To improve their efficacy and reduce the risk of carriage instead of simply protecting against Hib's invasive forms, conjugate vaccines were developed in the 1980s and became part of the national immunisation systems of the United States (1985), Finland (1986) and the United Kingdom (1992), resulting in dramatic decreases in reported Hib cases in each country.

Streptococcus pneumoniae (Pnc) was identified as an organism by Louis Pasteur and George Stenberg in 1881. Similarly to Hib, Pnc colonises the human nasopharynx, but its polysaccharide capsule and its more than ninety different strains made vaccine development difficult. Conjugate vaccine programmes have been introduced in the United States and in the United Kingdom, but European Union (EU) vaccination strategies have yet to be refined.[22] Both bacteria cause meningitis, with the latter being the leading cause of the disease. The World Health Organization (WHO) report that Hib alone caused approximately 386,000 deaths annually, as well as 3 million serious cases of disease among children aged 4–18 years worldwide (WHO, 2005).

These public health facts were the starting point for the Helsinki model building, and were translated into the following Hib study research questions in their 1994 Research Plan:

> Does vaccination alter the age distribution of Hib disease and incidence?
> Does natural immunity vanish from the general population, which would indicate the need for revaccination?
> How high must the vaccination coverage be in order to prevent the transmission of Hib disease in the population?

3.2 The Family Tree of the Goodnight Kiss Model

The GNKM (M1), first published in Auranen et al. (1996), was the parent of the facts in this case. It was an individual-based model built to describe

---

[21] A detailed story of understanding population-level Hib transmission is presented in Mansnerus (2009a).

[22] O'Brien et al. (2003), Pebody et al. (2005), Noakes et al. (2006).

asymptomatic Hib infection in a family with small children, and was designed to estimate family and community transmission rates simultaneously by studying the spread of Hib via goodnight kisses between family members with small children. It was fitted with datasets collected in Finland 1985–1986 and the United Kingdom 1991–1992, just before these countries introduced their Hib immunisation programmes. The model itself was the first of a set of Helsinki project models and represents the beginning of the modelling collaboration.[23] This model (M1) produced a set of facts (F1–8) that travelled to other Hib and Pnc studies, were referenced by different research groups and even ended up in a WHO vaccination policy report promoting Pnc vaccination. This case study records in detail how the facts from this single model travelled and facilitated the emergence of these two research networks.

Figure 14.2 illustrates the Helsinki models' 'family tree,' showing how research networks emerge from the travelling of facts first established in a single model and giving the names of the models (M1–M10) that adopt and use the facts (F1–12) and the location of the modelling group or the origin of the report where the fact was used. The figure also records different types of ties associated with facts' different journeys: As before, dotted lines indicate dissemination, dashed lines communication and solid lines cross-fertilisation.

Figure 14.2 shows how facts journeyed over an extended time from the parent GNKM model (M1) to two sets of models built to examine Hib and Pnc bacteria (the latter shaded in grey). The figure shows that between 1996 and 2004, the Helsinki group built a set of Hib (M1, 2, 4 and M8) and Pnc (M3) models that adopted facts (F1–F8) established in the GNKM model.[24] It also shows that the M3 Pnc model itself established a set of facts (F9–F12) and that both sets of facts journeyed outward to be adopted by a wider community. The following sections examine these journeys in more detail.

3.3 Journeys Towards Broader Research Networks

The facts established in the 1996 GNKM model (M1) were adopted by three subsequent Helsinki Hib models (M2, M4 and M8). The estimate for the

---

[23] Discussed in Mattila (2006a, c).

[24] Figure 14.2 illustrates only these five models, but in fact, the group built ten altogether between 1996 and 2004. References to models illustrated in Figure 14.2 are M1 (Auranen et al., 1996); M2 (Auranen et al., 1999); M3 (Auranen et al., 2000); M4 (Auranen et al., 2004); M5 (Coen et al., 1998); M6 (O'Brien et al., 2003); M7 (McVernon et al., 2004); M8 (Leino et al., 2004); M9 (Eerola et al., 2003) and M10 (Cauchemez et al., 2006b).

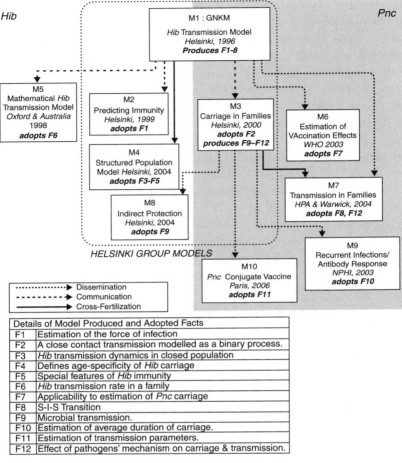

Figure 14.2. The family tree of the Goodnight Kiss Model.
From: © E. Mansnerus & J. Morgan.

force of infection fact (F1) can be described as *communicating* to M2 (studying predicting immunity) as the adoption of the estimate required returning to the parent GNKM model to determine how it should best be calculated in its novel context. The F3–F5 cluster of Hib transmission facts (concerning age specificity of carriage, duration of immunity and dynamics of transmission) were also fed into a model (M4) that simulated Hib transmission dynamics in a structured population This journey can be labelled as *cross-fertilisation* (productive of new outcome facts), as their adoption and use by the M4 model resulted in outcomes in the form of further simulation-based facts.

Facts about Hib transmission were also adopted into models studying Pnc transmission (M3, M6, M7, M9 and M10), which was of current interest in 2003, as conjugate vaccines had been launched onto the market and many EU countries were in the process of shaping their national vaccination policies,[25] and searching for resources to assist studying Pnc vaccine effects and mass vaccination estimations. Infected individuals do not remain immune, but become susceptible again over time so that conjugate vaccines – used in both cases to reduce carriage – can have unwanted long-term effects on decreasing natural immunity in a population.

The M3 model on Pnc carriage in families produced and then *disseminated* and *cross-fertilised* further facts (F9–12), as well as spreading some associated templates that reported data augmentation methods.[26] Many of these facts also travelled beyond the Helsinki group, linking with other models and helping create a wider research network. F6 (concerning transmission rates) first established in M1 was adopted into the M5 model published by researchers from Oxford and Australia, while the M7 model (on transmission in families) built by a UK Health Protection Agency/Warwick University group utilised the M1-produced fact F8 (about the S-I-S pattern of Hib infection) in modelling a similar dynamics structure for Pnc.[27]

This is an excellent example of how travelling facts can serve to set up a new research area. The similarities between the two pathogens meant that Hib-related facts also could be adopted in Pnc models, so facts F1–6 (especially the F3–5 cluster) provided a stepping stone for researchers modelling Pnc carriage in families. The M3 model output facts F9 and F12 (concerning microbiological transmission) made *dissemination* and *cross-fertilisation* journeys to be adopted into two further Pnc models (M7 and M9), which studied questions related to immunity and vaccine efficacy, respectively. The model-produced fact F6 about Hib disease transmission, which was applicable to problems of estimating vaccination effects, was adopted by the M6 WHO model to develop methods of detecting Pnc carriage in families. Facts about transmission, carriage and estimations of vaccination effects (F9–F12) were also adopted in models studying these effects in the UK (M7) and France (M9).

Facts were not the only useful thing that travelled. Some techniques documented in the Pnc model (M3) were designed to overcome problems

---

[25] Cf. Pebody et al. (2005)

[26] In particular, MCMC methods.

[27] Both Hib and Pnc follow the same S-I-S 'Susceptible-Infected-Susceptible' pattern, in which infection does not convert to a permanent immunity in an individual.

of missing data[28] (available data are more likely to come from earlier surveillance studies, and to have been collected for a particular purpose, so
may be missing some information needed for the new modelling exercise).
These techniques were used to support data augmentation techniques in
Pnc and Hib models targeted at estimation purposes, leading, for example,
to the development of latent Markov chain Monte Carlo (MCMC) models[29]
and MCMC sampling methods.[30] Techniques also sometimes travelled
(to, for instance, M10) via anecdotal observations that they had been
shown to be particularly efficient in addressing specific questions in their
production site.

This story shows a single model acting as the source of travelling facts
and their different journeys' ties to other modelling networks, widening the
production and distribution of factual knowledge, and gives a perspective
on how research networks often rest on such circulated 'factual' knowledge
(and their partner computational methods, algorithms and techniques)
prior to formal collaboration ties. We now focus in more detail on the
changing characteristics attributed to travelling facts by different communities and the functional roles they take in their new environments. Both
of these aspects allow us to explore in greater detail the dynamic, changing
nature of factual knowledge.

4. Identifying Facts, their Characteristics and Functionality

In order to understand what happens to facts after their original production
in epidemiological models, we have studied how the journeys of facts carry
information, but also facilitate the formation of research networks. Some of
these model-produced facts become adopted into receiving communities,
where they are identified and accepted according to their characteristics.

4.4 The Identities of Facts

Do facts have a character – a personality? They are commonly held to be
reliable and usable pieces of information, but what happens to them after

---

[28] MCMC data augmentation techniques, for example, to Bartolucci (2006) and to
Cauchemez et al. (2006a).

[29] Markov chain models are standard formulae used to describe random, unpredictable
processes that are 'memoryless', that is, 'the future is independent of the past given the
present.'

[30] MCMC sampling methods are a class of algorithms for sampling from probability
distributions.

their original construction – do they change when they travel into different modelling communities? Character is oftentimes linked with the innate nature of things or people – 'she's quite a character' – implying there is something special – both unique and immutable – in the person herself. However, we may also consider 'character' as something that is given by the community – as a socially defined property of people and of things, and this shift in perspective helps us be more precise about what we mean by the 'characteristics' of facts. In our case, the modelling communities do not, per se, characterise facts. It is how they are adopted into the receiving communities, used and interpreted there, that give us the grounds to analyse their characteristics. Our focus is on the changes we observe to the characteristics of facts in the course of their various modes of travelling.

Could we consider facts in terms of their social recognition – as, metaphorically, having a social identity? Such a concept would be based on an assumption that modellers, in their practice of building or applying models, identify facts for what they are and then *characterise* them according to their usefulness and applicability, recognising their knowledge claims and re-interpreting, adopting, shaping (or even ignoring) them in the course of their work. As with individuals, facts' social identities rest on a dialogical relationship with their environment: The notion of social identity relies on a socio-cultural understanding, and implies a way of behaving defined by relationships formed with a community.[31] The 'social identity of a fact' should be understood as a similar concept, which underlines the importance of how the receiving community recognises and adopts, and then interprets, uses and applies, factual claims. It is these different practices, these different 'attitudes' towards facts, that can allow us to observe in detail the changes they undergo in travelling to new contexts – and, in the end, tell us more about facts themselves.

My main tactic in the following sections has been to provide a snapshot – to 'freeze the moment' – so as to allow us to observe and analyse clear examples of facts' characteristics and functional roles, to tell the story of what happens to facts on their journeys. To rigorously include *all* the relevant information about these travelling facts would complicate the story. Some facts and models are represented in Figure 14.2 – but others are revealed in the analysis of a Model on the dynamics of natural immunity (DNIM – Leino et al. 2000). This model studies how

---

[31] For example, following Vygotskian ideas of child development, it is in a dialogical relation with the community that a child develops skills of using tools, material and symbolic means (language) and thus becomes part of that community (Vygotsky 1978).

vaccinations affect population-level immunity and (given that they also reduce carriage) whether they may actually weaken the natural immunity in the population. The DNIM model was the third model chosen to investigate details of facts' travels, but space prevents us from presenting its analysis to the same level of detail as the other two. (Thus, in what follows, models and facts that are included in Figure 14.2 are numbered to aid reader reference – where no numbers are noted, they refer to models/ facts not so illustrated.)

4.5 The Characteristics of Facts

As facts travel through a research environment, they become identified, and some are adopted into new model contexts. As the new model 'works' with them, their characteristics become revealed, and we can categorise these characteristics with simple labels, such as *stubborn, flexible, enriched* or *simplified*, as summarised in Table 14.1.

In our case, the F3–5 cluster of Hib transmission facts are exemplary representatives of *stubborn* facts, that is, those that are comparatively resistant to change during their travels. Epidemiological studies[32] show that understanding Hib transmission dynamics is a rather challenging issue, as the details and specificities of not only the pathogen but also its circulation in specific age groups, its capability to hide in asymptomatic carriers and its incapability to enforce permanent immunity must be properly appreciated.

Due to these challenges, this cluster of facts became the major source in those models used to estimate vaccination effects, to optimise herd immunity thresholds (M8) and to produce evidence for recommendations (M6) for implementing expensive conjugate vaccination programmes. The characteristic of *stubbornness* describes this cluster of facts as unchangeable, inflexible and, we might even say, robust. These stubborn facts, if you like, are good at communicating, uncompromisingly 'shouting' their main message out loud: 'This is how we are!' This means, for example, that the structure of the adopting model may need to be adjusted to represent the information carried by a *stubborn* fact.

The 'opposite' characteristic is *flexibility* – where facts can be easily accommodated to new environments. Model structure is an important part of model design practice. In the case we have presented, knowledge about infection dynamics, population structures and transitions among the

---

[32] Explored in Mattila (2008).

Table 14.1. *Summary of the analysis of the characterisations of travelling facts*

Characterisation of fact	Example	Demonstration of characteristic
Stubborn* A fact that is *resistant* to change May require some auxiliary measures to be operationalised or quantified in a model	'Transmission of Hib occurs through asymptomatic carriers. Most episodes of Hib carriage pass without clinical symptoms, and only in rare cases does carriage proceed to invasive disease.' (Auranen et al. 2004)	This fact relied on general epidemiological knowledge on dynamics of Hib transmission, and can be regarded as a 'textbook fact.' It supported the cluster of factual claims (F3–5) about Hib transmission produced in the Goodnight Kiss model (M1)
Flexible A fact that can be accommodated well into a new environment. Such facts can 'change their colour' easily or frequently, like chameleons	'[as in Auranen et al. 1996], we set the transition from C to S to be dependent on a constant recovery rate.' (Melegaro et al. 2004)	This fact – the SIS-structure (F8) – about the dynamics of transmission patterns of immunity against Hib infection, was easily adopted and modified to fit into the M7 Pnc study. (Pnc follows similar but not identical immunity dynamics as Hib)
Enriched A fact that gathers meaning as it is employed in its new context	'In the previous study on Hib carriage in families (1996), the force of infection is probably related to [the] different nature of data. In that article, data on antibodies was not included.' (Auranen 1999)	This fact (F1) – an estimate of the force of infection – was enriched in its destination model (M2) with data about antibodies. The importance of estimating the force of infection in relation to all the transmission dynamics details was discussed (carriage, antibody levels, cross-reactive bacteria)
Simplified A fact that becomes slimmed down so that only its core claim or key message remains	'For example, Hib transmission rate is thought to be greater within families whose members have experienced Hib disease.' (McVernon et al. 2004)	This fact – originally a sophisticated numerical estimate for a transmission rate in a family – was simplified into the factual claim 'thought to be greater' in a model on trends in Hib infections

Note: This analysis uses the models studied in Figure 14.2, but extends the analysis to a model published in Leino et al. (2000 – see earlier). Only the models and facts represented in the figure are given their corresponding numbers in the table and text.

*Also used by Daston (1994).

different sub-groups of population – namely between those who are susceptible to the infection, S, and those who become infected, I – needed to be
incorporated. The structure also fed into quantification efforts, so each tran-
sitional step was denoted with parameters leading to an estimation process.
The S-I-S-structure (which carries facts, e.g., F8) about the Hib transmission
pattern and the levels of immunity caused by the infection) was adopted by
a group of models (M3, 6–10) that studied the transmission of Pnc:

> As in Auranen [et al.,1996], we set the transition ..... to be dependent on a con-
> stant recovery rate. (M7, Melegaro et al. 2004)

This fact about constant recovery rate, which was first established in the M1
Hib model, can be seen as *flexible*, in that it was adaptable to being modified and adjusted to accommodate to Pnc transmission models (which is
thought to follow a similar epidemiological S-I-S mechanism in not converting to permanent immunity). The adjustment (which Melegaro et al.
(2004) define as "transition to be dependent on constant recovery rate")
helped solve effectively a model parameterisation problem. The ease with
which the fact could be adjusted to make the transition from the Hib to the
Pnc context confirms its characterisation as *flexible*.

These characteristics – *stubborn* and *flexible* – are related to the environment: The facts either resist relocation or accommodate themselves well.
But we also observed that some facts can be characterised in terms of the
information they carry with them on their travels and the way in which they
become *enriched* or *simplified* in their new context. Estimation of the force
of infection (the rate of infectivity, F1) is an example of an *enriched* fact. At
its origin (in the GNKM (M1)), it was a clearly defined, model-produced
fact carrying a numerical value of the estimated infection rate in a family (a closed population). But in use in other Helsinki group models (e.g.,
M2–4), its information content became 'enriched' with other elements –
information about the impact of antibody levels, cross-reactive bacteria and
other transmission dynamics details were added into it – giving its factual
content greater weight.

To give two other examples of this enriching characteristic: A fact that
reliably predicted trends (decreases or increases) in natural immunity was
adopted in the DNIM immunity model that examined a long-term persistence of immunity after vaccinations. This fact was enriched in a vaccination model that studied the duration of immunity and predicted antibody
resistance after Pnc polysaccharide vaccination.[33] In a similar case, a Hib

---

[33] Mäkelä et al. (2003).

model-based prediction about the decline rate of the number of infections was taken as a fact into a model about Pnc carriage,[34] and was also reinforced by observed data trends in Hib infections studied in England and Wales (M7).[35] In both these cases, the decline rate became an enriched fact when taken into a new context and augmented with information from the existing literature and datasets.

An example of a *simplified* fact is a model-based numerical estimate for a Hib transmission rate in a family (F6). Even though this fact was originally highly sophisticated and the outcome of a transmission model (M1), it was only used as a comparative reference point in McVernon et al. (2004), losing the connection to the numerical estimate it had in the original model. It was slimmed down and simplified, yet still usable when circulated to a new context. Another example of a fact being used in simplified form as a 'reference point' was when (F7) (about the applicability of modelling in estimating Pnc carriage rates) was evaluated in a WHO report (M6).

This labelling of facts' characteristics – as *stubborn, flexible, enriched* or *simplified* – offers a new dynamic to understand what happens to facts when they journey between models, in this case, from infectious disease epidemiology to non-communicable diseases. First, *stubborn* facts resist change, while the *flexible* 'chameleons' can 'change their colour' and form in order to be accommodated well into new models. Second, on their travels, facts may have their information content *enriched* or *simplified* in their adoptive contexts. Through their capacity to be accommodated into new environments, yet carrying their stable core of information, a degree of their central integrity, they prove useful in advancing model-based knowledge.

4.6 The Changing Functional Roles Facts Play

We have explored how modelling communities characterise facts, but to elaborate a broader and more dynamic account of evidence, we need to become familiar with what kinds of different functional roles they play when they journey beyond their production sites to be adopted into a new research model. These functional roles are, in a way, the other side of the coin – so looking at the process from this other angle, we can see *what adopting communities do with* facts, and begin to define and conceptualise the different functional roles they might play.

---

[34] Leino et al. (2001).
[35] McVernon et al. (2004).

We have discussed how, when facts spread across modelling communities – either as pieces of already existing knowledge (e.g., derived from textbooks) or as facts produced from models in previous studies – they may then be adopted into new models as various assumptions, estimates and parameters. This process is an iterative exercise, in which a modeller seeks to find the balance between a realistic-enough description of the phenomenon, the amount of information available and the sampling scheme of the data.[36] Once the facts have travelled into the new modelling environment, the inferences drawn from them are reasonably stable. Models' needs for input facts (e.g., estimates, parameter values and data) suggests a closer look at the different roles facts may play in their new environments and how they actually function within and across models. Table 14.2 describes the functional roles fulfilled by model-adopted facts observed in our case and analyses some examples. Although there may be others that could be observed in different contexts, we can identify these adopted facts as functioning in three such 'new' roles in our case: They may be used as *brokers* to open new areas of research, as *mediators* to reconcile and mediate different approaches and as *containers* for storing information.

The *fact-as-broker* functions by helping to open up and support new research areas, expanding and challenging aspects of a given research topic, as it travels across contexts (such as from one pathogen to another). In this functional role, facts can indicate the need for new explorations, new studies or perhaps new applications by, for example, communicating transmission rates or facts embedded in model structures into new production contexts. The major output of the GNKM (M1), was to establish the fact of the Hib transmission rate in a closed population of family members in relation to the size and age structure (F6), which was later adopted in a Pnc model (M7) as a way to understand the transmission of the Pnc pathogen within a household.

In this case, the way in which GNKM communicated the 'facts of transmission' (F3–5) actually provided the basic transmission rate (F6). The use of this information in the M7 Pnc model shows it fulfilling the function of *fact-as-broker* in stimulating more research, more specifications and negotiating a new space by doing so. The way in which this *broker* fact operated in its new domain was not by simply bringing in a transmission rate estimate to be added into the set of model parameters. The fact also carried with it a message about the quality of the estimate, recommending

---

[36] Auranen et al. (1999, 16)

Table 14.2. *Functional roles of travelling facts: their definitions and examples*

Functional role and definition	Examples of facts' functionality
Broker A fact that is used to create and negotiate a space, to open new lines of research, ask new questions, expand ongoing processes	A fact (F6) showing Hib transmission rate in a closed population, taking into account the basic population dynamics (structure of the family, children's relatedness in peer groups) The fact functions as a *broker*, since it opens new research in Pnc studies in a model by giving the direction (transmission in a family) and the estimated rate 'Following the work by Auranen and colleagues, the model considers transmission of Pnc within the household.' (Melegaro et al. 2004, 435)
Mediator A fact that is used to reconcile different approaches, techniques, methods, datasets, parameter values; functions as an intervening fact that enables reconciliation	A fact (F9) describing the indirect effects from Hib conjugate vaccines (reduction in carriage and boosting of immunity levels), showing that conjugate vaccines are able to reduce colonisation of Hib in the nasopharynx The fact functions as a *mediator* since it bridges the gap between Hib and Pnc studies, and mediates between childhood and adulthood studies 'How much vaccine coverage (for Pnc) is needed for indirect effects remains a key question. A model evaluating this in Hib conjugate vaccine showed that much of the decline in invasive disease could be attributed to indirect effects of the vaccine, even at relatively low levels of vaccine coverage.' (Lexau et al. 2005)
Container A fact that is used to enclose and contain information (such as estimates, etc.), which can be picked up or referred to subsequently as a new input	A fact that predicts the decline rate of natural immunity of Hib It functioned as a *container* in McVernon et al. (2004), *storing information*, in this case, the numerical value of the decline rate 'Our data shows that the reduction in opportunities for boosting natural immunity has resulted in a decline in specific Hib antibody titres among adults.' (McVernon et al. 2004)

its suitability for supplying similar help in opening up the manifold phenomenon of Pnc population-level transmission. Thus, it not only brings in a reference point from previous work to open up a new research domain, it also seems to bring 'encouragement' from its producer domain to support new efforts.

The *fact-as-mediator* can be defined by its functioning to enhance reconciliation between different approaches, problems and domains. In our case, *mediators* were facts that increased the reconciliation between the epidemiological data derived from surveys or experimental settings and those models that adopted them.[37] As the example in the table notes, a fact that described the effects of conjugate vaccines for Hib (unnumbered here, as it does not appear in Figure 14.2) substituted for a lack of knowledge about polysaccharide vaccines' effects on immunity when adopted by a different (again unnumbered) model dealing with Pnc in adults.[38] Thus, it can be said to have mediated – built bridges – between different epidemiological studies between these two groups – the fact is shown as being able to function not only between different types of pathogens (Hib and Pnc), but also between different groups in the studied population (children and adults).

The *fact-as-container*[39] function shows how facts can store information, which at times may enable it to carry 'forgotten' pieces of knowledge, which might be valuable in a new context. The dynamics of natural immunity model (DNIM – introduced earlier) provided a source model for studies of the impact of conjugate vaccines and the risk that they might reduce natural immunity in a population by establishing how such vaccination affected the number of sub-clinical infections occurring between vaccinations. This fact relies on knowledge from Hib studies that conjugate vaccines were effective in reducing carriage of Hib in vaccinated populations, which may also have resulted in the waning of natural immunity among the unvaccinated. (Because conjugates both prevent infection from and reduce the carriage of the bacteria, vaccination affects the levels of immunity in the population. Reducing the circulation of the bacteria may result in a weaker level of natural immunity, since our immune systems are strengthened if they are 'challenged' by a certain level of circulating bacteria.) As noted in Table 14.2, this fact predicts the decline rate of natural immunity to the Hib pathogen. This prediction was based on follow-up measurements of Hib antibody data in Finland (Auranen 1999),[40] and the specific estimate for the decline rate was

---

[37] For example, observations from serological data.

[38] Polysaccharide (PS) vaccines against infections caused by *S. pneumoniae* are given to immune-compromised, elderly or frail individuals. Pnc conjugate vaccines are targeted to children, since they are reducing the carriage of the bacteria, which tends to circulate among infants.

[39] Cf. Leonelli's (this volume) study of databases and the circulation of 'small facts' and Ankeny's (this volume) observation on case studies and their capacity to carry facts.

[40] The data were gathered during a polysaccharide vaccine efficacy trial in the 1970s.

then adopted and used as an input by McVernon et al. (2004) into a model that studied the impact of cross-reactive bacteria.[41] Here, the natural immunity estimate stored facts about immunity rates and other details that influenced the process and carried them forward to 'remind' the new model of these details. In a way (as a *container*), the fact carried information from beyond the model, since it 're-collected and recycled' data about the impact of polysaccharide vaccines, which had been the primary input data for the M3 model.

To sum up, analysing the functional roles of facts underlines two aspects of model-based evidence. First, it shows that the facts are rarely taken into the new contexts without having some effect: opening up further research, providing mediating solutions, which are reliable because they have been tested and acknowledged in other domains, or containing information, parameter values or estimates that are useable in building the new model. In some examples, the origin of the fact was clearly expressed and openly linked with the new groups' work; in other examples, the link was weaker, merely an acknowledgement. Second, by elaborating the *functions* of facts, we learn more about ways in which they facilitate the crucial links between different models.

Modelling is central to scientific investigations – but composing a reliable model is not a simple task. In order to gain such a model, and to be able to produce evidence for the chosen research questions, one needs to master the techniques, develop suitable algorithms and understand what the available data contains, and what may be missing and how model-adopted facts might need to be adjusted to address a new field or domain. This iterative process of modelling can be greatly eased, where modellers can rely on estimates established in other studies, or compare model-derived rates with those that have been empirically validated. A reliable model can establish factual evidence that can be used in further applications, and models that are supported by previously established 'facts' can increase the chance of these journeying facts being 'fruitful' and yielding outcomes that can be beneficial for further generations of models.

Considering the different ways in which facts journey beyond their original models and into new production sites, and understanding how they are identified and characterised there and what kinds of functional roles they fulfil in their subsequent uses and applications, is all part of the extended study of modelling and – beyond that – of how facts function in the scientific world.

---

[41] CR bacteria circulate in population and boost immunity.

5. Conclusions: Productive Journeys

What can productive journeys tell us about facts and evidence? Evidence for medical science and clinical practice is built by multiple methods and derived from various sources of knowledge: experiments (including randomised control trials), surveillance studies, bioinformatics and modelling. Computer-based modelling and simulation techniques have gained a particular importance in infectious disease epidemiology due to their explanatory and predictive capacities, and their capability to address indirect, population-level effects such as herd immunity. Modelling techniques form a set of sophisticated investigative instruments enabling researchers to try to keep up with (or even get ahead of) emerging infections such as pandemics.[42]

What do we understand by evidence? And how do facts relate to evidence? Are facts "evidence in potentia," as Daston (1994) suggests? Evelyn Fox Keller (2002) discusses two different senses: evidence *for* "a theory, argument or hypothesis" and evidence *of* "a phenomenon." Let us briefly re-familiarise ourselves with this distinction. 'Evidence *of*' refers to the common 'detective story-type' elements (a particular item '*a*' is evidence of person '*x*' being in a particular place '*y*' at a particular time '*z*'.....). But 'evidence *for*' points forwards, and I suggest, in two directions (rather than the one she suggests): evidence for (theoretical) claims; or evidence for *use* – for *action*. It may seem that evidence *of* something is nurtured in the 'production' domain and it turns into evidence *for* something when entering a 'use' domain.

Our account of facts shows them being used *both* as evidence *of* a phenomenon (like microbial transmission) and as evidence *for* further theoretical claims, such as a transmission pattern in a population, and for *action* in terms of assisting (for instance) a decision-making process for identifying optimal vaccination coverage. In a similar way, through examining facts' functional roles as brokers, mediators or containers, we subscribe to a more dynamic account of evidence in which facts aren't just evidence for something, but can be used to 'do things.'

What happens to facts on their travels? Despite the conventional view of facts as a solid, monolithic body of knowledge, perhaps a constellation of facts, our story tells us something different – it tells about how facts have changed and begun to become evidence *for*.

First, there are different modes of travelling: Facts may be sown as seed (i.e., disseminated). In this mode, they are not necessarily making an impact

---

[42] Cf. Mansnerus (2010) on explanatory and predictive functions of models.

on an application domain, and may be just acknowledged as they are. Facts that were communicated across different models engaged in bi-directional communication between the models: Adopting the fact entailed further communication with the originating (source) model. Cross-fertilisation journeys see facts as not only being adopted into a new research domain, but also as bearing fruit, producing new facts, new knowledge. Evidence *of* something now becomes evidence *for* something in a rich way – evidence for further claims about infection, transmission dynamics or immunity, or it finds its way to action – as evidence supporting vaccination policies, perhaps.

Second, researchers use facts – to estimate, parameterise and evaluate their models – as their research base for new facts. But they also use facts to connect, to establish new research collaborations – as if a fact carries with it a trace of the research model that originally produced it into its next 'life.' And those who use them think, 'Well, that's sound, plausible, usable, reliable enough for my purposes, it's come from a model that has been built intelligently –I can use it.'

So, how do our facts relate to evidence? It seems to me that facts can stand on the shoulders of previous achievements (those that established the information in the first place) and go on, in turn, to produce new 'facts' to address (in our case) broader public health concerns. Where receiving communities characterise facts and hence introduce changes, or where facts adopt new functional roles in those contexts, we see facts in a dynamic role. Now evidence seems to be a texture of facts – handmade, characteris-able, functioning, connecting – actively utilised by those who first made them, and again by those who use them – maybe somewhat differently – to do something more. There are re-interpretations, interventions and actions that rely on the solidity of facts. This chapter emphasises how facts are embedded both in their producing and receiving communities, and suggests that modelled knowledge brings an interactional, experiential flavour to the ways in which facts relate to bodies of evidence.

Acknowledgements

This study is conducted in a Leverhulme Trust/ESRC funded project 'The Nature of Evidence: How Well Do "Facts" Travel?' (grant F/07004/Z) at the Economic History Department, LSE. I wish to thank Mary Morgan, Peter Howlett, Naomi Oreskes, Lambert Schneider, Helen Lambert, Julia Mensink, Jonathan Morgan and my colleagues in the research project for their feedback on earlier versions of this paper. This research was presented in the Biennial Conference of the Society for Philosophy of Science in

Practice (SPSP 2007). I thank Martina Merz and Rachel Ankeny for their comments on that occasion.

Bibliography

Auranen, K. (2000). Back-Calculating the Age-Specificity of Recurrent Subclinical *Haemophilus influenzae* Type B Infection. *Statistics in Medicine*, 19, 281–96.

——— (1999). *On Bayesian Modelling of Recurrent Infections*. Thesis publication, University of Helsinki, Helsinki.

Auranen, K., Arjas, E., Leino, T., & Takala, A. (2000). Transmission of Pneumococcal Carriage in Families: A Latent Markov Process Model for Binary Longitudinal Data. *Journal of the American Statistical Association*, 95(452), 1044–53.

Auranen, K., Eichner, M., Käyhty, H., Takala, A., & Arjas, E. (1999). A Hierarchical Bayesian Model to Predict the Duration of Immunity to Hib. *Biometrics*, 55(4), 1306–14.

Auranen, K., Eichner, M., Leino, T., Takala, A., Mäkelä, P. H., & Takala, T. (2004). Modelling Transmission, Immunity and Disease of *Haemophilus influenzae* Type B in a Structured Population. *Epidemiology and Infection*, 132(5), 947–57.

Auranen, K., Ranta, J., Takala, A., & Arjas, E. (1996). A Statistical Model of Transmission of Hib Bacteria in a Family. *Statistics in Medicine*, 15, 2235–52, 2235.

Bartolucci, F. (2006). Likelihood Inference for a Class of Latent Markov Models under Linear Hypothesis on the Transition Probabilities. *Journal of Royal Statistical Society B*, 68(Part 2), 155–78.

Becker, Howard S. (2007). *Telling about Society*. Chicago: University of Chicago Press.

Cauchemez, S., Temime, L., Guillemot, D., Varon, E., Valleron, A.-J., Thomas, G., et al. (2006a). Investigating Heterogeneity in Pneumococcal Transmission: A Bayesian MCMC Approach Applied to a Follow-up of Schools. *Journal of the American Statistical Association*, 101, 475.

Cauchemez, S., Temime, L., Valleron, A.-J., Varon, E., Thomas, G., Guillemot, D., et al. (2006b). Pneumoniae Transmission According to Inclusion in Conjugate Vaccines: Bayesian Analysis of a Longitudinal Follow-up in Schools. *BMC Infectious Diseases*, 6, 14.

Coen, P. G., Heath, P. T., Barbour, M. L., & Garnett, G. P. (1998). Mathematical Models of *Haemophilus influenzae* Type B. *Epidemiology and Infection*, 120(3), 281–95.

Cooper, B., & Lipstich, M. (2004). The Analysis of Hospital Infection Data Using Hidden Markov Models. *Biostatistics*, 5(2), 223–37.

Eerola, M., Gasparra, D., Mäkelä, P. H., Linden, H., & Andreev, A. (2003). Joint Modelling of Recurrent Infections and Antibody Response to Bayesian Data Augmentation. *Scandinavian Journal of Statistics*, 30, 677–98.

Daston, L. ([1991] 1994). Marvelous Facts and Miraculous Evidence in Early Modern Europe. In Chandler, J., Davidson, A., & Harootunian, H. (eds.) *Questions of Evidence. Proof, Practice, and Persuasion across the Disciplines*. Chicago: The Chicago University Press.

Fox Keller, E. (2002). *Making Sense of Life. Explaining Biological Development with Models, Metaphors and Machines*. Cambridge, MA: Harvard University Press.

Howlett, P. (2008). *Travelling in Social Science Community: Assessing the Impact of Indian Green Revolution Across Disciplines.* Working papers on The Nature of Evidence: How Well Do Facts Travel? Number 24/08, Department of Economic History, LSE.

Humphreys, P. (2004). *Extending Ourselves. Computational Science, Empiricism, and Scientific Method.* Oxford: Oxford University Press.

Jackson, B., Thomas, A., Carroll, K., Adler, F., & Samore, M. (2005). Use of Strain Typing Data to Estimate Bacterial Transmission Rates in Healthcare Settings. *Infection Control and hospital epidemiology*, 26(7), 638–45.

Jovchelovitch, Sandra (2007). *Knowledge in Context: Representations, Community and Culture.* London: Routledge.

Knorr Cetina, K. (1981). *The Manufacture of Knowledge: An Essay on the Constructivist and Contextual Nature of Science.* Oxford: Pergamon Press.

Latour, B., & Woolgar, S. (1979/1986). *Laboratory Life: The Social Construction of Scientific Facts.* London: Sage.

Lexau, C., Lynfield, R., Danila, R., Pilishvili, T., Facklam, R., et al. (2005). Changing Epidemiology of Invasive Pneumococcal Disease among Older Adults in the Era of Pediatric Pneumococcal Conjugate Vaccine. *The Journal of the American Medical Association*, 294, 2043–51.

Leino, T., Auranen, K., Mäkelä, P. H., Käyhty, H., Ramsey, M., Slack, M., et al. (2002). *Haemophilus influenzae* Type B and Cross-Reactive Antigens in Natural Hib Infection Dynamics; Modelling in Two Populations. *Epidemiology and Infection*, 129, 73–83.

Leino, T., Auranen, K., Jokinen, J., Leinonen, M., Tervonen, P., & Takala A. (2001). Pneumococcal Carriage in Children during Their First Two Years; Important Role of Family Exposure. *The Pediatric Infectious Disease Journal*, 20, 1024–9.

Leino, T., Auranen, K., Mäkelä, P. H., & Takala, A. (2000). Dynamics of Natural Immunity Caused by Subclinical Infections, Case Study on *Haemophilus influenzae* Type B (Hib). *Epidemiology and Infection*, 125, 583–91.

Leino, T., Takala, T., Auranen, K., Mäkelä, P. H., & Takala, A. (2004). Indirect Protection Obtained by *Haemophilus influenzae* Type B Vaccination: Analysis in a Structured Population Model. *Epidemiology and Infection*, 132(5), 959–66.

McVernon, J., Trotter, C. L., Slack, M. P. E., & Ramsey, M. E. (2004). Trends in *Haemophilus influenzae* Type B Infections in Adults in England and Wales: Surveillance Study. *British Medical Journal*, 329, 655–8.

Mäkelä, P. H., Käyhty, H., Leino, T., Auranen, K., Peltola, H., Lindholm, N., et al. (2003). Long-Term Persistence of Immunity after Immunisation with *Haemophilus influenzae* Type B Conjugate Vaccine. *Vaccine*, 22, 287–92.

Mansnerus, E. (2009a). The Lives of Facts in Mathematical Models: a Story of Population-Level Disease Transmission of *Haemophilus influenzae* Type B Bacteria. *Biosocieties*, 4, 2–3.

(2009b). Acting with 'facts' in order to re-model vaccination policies: The case of MMR-vaccine in the UK 1988. Working papers on The Nature of Evidence: How Well Do Facts Travel? Number 37/08, Department of Economic History, LSE.

(2010). Explanatory and Predictive Functions of Simulation Modelling: Case: *Haemophilus influenzae* Type B Dynamic Transmission Models. In G. Gramelsberger (ed.), *From Science to Computational Sciences. Studies in the History of Computing and Its Influence on Today's Sciences.* Diaphenes: Zuerich (forthcoming).

Mattila, E. (2006a). Interdisciplinarity in the Making: Modelling Infectious Diseases. *Perspectives on Science: Historical, Philosophical, Sociological*, 13(4), 531–53.

(2006b). *Questions to Artificial Nature: A Philosophical Study of Interdisciplinary Models and Their Functions in Scientific Practice* (Vol. 14). Helsinki: University of Helsinki.

(2006c). Struggle between Specificity and Generality: How Do Infectious Disease Models Become a Simulation Platform. *Simulation: Pragmatic Constructions of Reality – Sociology of the Sciences Yearbook*, 25, 125–38.

(2008). The Lives of Facts: Understanding Disease Transmission through the Case of *Haemophilus influenzae* type b bacteria. Working papers on The Nature of Evidence: How Well Do Facts Travel? Number 26/08, Department of Economic History, LSE.

Melegaro, A., Gay, N., & GF, M. (2004). Estimating the Transmission Parameters of Pneumococcal Carriage in Households. *Epidemiology and Infection*, 132(3), 433–41.

Morgan, M., & Morrison, M. (1999). *Models as Mediators. Perspectives on Natural and Social Sciences*. Cambridge: Cambridge University Press.

Noakes, K., & Salisbury, D. (2006). Immunization Campaigns in the UK. In S. Plotkin (Ed.), *Mass Vaccination: Global Aspects – Progress and Obstacles*, 304 ed., Vol. 304, pp. 53–70). Berlin, Heidelberg: Springer Verlag.

O'Brien, K. L., & Nohynek, H. (2003). Report from a WHO Working Group: Standard Method for Detecting Upper Respiratory Carriage of Streptococcus Pneumoniae. *Pediatric Infectious Disease Journal*, 22(2), 133–40.

Pebody, R., Leino, T., Nohynek, H., Hellenbrand, W., Salmaso, S., & Ruutu, P. (2005). Pneumococcal Vaccination Policy in Europe. *Euro Surveillance*, 10(5), 174–8.

(1933). The Action of Type-Specific *Haemophilus influenzae* Antiserum. *The Journal of Experimental Medicine*, 58(6), 683–706.

Pittman, M. (1931). Variation and Type Specificity in the Bacterial Species *Haemophilus influenzae*. *The Journal of Experimental Medicine*, 53, 4, 471–92.

Rheinberger, H.-J. (1999). Experimental Systems. Historiality, Narration and Deconstruction. In M. Biagioli (Ed.), *The Science Studies Reader*, pp. 417–28. New York: Routledge.

Vygotsky, L. (1978). *Mind in Society: The Development of Higher Psychological Processes* (M. Cole, V. John-Steiner, S. Scribner & E. Souberman, Trans.). Cambridge, MA: Harvard University Press.

WHO (2005). *Haemophilus influenzae* Type B Bacteria, Hib. *Fact sheet no 294.*

FIFTEEN

THE FACTS OF LIFE
AND DEATH: A CASE OF
EXCEPTIONAL LONGEVITY

DAVID BOYD HAYCOCK

1. Introduction

In 1635, during a visit to his properties in Shropshire, Thomas Howard, fourteenth earl of Arundel and Earl Marshall of England, encountered a remarkable blind old man. His name was Thomas Parr (Figure 15.1), and according to common repute, he had been born in 1483. He was thus 152 years old. The principal contemporary source for our knowledge of Parr's life was the poem (and its prose preface), *The Old, Old, Very Old Man*, written by John Taylor and published shortly after Parr's death that same year. According to Taylor, Parr was the son of John Parr of Winnington, born in the parish of Alberbury, Shropshire, and he lived 152 years, 9 months 'and odd dayes.'[1] Howard took Parr to London, where he was introduced to the king, Charles I; his portrait was painted by Sir Peter Paul Rubens, and for a few weeks he was an object of considerable popular interest (Taylor records that at Coventry such 'multitudes' came to see Parr that he was almost suffocated by the crowds).[2] A few weeks after his arrival in London Parr died; he was autopsied, and his remains buried in Westminster Abbey. As I shall show in this chapter, Parr's remarkable longevity soon became established as a natural historical and gerontological 'fact.' As an almost entirely undisputed truth of potential human longevity, this fact would endure for well over two centuries. Indeed, it was cited as late as the early twentieth century as evidence for what was believed by one Nobel Laureate to be the unnaturally curtailed life spans of modern humans.[3]

---

[1] Taylor (1635), title page.
[2] Taylor (1635), un-paginated preface.
[3] See Metchnikoff (1910), pp. 86–9.

OLD PARR.

From the Original by Peter Paul Rubens in the possession of Uvedale Price Esq.

Figure 15.1. Thomas Parr, by George Powle, after Sir Peter Paul Rubens drypoint, late eighteenth century.
From: © National Portrait Gallery, London.

By examining Parr's case in detail and exploring the ways in which his age was established as fact, its penetration into popular consciousness and its endurance over such a long period, I will reveal some of the ways in which a fact can be constructed, how it can travel successfully (temporally, geographically and immutably) in a certain cultural and intellectual climate and how, when changes in that climate ultimately undermine what was once widely thought of as being a statement of truth, such a fact can then cease travelling (at least in so far as it is understood to be 'true fact'). We shall thus see the kinds of support that enable a fact to be established and become sufficiently stable to travel well, a support process that I label 'scaffolding.'

I shall begin my chapter by examining the contextual scaffolding that supported a widespread contemporary social and medical belief in the authenticity of Parr's great age; these include the early modern intellectual context; the status of witnesses and seventeenth-century conceptions of 'the matter of fact' and the complementary nature of further, apparently supportive, evidence. I will then look at how this particular fact – Parr's great age – travelled successfully over the next two centuries, before showing how the theoretical and cultural scaffolding that supported it was slowly dismantled over time and alternative scaffolding that failed to support the case for Parr's 152 years was eventually erected. Though my focus is on one individual fact (and the ways in which it was constructed, the ways in which it endured and the ways in which it was finally dismantled), I would suggest that it has broader epistemological relevance.[4]

2. Intellectual Context

The immediate and most important piece of scaffolding helping to establish Parr's seemingly remarkable longevity as a fact – and the one from which he is now so obviously divorced – is the intellectual context in which it first appeared. In the case of Thomas Parr, this was the contemporary belief in the potential for men and women to live very long lives. Without this initial element of scaffolding, Parr's case would have seemed even more remarkable – and even more improbable.

The ultimate source for the seventeenth-century belief in potential human longevity measured in centuries rather than decades was the Bible; this, in the early modern period, was still one of the most reliable authorities on most issues in natural history. As Genesis made clear, Adam and Eve had been created immortal. It was only by eating the fruit from the Tree of the Knowledge of Good and Evil that they had become corrupted and mortal. But although Adam and Eve had forsaken the possibility of immortality, they had still lived for many hundreds of years. According to Genesis, Adam lived to 930, Noah to 950 and Methuselah, eldest of all these Patriarchs, to the great age of 969.[5] As the London physician Everard Maynwaring pointed out in 1670 in 'the *Primitive* Age of the *World*, man's

---

[4] Parallels to this story can be found both in way that archaeologists construct scaffolds to support their interpretations of artefacts (see Wylie, this volume), while the account of how lots of particular bits come together to support a 'true story' of the construction of St. Paul's Cathedral can be found in Valeriani, (this volume).

[5] See Genesis, 5:5, 5:27 and 8:29.

life was accounted to be about 1000 *Years*: but after the *Flood*, the Life of Man was *abbreviated* half.' It had continued to fall, such that by the time of Moses it was 'commonly not exceeding 120 Years.' And '[n]ow the *Age* of Man is reduced to half that: 60 or 70 years we count upon.'[6] This was the famous lament of King David in the Book of Psalms: 'The days of our years are three score years and ten; and if by reason of strength they be fourscore years, yet is their strength labour and sorrow; for it is soon cut off, and we fly away.'[7]

Though the Bible appeared explicit that human life spans had fallen considerably, medieval and early modern philosophers and physicians contested the possible causes. The question was even asked by some whether it would be possible to considerably prolong human life. Roger Bacon in the Middle Ages suggested this possibility, and his namesake, Sir Francis Bacon, advanced it again in the early seventeenth century. In *The Historie Naturall and Experimentall, of Life and Death* (first published in Latin in 1623 and, translated into English, twice in 1637, only two years after Thomas Parr's death), Bacon explored the possibility of restoring mankind to that state of perfection he had enjoyed in the Garden of Eden. In Bacon's opinion, the prolongation of human life for hundreds of years was a real possibility; if the right means could be found, he believed it conceivable that human immortality could be restored.[8]

For Bacon, part of the process to prove that the prolongation of life was possible, and to counter the psalm's seemingly authoritative 'three score years and ten,' was the mass accumulation of individual facts. He thus filled many pages of the *Historie ... of Life and Death* with names of people, in both ancient and modern times, who had lived beyond eighty – some well beyond. His most extreme examples of modern longevity in the Western world included the Countess of Desmond, who was reported (probably by the learned scholar and adventurer, Sir Walter Ralegh) to have 'lived to an hundred and forty yeares of Age.' The inhabitants of the Barbary Mountains in North Africa, Bacon also noted, 'even at this day, they live, many times, to an Hundred and fifty yeares.'[9] Indeed, he wrote that he had himself met a man of over a hundred. This was an important example, as the establishment of 'matter of fact' (as we shall see) depended particularly upon

---

[6] Maynwaringe (1670), pp. 1 – 2.
[7] Psalm 90:9 – 10.
[8] See Bacon (1638) and Shapin (1998).
[9] Bacon (1638), pp. 244, 241. Bacon's source for Katherine Fitzgerald, Countess of Desmond, was probably Ralegh's *History of the World* (1614). It seems likely that the countess was at least in her nineties when she died in 1604. See Thoms (1873), pp. 95–104.

the testimony of reliable witnesses – and who could be more reliable than Bacon himself?[10]

Establishing the maximum human life span was not Bacon's intention. By collecting individual examples he wished simply to illustrate that maximum life span was uncertain – and certainly far in excess of what most seventeenth-century Englishmen and women ever attained. Bacon thus showed that it was no idle ambition to attempt the vast prolongation of life – the project he then laid out in greater detail in the rest of his book, as well as elsewhere in his philosophical writings.[11] Further contemporary evidence indicates that this was not an idiosyncratic speculation. In his hugely popular *Directions for Health, Naturall and Artificiall* (which was first published in 1600, but went through numerous further editions), the Oxford graduate and writer Sir William Vaughan expressed his opinion that 'To live forever, and to become immortall here on earth, is a thing impossible: but to prolong a man's life free from violent sicknesses, and to keep the humors of the body in a temperate state, I verily believe it may be done.'[12] In an era when, as Mary Abbott has observed, old age 'was much less clearly defined' and birth certificates nonexistent, a mass of accumulated facts from contemporary England appeared to support Bacon's case, and numerous examples can be given of remarkable examples of longevity collected by reliable witnesses.[13] These were a further form of scaffolding, helping to support the fact of Parr's great age.

The sources for other longed-lived men and women were authoritative. In Oxford, for example, the antiquary and sometime librarian of the Bodleian Library, Anthony Wood, noted the deaths of two local women, both aged 104, in 1679 and 1680.[14] In 1681, in the same city, the physician and philosopher John Locke recorded a long conversation with an old woman who told him that she was 108 and that her grandmother had lived to 111.[15] The Oxford natural historian, Robert Plot, remarked in his *History of Staffordshire* (1686) that James Sands had died in that county in 1588, aged 140.[16] And a few years after Plot's account, Sir William Temple – English ambassador to The Hague and a scholarly man – wrote of meeting a beggar at a Staffordshire inn who professed to be 124; this old man

[10] Bacon (1638), pp. 134–5, p. 159.
[11] See Shapin (1998), pp. 21–50; and Haycock (2008), chapters 1 and 2.
[12] Vaughan (1633), p. 153.
[13] Abbott (1996), p. 133.
[14] Wood (1892–4), 2.461 and 2.476.
[15] Laslett (1999), p. 23.
[16] Cited in Steele (1688), p. 16.

told Temple that he ate milk, bread and cheese, and meat only rarely, and drank mostly water.[17] Temple would also note that according to travellers' accounts, natives of Brazil were said 'to have lived two hundred, some three hundred Years.'[18] In 1633 the Northampton doctor James Hart, in a book attacking the Paracelsian suggestion that it might be possible to prolong human life almost indefinitely, declared nonetheless that to 'attaine to 100 is no wonder, having my selfe knowne some of both sexes.'[19]

The most famous corroborating example of Parr's great age was a man even longer lived. In 1670, Henry Jenkins of Ellerton-upon-Swale, North Yorkshire, died at the remarkable age of 169. Like Bayles, later Jenkins had an historic event – as well as a number of other seemingly verifiable anecdotes – by which the authenticity of his great age could be verified: He claimed to have carried arrows to the battle of Flodden Field in 1513, when (he said) he was a boy of twelve. According to an account forwarded to the Royal Society by the physician Dr. Tancred Robinson, which was published in the *Philosophical Transactions* for 1695 – 1697, four or five old men of Ellerton, 'that were reputed all of them to be an Hundred Years Old' or so, all agreed Jenkins 'was an elderly Man ever since they knew him; for he was born in another Parish, and before any Register were in Churches.' Therefore, Jenkins was, Dr. Robinson noted, 'the oldest Man born upon the Ruines of the *Postdiluvian* World.'[20]

The case of Jenkins is interesting, as it appears to rest on a remembered event – and such memories were important ways of establishing dates and ages amongst largely oral communities. These many examples – which are by no means exhaustive – help us understand the clear context for Thomas Howard's discovery of Thomas Parr. John Taylor wrote ambiguously in his 1635 poem of 'Records, and true Certificate' that had been provided by 'Relations' from Shropshire, detailing Parr's long life. According to Taylor, Parr had lived his first 17 years at home, and then 18 with his master, before inheriting the remaining 4 years of a 21-year lease on a property held by his deceased father; this had been renewed three times (63 years), before being granted to Parr for life – and Parr had held this lease another half-century: making, in all, 152 years.[21]

---

[17] Temple (1701), p. 112.
[18] Temple (1701), p. 112.
[19] Hart (1633), pp. 7–8; even Hart still placed the upper reaches of the human life span at 120 years.
[20] Robinson (1695–1697), p. 268.
[21] Taylor (1635), B3: It is not clear if by 'Relations,' Taylor means relatives, or simply information that has been related to him.

It must be noted, however, that there were sceptics at the time, who pointed out that there was no way of actually *proving* Parr's great age (and thus undermining any factual claim). As John Taylor observed:

> Some may object, that they will not believe
> His Age to be so much, for none can give
> Account thereof, Time being past so far,
> And at his Birth there was no Register.[22]

Parish priests had first been instructed to collect and keep register of births, marriages and deaths in 1537, but these were not always well kept – and besides, if Parr was as old as was claimed, his birth (supposedly in 1483) preceded this date. Modern historical research *has* unearthed a document from 1588 that confirms Thomas Parr was a married man in that year; assuming that this was his first wife (Parr married twice) and that he had married her in 1588 at the age of twenty, he would have been *at least* sixty-seven in 1635.[23] If so, he would have been at least in his mid-to-late sixties, but he may even have been in his seventies, eighties or nineties. There is simply no way of knowing. That latter figure, though high, is not impossible: There were verifiable seventeenth-century nonagenarians, such as the philosopher Thomas Hobbes (whose own birth was dated by the arrival of the Spanish Armada – the shock allegedly making his mother enter premature labour[24]).

As the clergyman Thomas Fuller sagaciously observed in 1647, 'many old men use to set the clock of their age too fast when once past seventie; and growing ten yeares in a twelve-moneth, are presently fourscore, yea, within a yeare or two after, climbe up to an hundred.'[25] Exaggerating their age was one way for the old to draw attention to themselves, for as Keith Thomas has suggested, they were becoming increasingly marginalised in this period – especially if they were poor.[26] But, based on Parr's various historical recollections and anecdotes, John Taylor was confident that the old man *was* as aged as he claimed. Bacon, along with a number of other seventeenth-century writers, attempted to explain ways in which life could thus be prolonged, and many focused on the importance of diet, hard outdoor work and temperance. Taylor's explanation for Parr's longevity – that

---

[22] Taylor (1635), D4 verso.
[23] See Keith Thomas's article 'Thomas Parr,' in *The Oxford Dictionary of National Biography* (Oxford: Oxford University Press, 2004), on-line edition.
[24] See John Aubrey's *Brief Lives*.
[25] Fuller, (1647), p. 261.
[26] See Thomas (1976).

he had prolonged his life by eating good Shropshire butter and garlic, enjoying clean air and hard outdoor labour, whilst avoiding wine, tobacco and the pox[27] – thus fitted very comfortably with contemporary notions of how a long life might be achieved. Again, this was no doubt of considerable importance in understanding his great age as fact.

All the aforementioned examples of remarkable longevity were important. They helped to establish the context in which substantial life spans could be accepted and believed as mutually supportive facts. It is possible, of course, that some of these writers were inspired in part to seek out such cases by their own reading or awareness of Bacon's *Historie of Life and Death*. But this, too, helps illustrate the way in which a fact, when established on the work of an influential author, may become embedded and disseminated.

3. Witnesses and 'Matters of Fact'

The status and reliability of witnesses has been seen as being particularly important in the creation, acceptance and conveying of a 'fact'. The historian Barbara Shapiro has made the most thorough investigation of this constructivist, judicial approach to knowledge formation, in particular in her 1994 essay, 'The concept "fact": Legal origins and cultural diffusion',[28] and her subsequent book, *A Culture of Fact: England, 1550–1720* (2000). Shapiro shows how law and history were intermingled first in France, and then in England; whilst we today might consider human life span a scientific subject, in the classical and early modern era, when knowledge was going through the process of categorization, this subject was the preserve of historians and philosophers. Thus, when Thomas Hobbes wrote, as Shapiro explains, 'that seeing "a fact done" and remembering it "done" together produced the "Register of Knowledge of Fact … called History."'[29]

History, like the law, depended upon witnesses, testament and documentary evidence, and the notion of the 'matter of fact' has been seen as a key idea in the development of natural philosophy (what we have now come to call science) in seventeenth-century England.[30] As Steven Shapin has written in his important study, *A Social History of Truth*:

> An experience, even of an experimental performance, that was witnessed by one man alone was not a matter of fact. If that witness could be extended to many,

[27] Taylor (1635).

[28] Shapiro (1994), pp. 1–26.

[29] Shapiro (2000), p. 37.

[30] See in particular Shapin (1984), pp. 481–520; Shapin and Shaffer (1985), and Poovey (1998).

and in principle to all men, then the result could be constituted as a matter of fact. In this way, the matter of fact was at once an epistemological and a social category.[31]

Thus, a fact can be put on trial, tested and cross-examined; expert witnesses can be called; a 'jury' can reach a decision and a judgment can be made upon it. It is that process that we have seen occurring with Parr – the jury, in effect, being the court of public opinion, which widely accepted and embraced Parr's longevity.

But, as in any miscarriage of justice, the final verdict that a fact is to be believed and accepted does not, of course, of necessity actually make it true. It may at the time of first judgment be found to be 'beyond reasonable doubt' – but if the witnesses change, or if the evidence is challenged, a judgment may in time be overturned. 'The testimony of the poor and the mean in general,' writes Shapin, 'was widely deemed by gentle society to be unreliable, even in legal settings where testimony was taken on oath.'[32] Far more reliable was the word of a gentleman – or better still, an aristocrat. One of the key figures Shapin has used in his discussion of the establishment of fact in the seventeenth century is Robert Boyle. Shapin quotes Boyle's opinion that 'the two grand requisites, of a witness [are] the knowledge he has of the things he delivers, and his faithfulness in truly delivering what he knows.'[33]

In the case of Parr, we already have two reliable (i.e., socially upstanding and prominent) 'witnesses' to the fact of this old man's great age. The first – though he knew nothing about Parr – was Bacon, a highly respected philosopher and former Chancellor of England, who helped to establish a supporting context in which old age was credible and – indeed – desirable. The second was Thomas Howard, who discovered Parr and brought him to London.[34] But the third, and final 'witness,' was one with both knowledge and faithfulness: the royal physician, Dr William Harvey.

Following Parr's death, and on the command of Charles I, Harvey conducted an autopsy. This itself was a relatively new practice in England, and the posthumous medical examination of a commoner who had died of natural causes (as opposed to having been executed and his body donated to science) would have been highly unusual.[35] The choice of anatomist was also important. Harvey was famous throughout Europe for his discovery

---

[31] Shapin (1994), pp. 483–4, my emphasis.
[32] Shapin (1994), p. 93.
[33] Boyle, *The Christian Virtuoso* (1690), quoted in Shapin (1984), p. 489.
[34] It is worth noting that Howard had known Bacon. Indeed, after taking suddenly ill in April 1626, Bacon had died in Howard's Highgate house.
[35] See Harley (1994), pp. 10–11.

of the circulation of the blood, as explicated in *De motu cordis* (1628), and
he was highly respected for his skill, knowledge and experience. Harvey's
examination of Parr's 'organs of generation' confirmed the sensational
report that even at the age of 120, the old man had had intercourse with his
wife. Nor did Harvey find any great signs of ageing in the old man's other
organs; having examined the stomach and intestines, the great physician
deduced that by 'living frugally and roughly, and without cares, in hum-
ble circumstances,' Parr had thus 'prolonged his life.' Indeed, Harvey found
that 'all the internal organs seemed so sound that had he changed nothing
of the routine of his former way of living, in all probability he would have
delayed his death a little longer.' Harvey blamed what actually appeared to
be Parr's *premature* death on the smoky atmosphere of London compared
to the fresh country air of Shropshire, compounded by the old man's sudden
change of diet to one more rich and varied than those plain foodstuffs to
which he had been long accustomed. Notably (though this would be widely
overlooked), Harvey attempted no explicit confirmation of Parr's supposed
age. He merely conducted an autopsy, from which he concluded that Parr
could have lived longer than he had. His autopsy report would be given
additional gravitas through publication in the august pages of the Royal
Society of London's *Philosophical Transactions* in 1668.[36] Completely unwittingly, Harvey seemingly gave Parr's longevity a medical stamp of scientific
approval. As we shall see, the fact of Parr's great age would, to use one of our
metaphors, ride successfully on the back of Harvey's great name.[37]

4. The Fact Travels

Taken in isolation (as it often can be today), Thomas Parr's 152 years looks
preposterous. But associated alongside a long series of other supporting
facts – about how the world was thought to work; the accepted long lives
of the Patriarchs; complementary evidence of other long livers; theories
on how long life could be achieved and seemingly reliable and upstanding
witnesses verifying this evidence – it does make sense. (This is true even
though for some critics at the time, and afterwards, it remained contentious; these voices were overwhelmed and effectively 'outvoted.' The jury
gave a majority verdict, and their decision held.) With such relatively secure
factual scaffolding based upon interlocking conceptual, epistemological

---

[36] Keynes (1966), p. 224.
[37] See Howlett and Velkar (this volume) for another account in which expertise of science
      and of experience are critical companions to get facts to travel well.

and social elements, as well as upon complementary evidence, Parr's long life had the stability and authority to travel very successfully, unchanged, down some 250 years of history. For this is exactly what it did do, as we shall now see.

One reason that the fact of Parr's great age travelled so well was audience receptivity. This was a fact to be welcomed, since it suggested that human life could be advanced considerably; it also appealed to the common human interest in the bizarre, the extraordinary and the extreme (the continuing modern fascination with 'record breaking' facts testifies to this.) Another important, related factor was sheer repetition. Parr's case was worthy of remark and reiteration. His face appeared in portraits, prints and even pub signs; his story was reproduced in endless popular tracts of local lore, curiosities and remarkable tales and in county histories over the next two centuries; his Shropshire house even became a tourist destination, and an image of it survives in an early lanternslide.[38] But whilst accepted (and almost venerated) at a popular level, thanks largely to Harvey's autopsy, Parr's case also retained credibility in scientific and medical circles.

An early example of the fact of Parr's life and death travelling occurs in 1661, when John Evelyn – a scholarly man who had studied at Oxford and had toured the continent, and was involved in the foundation of the Royal Society – happily used Parr's seemingly untimely death as clear evidence of the harmfulness of London's polluted, smoky air.[39] It appeared again in a scholarly context in 1706, when the post-mortem of another old Englishman, by another prominent anatomist, added further credibility to Parr's great age, lending additional strength to the scaffolding and providing a further example of its successful travel. In April 1706, an elderly, emaciated button-maker named John Bayles died in Northampton – supposedly in his 130th year. Whilst still alive, Bayles had come to the attention of a local doctor, James Keill (1673–1719). Keill was no ordinary country practitioner, however. A graduate of the University of Edinburgh, he had studied medicine at Leiden and had lectured at Oxford and Cambridge universities before settling in Northampton. When Bayles died, Keill undertook an autopsy, sending his report to the Royal Society, who published it in *Philosophical Transactions*. Keill observed that there was 'no Register so old in the Parish' where Bayles lived by which to date his birth, but the oldest

---

[38] 'Old Parr's Cottage, Winington', Shrewsbury Museum Service (SHYMS: P/2005/1604). The role played by fictions in getting facts to travel well is treated also by Adams, while the importance of visual images in facts' transmission, within an expert, not public, community, is discussed by Merz (both this volume).

[39] Evelyn (1661) p. 21.

locals – 'of which some are 100, others 90, and others above 80 years' – all agreed that Bayles had 'been old when they were young.' Keill recorded that though their accounts differed 'much from one another,' they concurred 'that he was at least 120 years' old. Bayles' claim that as a twelve-year-old he had been at Tilbury Camp, where Queen Elizabeth had addressed her army before the Spanish Armada in 1588, was proof enough for Keill that Bayles 'must have been 130 when he dyed.'[40]

As Keill noted, his autopsy 'agrees with that given of old *Parre* by the famous *Harvey* in most particulars.'[41] Keill attributed both men's longevity to the size and strength of their heart and lungs – though he noted that his conclusions would only be confirmed by more 'Dissections of old persons, and these are not numerous enough to ground any thing certain upon.'[42] The editor of the *Philosophical Transactions* may have been more sceptical than Keill: He noted that Bayles was only 'reputed to have been 130 years old.' In a sense, Keill's testimony was self-fulfilling. Bacon had helped create an intellectual context in which old age was a subject of philosophical and medical interest, and the case of Parr, with Harvey's 'supporting' autopsy, had provided a 'factual' example. Keill merely provided further support – additional, strengthening scaffolding.

And the fact continued to travel. In 1713, the respected natural philosopher and clergyman William Derham accepted the great ages of, among others, Methuselah and Thomas Parr in his oft-published Boyle Lectures, *Physico-Theology, or, A Demonstration of the Being and Attributes of God from His Works of Creation.*[43] In 1744 in his *Guide to Health through the Various Stages of Life*, Dr. Bernard Lynch dismissed the claims made by chemists that they could prolong human life, but accepted that the Patriarchs had been long-lived, and that there had been exceptional modern cases of longevity. He noted the long life of Parr – as well as a more recent case: a woman named Margaret Paten from Paisley in Scotland, who had died at St. Margaret's workhouse, Westminster, on 26 June 1739, aged 138. But it is important to note that Lynch considered these examples to be anomalies; he assured his readers that the natural human life span since the Flood was a mere eighty years.[44]

The Enlightenment saw renewed interest in the prolongation of life, providing a context in which belief in Parr's great age could continue to

---

[40] Keill (1706–7), p. 2, 249.
[41] Ibid.
[42] Ibid., 2, p. 252.
[43] Derham (1713).
[44] Lynch (1744), p. 30, pp. 47–8.

flourish (Figure 15.1). Thus, in 1787, a teenage Napoleon Bonaparte wrote to the Swiss physician Samuel Auguste Tissot, seeking a cure for his seventy-year-old uncle's gout: '[P]eople live to a hundred years and more,' young Bonaparte explained, 'and my uncle, by his constitution, ought to be of the small number of those privileged.'[45] In the 1790s, both the French philosopher the Marquis de Condorcet and the English philosopher William Godwin envisaged a future in which mankind would have achieved effective physical immortality.[46] Though their arguments were demolished in Thomas Malthus's 1798 *Essay on the Principle of Population, as It Affects the Future Improvement of Society*, belief in the possibility of long life – even if it was uncommon – clearly penetrated popular consciousness and kept afloat the belief that a long life such as Parr's was both reasonable and possible.

Such ideas were also promulgated by encyclopaedias of the long-lived. In Denmark, the civil servant Bolle Willum Luxdorph gathered information on the very aged, and in 1783 he published a book containing 254 examples of men and women aged from 80 to over 180.[47] They included Thomas Parr and John Bayles, as well as numerous Continental examples, including an Hungarian peasant, Petracz Czartan, who had died in 1724, supposedly at the age of 185.[48] His story, like 'Old' Parr's, soon spread around Europe. And like Parr's, it was faithfully accepted by many, recurring in numerous subsequent accounts. Again, these other instances maintained the scaffolding that supported the fact of Parr's remarkable age, as did a catalogue similar to Luxdorph's published in England in 1799. Its author, James Easton of Salisbury, produced a whole directory containing the names, ages, places of residence, years of decease and brief biographies of numerous men and women who, over the past 1,733 years, had lived for a century or more. Easton noted that he had 'most scrupulously refused admittance' of any account the authenticity of which he 'had the smallest doubt'. His list included Parr, whose supposedly 'premature' death suggested to Easton that a modern human life span of 200 years was not impossible.[49]

[45] Napoleon Bonaparte to Tissot, 1 April 1787, quoted in Troyansky (1989), p. 120.
[46] See Condorcet's *Outlines of an Historical View of the Progress of the Human Mind* (1796), and Godwin's *An Enquiry Concerning Political Justice, and Its Influence on General Virtue and Happiness* (1793).
[47] Published in Latin as Index tabularum pictarum et cælatarum qvæ longævos representant. See Petersen and Jeune (1999).
[48] Jeune and Vaupel (1999).
[49] Easton (1799), pp. xviii–xix.

In 1797, the German doctor Christoph Wilhelm Hufeland (1762–1836) published *The Art of Prolonging the Life of Man*, a work that acknowledged Francis Bacon's *Historie of Life and Death* as its model. The most long-lived moderns, Hufeland suggested, were those who laboured in the open air, such as farmers, soldiers and sailors. Experience, Hufeland observed, 'incontestably tells us' that in these circumstance a life of 150 or even 160 years was quite possible. Recent examples included a Cornishman who had died in 1757 in his 144th year; a labourer in Holstein who had died in 1792 in his 103rd year and an old soldier in Prussia who had died as recently as 1792, in his 112th year. But Hufeland's star example was Parr. Indeed, for Hufeland, as for so many others, Harvey's autopsy of Parr 'proves that, even at this age, the state of the internal organs may be so perfect and sound that one might certainly live some time longer.' Hufeland concluded that it was possible 'with the greatest probability' to 'assert, that the organization and vital power of man are able to support a duration and activity of 200 years.'[50]

Early in the nineteenth century the wealthy Scottish politician Sir John Sinclair (1754–1835) published numerous editions of his multi-volume *Code of Health and Longevity: Or, A General View of the Rules and Principles Calculated for the Preservation of Health, and the Attainment of Long Life.* Sinclair – who though well educated and highly inquisitive, had no medical experience – saw no evidence that the human frame had been built to last a long time. Rather, he observed that the human body clearly 'cannot possibly retain life for ever.'[51] Yet even Sinclair believed contemporary life spans were not what they had once been. He noted Hufeland's observation that in patriarchal times, a year was measured as three months, not twelve, which would make Methuselah a more believable 240. Inevitably, Sinclair's main piece of evidence was what he called the first known 'anatomical account drawn up of the dissection of any old person' – that is, Harvey's dissection of Thomas Parr.[52]

The promise that long lives might be true added strength to those who wanted to believe in such facts – wishful thinking made scepticism of such claims undesirable, and a willingness to accept further exaggerated claims of longevity resulted in a self-perpetuating loop of information. From Bacon and Harvey to the Enlightenment philosophers and encyclopaedia writers, the scaffolding that maintained the 'fact' of Parr's great age had been maintained largely without question or criticism.

[50] Ibid., pp. 70–7, p. 87.
[51] Sinclair (1816), Appendix, 5.
[52] Ibid. Appendix, 6: His wording here is clearly copied from Hufeland. See also Dr. Barnard van Oven, who cited Parr's great age in the preface to *On the Decline of Life in Health and Disease* (1853), pp. xiii–xiv.

5. Undermining the Fact

Further scientific attempts to understand longevity in the early nineteenth century failed to examine certain key facts – such as Parr's 152 years – in detail, despite new information becoming available that might have brought them into doubt. Sir Henry Holland, physician-in-ordinary to Queen Victoria, published a long and influential essay on longevity in the *Edinburgh Review* in January 1857. Modern statistical records of mortality, Holland pointed out, 'utterly' refuted the argument that had recently been made by Pierre Marie Flourens, perpetual secretary to the Academy of Science and professor of comparative physiology at the Museum of Natural History in Paris. In *On Human Longevity and the Amount of Life upon the Globe* (1854; English translation, 1855), Flourens had asserted that 100 years was the maximum natural human life span. But he had added that it was a 'fact' or 'law' that in 'extraordinary' circumstances human life could be prolonged to at least 150 years, if not even to two centuries. An 'extreme' example was Thomas Parr's 152 years. This, Flourens noted, was significant 'because it cannot be disputed [as] it has the testimony of Harvey.' An 'extraordinary life' of a century and a half, he asserted, was 'the prospect science holds out to man.'[53]

In response to Flourens, however, Sir Henry Holland asserted, 'A hundred years is not, and has never been, the natural or normal age of man.'[54] What mattered in understanding the truth of this, he explained, was the law of averages:

> The law of averages, indeed, has acquired of late a wonderful extension and generality of use; attaining results, from the progressive multiplication of facts, which are ever more nearly approaching to the fixedness and certainty of mathematical formulæ. Every single observation, and every new fact added, comes into contributing to these resulting truths. Phenomena, seemingly the most insulated, and anomalies the most inexplicable, are thus submitted to laws which control and govern the whole.[55]

In terms of undermining the scaffolding that had supported Parr so well for so long, this sounds promising. Yet it did not mean that Holland dismissed the idea that some humans *could* and *did* live as long as – or even longer than – a century. Whilst he felt that Bacon's example of the Countess of Desmond was doubtful, he could not reject 'the evidence as to the 152

---

[53] Flourens (1855), pp. 54, 75.
[54] Holland's essay, originally published in 1857, was reprinted in his collected *Essays on Scientific and Others Subjects*: see Holland (1862), p. 105, 110, 115.
[55] Holland (1862), p. 113.

years of Thomas Parr's life, accredited as it is by the testimony of Harvey.' In fact, Holland observed that there were other instances 'of this extraordinary kind … fully admitted by some of the most eminent physiologists.' For Holland, there was 'sufficient proof of the occasional prolongation of life to periods of from 110, to 130, or 140 years.' Such anomalies occurred in human height and weight, he observed, so why not also in age? Indeed, he noted that there were anomalies 'either of excess or deficiency' occurring 'in every part of the physical structure of man,' as in the rest of nature.[56] Why should life span be any different?

Holland's holding on to the evidence of examples such as Parr shows just how strong the scaffolding was around this fact and how deeply embedded it was in both social and medical practice. But it was the statisticians, with the slow accumulation of facts described by Holland, who started to undermine the fortress of century-and-a-half-long life spans. One of the first was the mathematician Edmond Halley, a prominent fellow of the Royal Society. Using records from the Silesian town of Breslau, in 1694 Halley made the first advancement upon John Graunt's famous 1662 study of London's bills of mortality (a work that had actually been inspired by Bacon's *Historie of Life and Death*). Graunt had calculated that 36 per cent of those born in the capital died before they were six years old, whilst only around 7 per cent 'die of *Age*' – that is, over 70 years old.[57] Halley now showed that in Breslau the average life expectancy at birth in the period 1687 to 1691 was a mere 33.5 years. This rose to 41.55 at five years old, but then declined slowly after that. At eighty, life expectancy was a little less than six more years. No one lived beyond 100. Though flawed in its conception, Halley's calculations were the first statistical attempt to compute the *average* span of human life – and he was dismissive of any vision of its vast prolongation.[58] The mathematician Abraham de Moivre would make further advances in the new science of statistics, in particular in *Doctrine of Chances* (1718) and *Annuities on Lives* (1725). Notably, de Moivre reckoned that *nobody* lived beyond eighty-six years. This is what the most reliable evidence told him, and in the preface to *Annuities on Lives*, he explained that he was 'no more moved' to change his mind by some recent observations 'that Life is carried to 90, 95, and even to 100 years,' than he was 'by the Examples of *Parr*, or [Henry] *Jenkins*.'[59]

But, as the London mathematician actuary Benjamin Gompertz subsequently showed, statistics could not actually prove any maximum life

[56] Ibid., p. 110–2.
[57] Graunt (1662), p. 15, 18.
[58] Halley (1694), p. 655; see also Cassedy (1973).
[59] Moivre (1743), p. x.

span – they only indicated probability.[60] Gompertz realised that what he called a 'law of geometrical progression' pervaded 'large portions of different tables of mortality,' and that this could actually be expressed in a simple equation.[61] He published his work, based on the Carlisle and Northampton tables and Deparcieux's observations, in an important essay that appeared in *Philosophical Transactions* in 1825. It established what, with further revisions, would become known as the Gompertz law or the Gompertz curve: In essence, Gompertz's equation showed that from sexual maturity to old age, human death rates rise exponentially. Put in its modern form, for every eight years that we live, our statistical chance of dying doubles.[62] Most significantly, Gompertz suggested that the law of mortality he had identified 'would indeed make it appear that there was *no positive limit to a person's age*' (my italics). But Gompertz added that 'it would be easy … to show that a very limited age might be assumed to which it would be extremely improbable that any one should have been known to attain.' From Milne's Carlisle tables, he saw that from the age of 92 to 99, a quarter of the nonagenarians living at the commencement of each year died. Thus, the odds would be above a million to one 'that out of three million persons, whom history might name to have reached the age of 92, not one would have attained to the age of 192.' Nevertheless, though the projection of his equation makes it *highly unlikely* that anyone would live to nearly two centuries, it did not make it *impossible*: Whilst a quarter of the number may die each year, and though this number would eventually fall below one, as a fraction of a fraction, it would *never* actually reach zero.[63]

The 'limit to the possible duration of life,' Gompertz thus pointed out, 'is a subject not likely ever to be determined, even should it exist.' Indeed, he observed in his *Philosophical Transactions* article that 'the non-appearance on the page of history of a single circumstance of a person having arrived at a certain limited age, would not be the least proof of a limit of the age of man.' Furthermore, 'neither profane history nor modern experience could contradict the possibility of the great age of the patriarchs of the scripture.' Indeed, 'if any argument can be adduced to prove the necessary termination of life, it does not appear likely that the materials for such can in strict logic be gathered from the relation of history.'[64] So although mortality

---

[60] See Hooker (1965), pp. 202–12.
[61] Gompertz (1825), p. 517.
[62] Kirkwood (2000), p. 33.
[63] Gompertz (1825), p. 516.
[64] Ibid., pp. 516–7.

statistics showed the increasingly likelihood of death the older we become, they could never conclusively prove that an absolute age limit had eventually to be reached!

6. The Death of the Fact

In the end it took an historian, William Thoms (1803–1885), fellow of the Society of Antiquaries of London, deputy librarian at the House of Lords and founder (in 1849) of the esteemed literary journal *Notes and Queries*, to demolish the case for Parr's 152 years. Thoms was, in part, dependent upon the new context established by the statisticians noted earlier: They had helped to create the conditions in which it was possible to be more sceptical about claims for long life; furthermore, increasingly critical studies of the Bible had undermined its claim to authority and authenticity. Thomas Paine, for example, in *The Age of Reason*, had placed the Bible under close critical investigation, and had found its claims to be taken seriously as fact to be seriously wanting.

By the time Thoms came to examine the case of extreme human longevity, much of the scaffolding that had supported the 'authenticity' of Parr's great age for so long had either – very gradually – been dismantled or it had thoroughly collapsed altogether. The fact of Parr's age, and his remarkably longevous colleagues were thus left tottering, largely unsupported: They were still held up as facts, but it could not be long before they were toppled. Furthermore, it was not simply the case that the scaffolding supporting the 'fact' of Parr's old age (scaffolding that had also enabled it to travel so well) had been dismantled. In its place, alternative scaffolding about facts of longevity had gradually been constructed – a scaffolding that offered no hope of supporting Parr. With the construction of this new contextual and intellectual structure, the validity of Parr's great age could be questioned. The old scaffolding had depended upon witnesses, upon biblical examples, individual particular instances of longevity, the evidence of dissection, the reliability, repute and social standing of the witnesses and so on. By contrast, the new scaffolding, supporting a different notion of longevity, was gradually built up out of extensive records of numerous old people (and not a few particular extreme – and suspect – examples), the notion of normal life expectancy, the idea of averages and probabilities and so on.

Thoms identified sheer unreflective repetition as one of the chief supports for claims for extreme longevity, and complained of what he considered 'the unhesitating confidence and the frequency with which the public is told of instances of persons living to be a hundred years of age and upwards.'

This, he felt, 'so familiarises the mind to the belief that Centenarianism is a matter of every-day occurrence, that the idea of questioning the truth of any such statements never appears to have suggested itself.'[65] He was particularly struck by the 'child-like faith' with which even men of 'the highest eminence in medical science,' such as Sir Henry Holland, accepted 'without doubt or hesitation statements of the abnormal prolongation of human life.'[66] It was an article Thoms had published in *Notes and Queries* in 1862 that first spurred his interest in the subject. It contained details of what appeared to be a 'perfectly authentic' account of Mrs. Esther Strike, born at Wingfield, Berkshire, on 3 January 1659, and who had died at Cranbourne St. Peter's, Berkshire, on 22 February 1762. The author of the article, Sir George C. Lewis, stated that Mrs. Strike's case rested on 'certified extracts' confirming her baptism, and that no other 'well-authenticated case of a life exceeding 100 years has occurred,' either in the records of modern peerage and baronetage, or in the experience of life insurance companies. These facts suggested to him that human life 'under its ordinary conditions' was never prolonged much beyond a century.[67]

But what of all those commonplace reports in the press of centenarians? It was just such a report that spurred Thoms into action. In 1863, *The Times* announced that a Miss Mary Billinge had died in Liverpool on 20 December 1863, aged 112½ years old. Instead of accepting this fact at face value, Thoms decided to investigate. Whilst what he called 'persons of intelligence and position' accepted that Miss Billinge had indeed died in her 113th year, he took the hitherto almost unprecedented step of actually checking the records for himself. Though Henry Holland had expressed doubts about many claims of long lives, the only person who seems to have attempted to actually check the records earlier than Thoms were Sir George Lewis and Bolle Luxdorph. Yet baptisms, burials and marriages had been registered in England and Wales since 1537. As Thoms's investigation of the appropriate parish records eventually revealed, the Mary Billinge whose death in 1863 had been reported in *The Times* was *not* the same Mary Billinge, daughter of William Billinge, who had been born on the 24 May 1751. She was, in fact, that woman's *niece* – the spinster daughter of the other Mary's brother. This Mary had been born on 6 November 1772. So she was 91 years old, not 113 at all. Old, yes; but some way off even a century.

---

[65] Thoms (1873), p. 1.

[66] Ibid., p. 8.

[67] George C. Lewis, 'Centenarians,' *Notes & Queries*, 3rd series, volume 1, pp. 281–2; see Thoms (1879), pp. xiv–xv.

Thoms published the results of his researches in 1873 in the appropriately titled *Human Longevity: Its Facts and Fictions*. He eventually wished he had never bothered. As he observed in the second edition, he had become noted in the press 'as the ardent apostle of the strange scepticism that nobody exists to over 100 years.' He advised anyone who had 'the slightest desire to live in peace and quietness' not 'under any circumstances, to enter upon the chivalrous task of trying to correct a popular error.' The public response to his book had been such, he wrote, that he considered himself 'one of the best abused men in England.'[68] Thoms's experience shows how deeply ingrained the belief in long life had become. It was almost as if he was personally removing the chance of living to 150. Yet, given 'the number of works which have been written on Old Age, the means of attaining it, and other matters connected with the duration of human life,' Thoms was surprised that his book was 'the first in which the important question, What is the average extension of human life? has ever been tried by the logic of facts.'[69]

Among the many claims he investigated were what he called the 'stock cases' of Thomas Parr, Henry Jenkins and the Countess of Desmond. For so long, these three had been cornerstones in the medical debate on human longevity. But as Thoms demonstrated, there was not one shred of historical evidence in support of *any* of them. Notably, he could not actually prove that Parr was not 152 years old when he died: Given the available evidence, the proof of such a negative was impossible; however, he effectively illustrated how the scaffolding that helped to support such a claim was simply untenable. The longest proven human life Thoms found, based on marriage and baptismal records, was that of a Martha Lawrence. Born on 9 August 1758, she had died on 17 February 1862, aged 103 years and 6 months old.[70] Without supporting facts, this would have to be taken as a maximum known life span. It was a fact; Parr's age was not even a conjecture; it was a claim whose authenticity rested entirely upon anecdote.

It is interesting to see how persistent a well-travelled fact can prove to be, and how facts established with the support of the old scaffolding continued to exist side by side with those facts supported by the new scaffolding. Although *Chamber's Journal* explained that Thoms's was 'a work which everybody who has not read it should read,'[71] many clearly did not. So

---

68 Thoms (1879), p. xii.
69 Ibid., pp. xviii–xix.
70 Thoms (1873), pp. 266–8.
71 Anon (1874), p. 684.

despite Thoms's dismissal of long lives much beyond a century, even as late as 1910, the Russian Nobel Prize–winning biologist Elie Metchnikoff cited the examples of both Drakenberg and the 'well-known instance of Thomas Parr' in his book, *The Prolongation of Life*, as supporting evidence for his belief that the 'natural' human life span was in the region of 150 years. The latter, he pointed out, rested 'on good authority', since William Harvey had autopsied Parr's body; that this was a non sequitur does not appear to have crossed his mind.[72] And whilst the medical investigation of old age was improving through the later part of the century – particularly with the work of Alfred Loomis in New York and J. M. Charcot in Paris from the 1860s to the 1880s – the press continued to indulge in stories of remarkable longevity. Thus, in 1874, the *Birmingham Morning News* reported the death of a Moslem man in Madras aged 143; in 1879, the *Telegraph* carried the story of a 'venerable gentleman' in Mexico aged 180 and John Burn Bailey (in 1888) published yet another catalogue of longevity, with the spurious mixed in with real nonagenarians such as Caroline Herschel (1750–1848).[73] At the same time as the Registrar General in 1885, using the latest statistics and death tables to support a new fact of old age, estimated that only one person in a million born in England could expect to live to 108,[74] in the preface to his *Modern Methuselahs*, Bailey declared that human life was capable 'of an extension to three times its present duration.'[75]

7. Conclusion: The 'False Fact'

As I have shown, even what eventually transpire to be 'false facts' can travel well when they are supported by an entrenched intellectual context, those taken to be reliable witnesses, complementary evidence and a willingness among both the professional classes and the general public to accept the truth of those facts based on those evidences and testimonies. The seemingly oxymoronic concept of a 'false fact' that this essay thus introduces is neither a new nor a contrived concept.[76] As Charles Darwin noted in the concluding chapter to *The Descent of Man* (1876): 'False facts are highly injurious to the progress of science, for they often endure long.'[77] A century

---

[72] Metchnikoff (1910), pp. 86–9.
[73] See Bailey (1888).
[74] Kirkwood (2000), p. 5, 43.
[75] Quoted in Rousseau (2001), p. 11.
[76] A parallel case of false facts that bears many of the same features is that of recent climate science; see Oreskes, this volume.
[77] Darwin (1989), p. 606.

earlier, in *Observations on the Duties and Offices of a Physician* (1770), Dr. John Gregory, professor of the Practice of Physic at Edinburgh University (where Darwin later studied medicine), had observed: 'An easiness of belief, in regard to particular facts, by admitting them upon weak authority, has corrupted every branch of natural knowledge, but none of them so much as medicine.'[78] As Gregory observed, '[T]here is frequently required the united labours of many to make a separation between facts that are fully and candidly represented, and such as are false or exaggerated.'[79] This was a not uncommon trope, and numerous examples could be cited from the period. Darwin was correct that these so-called 'false facts' could survive for a long time, and the example of Parr illustrates well how this may be the case.

The study of maximum human life spans has been beleaguered by an endless accumulation of such 'false facts.' As I have shown in this essay, a 'false fact' such as Thomas Parr's 152-year life can travel remarkably well, both in space and time, and across disciplines of knowledge. As the construct 'false fact' indicates, a fact is not of necessity something that is true; it is rather something that is taken to be true on the basis of current evidence in the context of a particular scaffolding of knowledge, ideas and beliefs that supports it. What I have shown are some of the methods by which a particular fact can be elevated to the status of truth, and how it can both travel and endure. I have also shown how difficult it can be for such a fact to be undermined and how such a fact, by processes of doubt and demolition, and by the erection of an alternative methodology of information, may decline to the status of merely being a factoid: 'an unprovable statement which has achieved unquestioning acceptance by frequent repetition.'[80] In due course, the factoid comes to be considered a 'false fact': something that was once thought to be true, but is now known to be false – even if the truth of its falseness cannot actually be proven, only inferred. It is a fact that Thomas Parr was born, and a fact that he died. He was buried in Westminster Abbey, where his supposed age at death is inscribed on his tombstone, and while his true age at death will probably remain forever unknown, it is one of the best travelled facts of old age.

Acknowledgements

The preliminary work for this essay was undertaken during an Ahmanson-Getty Research Fellowship at the William Andrews Clark Memorial Library,

---

[78] [Gregory] (1770), p. 137.
[79] Ibid., p. 138.
[80] *The Chambers Dictionary* (Edinburgh: Chambers Harrap Publishers, 1993).

University of California, Los Angeles, in 2004, and was completed during a Wellcome Trust Research Fellowship held in the Department of Economic History at the London School of Economics between 2005 and 2007. I am very grateful for the support also received from the LSE's Leverhulme Trust grant, 'The Nature of Evidence: How Well Do "Facts" Travel?' (grant F/07004/Z), and from all my colleagues at the LSE and the contributors to the accompanying workshops and seminars, who commented on various versions of this paper.

Bibliography

Abbott, Mary. 1996. *Life Cycles in England, 1650–1720: Cradle to Grave*. London: Routledge.

Anon. 1874. 'Prolongation of life.' *Chamber's Journal of Popular Literature, Science and Arts*, 565(October):684.

Bailey, John Burn. 1888. *Modern Methuselahs; Or, Short Biographical Sketches of a Few Advanced Nonagenarians or Actual Centenarians who were Distinguished in Art, Science, etc.* London: Chapman & Hall.

Bacon, Francis. 1638. *History Naturall and Experimentall, of Life and Death. Or of the Prolonging of Life*. London: printed by John Haviland for William Lee, and Humphrey Mosley.

Cassedy, James H. (ed.). 1973. *Mortality in Pre-Industrial Times: The Contemporary Verdict*. Farnborough: Gregg International Publishers Ltd.

Condorcet, Jean-Antoine-Nicolas deCaritat. 1796. *Outlines of an Historical View of the Progress of the Human Mind*. Dublin: printed by John Chambers.

Darwin, Charles. 1989. *The Descent of Man, and Selection in Relation to Sex*, in Paul H. Barrett and R. B. Freeman (eds.) *The Works of Charles Darwin*, vols. 21–22. London: William Pickering.

Derham, William. 1713. *Physico-Theology, or, A Demonstration of the Being and Attributes of God from his Works of Creation*. London: printed for W. and J. Innys.

Easton, James. 1799. *Human Longevity: Recording the Name, Age, Place of Residence, and Year, of the Decease of 1712 Persons, who Attained a Century, & Upwards, from AD 66 to 1799, Comprising a Period of 1733 Years*. Salisbury: printed and sold by James Easton, High-Street; sold also by John White, Horace's Head, Fleet St., London.

Evelyn, John. 1661. *Fumifugium: Or the Inconvenience of the Aer and Smoake of London Dissipated*. London: Gabriel Bedel and Thomas Collins.

Flourens, P. 1855. *On Human Longevity and the Amount of Life upon the Globe*, trans. from the French by Charles Martel. London: H. Baillière.

Fuller, Thomas. 1647. *The Historie of the Holy Warre*. London: printed by Tho. Busk.

Godwin, William. 1793. *An Enquiry Concerning Political Justice, and Its Influence on General Virtue and Happiness*. London: printed for G. G. J. and J. Robinson.

Gompertz, Benjamin. 1825. 'On the nature of the function expressive of the law of human mortality, and on a new mode of determining the value of life contingencies.' *Philosophical Transactions* 115:513–83.

Graunt, John. 1662. *Natural and Political Observations Mentioned in a Following Index, and Made Upon the Bills of Mortality. With Reference to the Government,*

Religion, Trade, Growth, Ayre, Diseases, and the Several Changes in the Said City. London: Printed by Tho: Roycroft, for John Martin, James Allestry, and Tho: Dicas, at the Sign of the Bell in St. Paul's Church-yard.

[Gregory, John.] 1770. *Observations on the Duties and Offices of a Physician; and on the Method of Prosecuting Enquiries in Philosophy.* London: printed for W. Strahan and T. Cadell.

Halley, Edmond. 1694. 'An estimate of the degrees of the mortality of mankind, drawn from curious tables of the births and funerals at the city of Breslaw; with an attempt to ascertain the Price of annuities upon lives.' *Philosophical Transactions* 596–610:654–6.

Harley, David. 1994. 'Political post-mortems and morbid anatomy in seventeenth-century England.' *The Society for the Social History of Medicine* 7:1–28.

Hart, James. 1633. Κλινικη *or Diet of the Diseased.* London: printed by John Beale, for Robert Allot.

Haycock, David Boyd. 2008. *Mortal Coil: A Short History of Living Longer.* London: Yale University Press.

Holland, Henry. 1862. 'Human longevity', in *Essays on Scientific and Others Subjects Contributed to the Edinburgh and Quarterly Reviews* (London: Longman, Green, Longman, Roberts & Green), 102–44.

Hooker, P. F. 1965. 'Benjamin Gompertz.' *Journal of the Institute of Actuaries* 91:202–12.

Jeune, Bernard, and Vaupel, James W. (eds.). 1999. *Validation of Exceptional Longevity.* Odense: Odense University Press.

Keill, James. 1706–1707. 'An account of the death and dissection of John Bayles, of Northampton, reputed to have been 130 years old.' *Philosophical Transactions* 25:2247–52.

Keynes, Geoffrey. 1966. *The Life of William Harvey.* Oxford: The Clarendon Press.

Kirkwood, Tom. 2000. *Time of Our Lives: Why Ageing is Neither Inevitable nor Necessary.* London: Phoenix.

Laslett, P. 1999. 'The bewildering history of the history of longevity', in Jeune and Vaupel (eds.), 23–40.

Lynch, Bernard, MD. 1744. *A Guide to Health Through the Various Stages of Life.* London: printed for the author.

[Malthus, Thomas.] 1798. *An Essay on the Principle of Population, as It Affects the Future Improvement of Society. With Remarks on the Speculations of Mr Godwin, M. Condorcet, and Other Writers.* London: printed for J. Johnson.

Maynwaringe, Edward. 1670. *Vita Sana & Longa. The Preservation of Health, and Prolongation of Life. Proposed and Proved. In the due observance of Remarkable Precautions. And daily practicable Rules, Relating to Body and Mind, compendiously abstracted from the Institutions and Law of Nature.* London: printed by J. D. Sold by the Bookseller.

Metchnikoff, Elie. 1910. *The Prolongation of Life: Optimistic Studies.* New revised edn, London: William Heinemann.

Moivre, Abraham de. 1718. *The Doctrine of Chances: Or, A Method of Calculating the Probability of Events in Play.* London: printed by W. Pearson, for the Author.

——— 1725. *Annuities Upon Lives: Or, The Valuation of Annuities Upon Any Number of Lives; As Also, old Reversions. To Which is Added, An Appendix Concerning the Expectations of Life, and Probabilities of Survivorship.* London: printed by W. P.

1743. *Annuities on Life: Second Edition, Plainer, Fuller, and More Correct than the Former*. London: printed for A. Miller.

Oven, Barnard van. 1853. *On the Decline of Life in Health and Disease, Being an Attempt to Investigate the Causes of Longevity, and the Best Means of Attaining a Healthful Old Age*. London: John Churchill.

Paine, Thomas. 1794. *The Age of Reason, Being an Investigation of True and Fabulous Theology*. Paris: printed by Barrois.

Petersen, L.-L. B., and Jeune, Bernard 1999. 'Age validation of centenarians in the Luxdorph gallery', in Jeune and Vaupel (eds.), 41–64.

Poovey, Mary. 1998. *A History of the Modern Fact: Problems of Knowledge in the Science of Wealth and Society*. Chicago: University of Chicago Press.

Robinson, Tancred. 1695–1697. 'A letter giving an account of one Henry Jenkins a Yorkshire man, who attained the age of 169 years, communicated by Dr Tancred Robinson, F. of the Coll. of Physitians, & R. S. with his remarks on it.' *Philosophical Transactions of the Royal Society* 19:266–8.

Rousseau, George S. 2001. 'Towards a geriatric enlightenment', in Kevin L. Cope (ed.), *1650–1850: Ideas, Aesthetics, and Inquiries in the Early Modern Era, Volume 6* (New York: AMS Press), 3–44.

Shapin, Steven. 1984. 'Pump and circumstance: Robert Boyle's literary technology', *Social Studies of Science* 14:481–520.

1994. *A Social History of Truth*. London: Chicago University Press.

1998. 'The philosopher and the chicken: On the dietetics of disembodied knowledge,' in Steven Shapin and Christopher Lawrence (eds.), *Science Incarnate: Historical Embodiments of Natural Knowledge* (Chicago and London: Chicago University Press), 21–50.

Shapin, Steven and Shaffer, Simon. 1985. *Leviathan and the Air Pump: Hobbes, Boyle and the Experimental Life*. Princeton: Princeton University Press.

Shapiro, Barbara. 1994. 'The concept "fact": Legal origins and cultural diffusion.' *Albion* 26:227–52.

2000. *A Culture of Fact: England, 1550–1720*. Ithaca and London: Cornell University Press.

Sinclair, John. 1816. *The Code of Health and Longevity: Or, A General View of the Rules and Principles Calculated for the Preservation of Health, and the Attainment of Long Life*. 3rd ed, London: printed for the author.

Steele, Richard. 1688. *A Discourse Concerning Old-Age, Tending to the Instruction, Caution and Comfort of Aged Persons*. London: printed by I. Astwood, for Thomas Parkhurst at the Bible and Three Crowns.

Taylor, John. 1635. *The Old, Old, Very Old Man: Or, the Age and Long Life of Thomas Parr*. London: printed for Henry Gosson, at his shop in London Bridge, neere to the Gate.

Temple, William. 1701. *Miscellanea. The Third Part*. London: printed for Benjamin Tooke.

Thomas, Keith. 1976. 'Age and authority in early modern England.' *Proceedings of the British Academy* 62:205–48.

Thoms, William J. 1873. *Human Longevity: Its Facts and Fictions*. London: John Murray.

1879. *The Longevity of Man*. London: Frederic Norgate.

Troyansky, David G. 1989. *Old Age in the Old Regime: Image and Experience in Eighteenth-Century France*. Ithaca and London: Cornell University Press.

Vaughan, William. 1633. *Directions for Health, Naturall and Artificiall: Derived from the Best Physicians, as well Moderne as Antient*. 7th ed, London: printed by Thomas Harper for John Harison.

Wood, Anthony. 1892–1894. *The Life and Times of Anthony Wood, Antiquary, of Oxford, 1632–1695, described by Himself. Collected from his Diaries and Other Papers by Andrew Clark, MA, Volumes 2 and 3*. Oxford: printed for the Oxford Historical Society.

SIXTEEN

THE LOVE LIFE OF A FACT

HEATHER SCHELL

1. Introduction

Once upon a time, a very masculine cluster of facts left their home for love. These facts first existed only as a story about competing sexual strategies. Their saga stretches back to a tryst between two prehistoric lovers, a rambling man and the pregnant woman he was tempted to abandon; their scarcely compatible reproductive needs left their mark, according to the story, in instincts that still drive much of gendered behaviour today. Ironically, though the story itself was about travel (the travel of gendered behaviour through time), the narrative form of these facts made their own travel cumbersome – the facts could only move slowly, from storyteller to storyteller. Then came a day when these facts were discovered by expert storytellers, romance writers, who realised that the facts could be used not simply to make a story but also to create a hero. They named him the Alpha Hero, and he gave the facts about sexual selection a second life in the world of romance. He was tall, sexy and powerful, a winner in the battle of the fittest; his main antagonist in each romance novel was a monogamous woman with romantic and reproductive interests antithetical to his own. As an avatar for the facts about sexual selection, he could speak for them, enact them, embody and even propagate them. This new hero helped his facts travel exceptionally well, better than many other facts in this volume. He was not merely a good companion but also a host; he not only maintained the facts' integrity while they ventured very far from their origins, but he also saw them safely through to the return trip: After years of expatriate living, the facts were rediscovered by their first storytellers and returned to their home discipline unscathed, hero in tow. This is their story.[1]

---

[1] I am treating the Alpha Hero facts as facts based on their acceptance as such within the communities I am discussing here; I do not intend to evaluate whether or not their status as facts is valid.

In this chapter, I will look at how the Alpha Hero was created to serve as a warrior in a battle of facts about the appeal of romance novels. This conflict engaged three communities: evolutionary psychologists, romance writers and, as their mutual antagonists, feminist scholars of popular culture. The original clash was precipitated when romance novelists discovered feminist scholarship on romance novels and objected strongly to what they found. Refuting the facts that the scholars had generated required finding new and better explanations for the appeal of romance. I will discuss the romance community's quest for persuasive explanations and explore why the story about the evolutionary basis of gendered behaviour – at first only one of many possibilities they entertained – attained the status of fact. A large part of the story's success derived from its finding an avatar, the Alpha Hero, a heroic figure who embodied the facts in an attractive, charismatic package who could speak for himself. Having an avatar enabled the sexual strategies facts[2] to expand from their role as a mere explanation for the genre's appeal: They could become an actual part of the romance storyline. Once in the storyline, they could be easily transmitted from one novel to the next, and they could move beyond the relatively narrow confines of the romance community's internal discussion into the homes of romance readers across the globe. Over time, the story's usefulness and appeal within the broader romance community was eventually played out as writers and readers began to question the factual status of the story. The Alpha Hero himself was relegated to a romance subgenre. This loss of status fortuitously coincided with the facts' rediscovery by the community who had originally generated them, and for whom they had never ceased to hold unquestioned authority: evolutionary psychology.

The round-trip journey raises interesting questions about what it means to travel well, especially given that the evolutionary psychologists seem not to have recognised the role and status enjoyed by the sexual strategies facts in the romance community. Indeed, both communities perceived themselves as the discoverers or creators of the Alpha Hero facts. My analysis suggests that the facts' integrity through their travels was not sustained by open cooperation between the two communities but instead maintained by their mutual antagonism to feminist critique, against which the sexual strategies facts appeared to be the best available defence. Despite his apparent suitability as a champion for this cause on both fronts, the Alpha Hero only worked as an avatar for the romance writers, not for the evolutionary

---

[2] My phrasing here is a nod to Buss and Schmitt's Sexual Strategies Theory, to which this essay will return later.

psychologists. The facts' ability to travel well required different conditions for each leg of the journey. For the romance writers, the facts' narrative predisposition made it possible to integrate them into pre-existing narrative explanations. For the evolutionary psychologists, the most important condition for the facts' travel was that their unauthorised excursion remain undetected; such camouflage was a necessity for the facts to retain scientific authority, which would have been undermined had the romance community's active participation in crafting and perpetuating them within the Alpha Hero been recognised.

Before moving into the history of the Alpha Hero facts, I want to raise one more point. These facts are a bridge between two fronts: "the truth about human nature" is on one side and "the truth about the appeal of romance novels" on the other. Depending on who is using these facts, they are grounded on one pole and aimed at elucidating the other pole. The evolutionary psychologists use the novels as evidence for their theory of human nature, and the novelists use that theory of human nature as the evidence that justifies their domineering hero. In that way, though the body of facts remains the same, their use changes.

2. The Journey From Hypothesis to Fact

Before the Alpha Hero became a fact about sexual strategies, the fact was a hypothesis and the term "alpha" referred to an animal. First, let's consider the term. "Alpha male" had long been used in animal ecology and zoology as a way of describing the dominant male in a group of hierarchal animals. Influential works in sociobiology, such as Lionel Tiger's *Men in Groups* (1969), E. O. Wilson's (1975) *Sociobiology* and Frans de Waal's *Chimpanzee Politics* (1982), suggested that there was much to be learned about human nature by studying the behaviour of other animals, particularly social mammals.[3] Although these authors do not refer to humans as alphas, their discussions of alpha males – among wolves, chimps and so forth – are explicitly intended to shed light on human behaviour. "Alpha male" in these works is often glossed with the very human term "leader."

Turning to the hypothesis, we find evolutionary psychologists in the 1980s suggesting that contemporary human gender relationships might be understood as a vestige of our ancestry, reflecting the sexual strategies most successfully used by past hominids to reproduce. For example,

---

[3] On a discussion of the travels of facts across meaningful comparisons in animal behaviour research, see Burkhardt, this volume.

David Buss and Michael Barnes, in "Preferences in Human Mate Selection" (1986), discovered a significant gender difference in the importance that people attribute to the three following characteristics in their mates: physical attractiveness, college education and good earning capacity (p. 568). These differences may be caused by culture ("structural powerlessness and sex role differentiation") or by biology ("cues to reproductive investment") (p. 569). Buss and Barnes explained the latter option with a fledgling version of the narrative that will eventually become our facts: "Individuals who have valued cues that discriminate mates most capable of reproductive investment from those less capable ... will be more represented genetically in the current generation than will individuals who have been indiscriminate or who have enacted preferences that do not correlate with reproductive advantage" (p. 569). The cue most tied to women's reproductive value is beauty, while the cue for men is most likely his success, especially (at least in our times) strong earning power (p. 569). Interestingly, the authors see the cultural and biological explanations as compatible. Buss's (1989) subsequent article on human mating strategy remains somewhat tentative, still explicitly marking the evolutionary explanation as a "hypothesis" in the article's title.

By 1993, this developing fact about the evolutionary basis of mate selection had accrued a longer, more complex narrative. Buss and Schmitt's tremendously influential "Sexual Strategies Theory: An Evolutionary Perspective on Human Mating" offered an elaborate development of Buss's earlier explanation of gender differences. The theory itself now had eleven premises and twenty-two predictions. Additions from the 1986 article included differentiations between short-term and long-term mating strategies, economic concepts about the minimum level of parental investment and concerns about fidelity and paternity. The article articulated all of the assumptions about human mating behaviour that remain axiomatic in the field of evolutionary psychology to this day, and also illustrated the essentially narrative nature of this fact. Not only did the authors need eleven premises in which to spell out their basic theory, but elucidating each point required the language of story-telling. Here, for example, is Buss and Schmitt's explanation of short-term mating strategies among men:

> The reproductive benefits that historically would have accrued to men who successfully pursued a short-term sexual strategy were direct: an increase in the number of offspring produced. A married man with two children, for example, could increase his reproductive success by a full 50% by one short-term copulation that resulted in insemination and birth. This benefit assumes, of course, that the child produced by such a brief union would have survived, which

would have depended in ancestral times on a woman's ability to secure relevant resources through other means (e.g., by herself, through kin, or through other men). (p. 207)

Note that what should in theory be a simple supporting fact about the numeric advantage of short-term mating strategies became a story about a cheating husband; a resourceful paramour, already, it appears, facing the social stigma of being unwed and now also pregnant; their bastard child and a supporting cast of unspecified kin and other men. Buss and Schmitt's argument demands merely that we recognise infidelity as potentially advantageous in evolutionary terms. A narrative example is not really necessary, though it may be that the facticity of this type of fact relies to some extent on narrative.[4] Meanwhile, the scientific writing meshes oddly with the romance. Buss and Schmitt attempted to make the narrative sound scientific by including a percentage, removing any quotidian language ("affair," "bastard," etc.), and carefully choosing words that sound clinical ("copulation," "insemination," "resources"). The one inconsistency comes from describing the fictional adulterer as "married," which too clearly draws our attention to the cultural underpinnings of the story; "mated" would probably have been a better choice.

The hypothesis about men and women's competing mating strategies developed rapidly in the early 1990s, prompting substantial media attention, numerous treatises geared towards a general audience, enthusiastic adoption within some fields (such as political science) and strong objections from others (principally from evolutionary biology and zoology, especially primatology). Within its home discipline, its status was not questioned, and it would soon become not just any fact but a foundational fact. I am not going to trace that transformation here; instead, I am going to follow our fact as it sets out on an unexpected journey: into the realm of the romance writing community.

3. The Crisis That Demanded the Fact

Before the early 1980s, there were not many facts about romance novels. By "romance novels" here, I mean not the "silly novels" condemned by George Eliot or the influential works by Jane Austen and Charlotte Brontë that would provide the template for the twentieth-century romance genre and its hero. Instead, I refer to a type of highly regimented, skilfully marketed, condensed love story, adhering to a standard format controlled primarily

---

[4] For another account of how scientists use narratives to make their facts travel, see Merz, this volume.

by a handful of large presses (Mills and Boon in the United Kingdom, Harlequin and Silhouette in Canada and the United States), with each publication having a shelf life of approximately one month. Romance novels had not received any of the attention that scholars had begun to direct towards other types of mass culture; there was thus no contention among academics about what these novels meant. Romance writers themselves weren't engaged in any collective soul-searching about the meaning of their work, either, in part because the conditions of their labour weren't such as to foster dialogue: Romance novels were written by hundreds of women working in isolation, without agents, connected individually to their publishing houses through correspondence and through the written guidelines to plot and character (i.e., the "formulas") to which prospective authors had to adhere. The facts about romance novels in the 1970s were limited to industry-generated data about sales and distribution.

That situation changed dramatically in the 1980s, for two reasons: romance writers organised, and scholars began to write about the genre and generate facts about what it meant. First, in 1980, Romance Writers of America (RWA) was founded; since then, it "has built itself into a self-enclosed counterweight to the dismissive attitude of the East Coast publishing and critical establishments as a trade organization focused on empowering its members" (Zaitchik 2003). In 1981, the founding of *Romantic Times* (*RT*), with its fiction reviews and conferences, generated the same unity among readers of romances. Most important for our purposes here is the way that such organisations provide the necessary conditions for the development of a collective intelligence, providing individual group members with access to information that might otherwise have remained esoteric and encouraging a consensus about ideas. The development of this collective intelligence was fostered by trade publications, by panels and word of mouth at conferences, by writing and reading groups and, after the arrival of the Internet, by websites and discussion boards. These conditions made it possible for romance writers to develop and disseminate facts about romance novels.

Then, in 1982, a young scholar named Tania Modleski published her doctoral dissertation, a psychoanalytic approach to examining the appeal of "mass-produced fantasies for women." A few essays on romance novels had previously been published, but Modleski's was the most nuanced analysis to date, as well as the most extensive.[5] *Loving with a Vengeance* was

---

[5] Ann Barr Snitow's "Mass Market Romance: Pornography for Women Is Different" (1979) and Ann Douglas's "Soft-Porn Culture" (1980), as their titles suggest, were discussing surface-level features of the romance novel; also, their argument applied more accurately to single-title bodice rippers than to the relatively chaste category romances of Harlequin.

intended as a response to the masculine bias of popular-culture scholars, who tended to extol the creative brilliance of the male-dominated genres of science fiction and mystery while denigrating or ignoring works with predominately female audiences (Modleski 1990, 1). At the same time, Modleski was uncomfortable with what she saw as the tendency of female critics to approach women's genres with "dismissiveness, hostility [or] a flippant kind of mockery" (p. 4). She devoted one chapter of her slim volume to analyzing Harlequin romances along the lines suggested by John Berger's (1973) *Ways of Seeing*, eventually concluding that reading these novels was a kind of "repetition compulsion" aimed at resolving the psychic conflict caused by a sexist society (p. 48). Further, the Harlequin "not only reflect[ed] the 'hysterical' state, but actually, to some extent, induce[d] it" (p. 49). She ended with the caution that our new understanding of the genre "should lead one less to condemn the novels than the conditions which have made them necessary" (p. 49). Two years later, her book was picked up by a larger press and released as a paperback. Within a few years of the original publication of her work, the other two important books in the field appeared: Janice Radway's *Reading the Romance* (1984) and Kay Mussell's *Fantasy and Reconciliation* (1984). These three works comprised the seminal scholarship in the field. No romance scholarship published since has had anything close to the influence of Modleski's and Radway's works.

Though a romance reader herself, Modleski probably never imagined that the romance community would run across her dissertation. It is not written with that audience in mind. In fact, a quick perusal of *Loving with a Vengeance* reveals that romance writers themselves were mostly absent from Modleski's discussion, perhaps because the narrow parameters of Harlequin's romance formula at the time may have seemed to preclude much of a creative role for individual writers. More importantly, Modleski aimed to explain the romance readers, not the writers. Yet romance writers did indeed discover her work. Modleski's ideas also showed up in the syllabi of college courses, in which some future romance writers were surely enrolled. Moreover, a number of romance writers were academics. And once some romance writers knew her work, her theory quickly became common knowledge among the RWA, where it was not well received. Romance writers had organised into a group of professionals who felt pride and ownership in their written work. Modleski treated romance novels as a monolithic text, not as individual works with varying merit; although a useful approach for a theorist looking for the big picture, it might well appear dismissive to members of the community. Modleski's analysis not only left the writers out of the picture, but also suggested that they should not be feeling any pride: Their writing might actually be harming women.

Thus, if Modleski's argument were accepted as true, it controverted the romance writing community's developing ethos as a profession dedicated towards bringing women happiness. There was no easy way to adapt the genre to address Modleski's concerns.[6] The classic plot of the romance novel works with extraordinary utility to provide the conflict and resolution we expect from a gripping story while using a minimal cast of two major characters and a few minor characters.[7] Conflict in this efficient genre often stems from having the love interest serve simultaneously as the villian.[8] A worthy antagonist, one who poses an adequate challenge to the heroine, needs to be powerful, unaffectionate if not hostile, with goals in opposition to the heroine's. Further, readers will be more impressed with the heroine's development, as well as anxious about the outcome of the contest, if the heroine herself doesn't seem quite up to the challenge; this effect can be reliably achieved by making the heroine young, inexperienced and vulnerable. The more extreme the power differential, the more tense the plot. This narrative dynamic works in any genre. In the hands of a competent storyteller, we would be similarly caught up in the uneven contest between a villain and the small orphan child whom he has decided to target. If the small orphan child happens to be above the age of legal consent and finds herself unwillingly attracted to the villain, who happens to be fiendishly sexy, then really all we need to make this a compelling romance is for the villain to reveal one redeeming feature, making it possible for us to like him.[9]

Entertaining Modleski's critique would have undermined the unselfconscious confidence with which an author could employ the genre's conventions. Savvy romance writers could no longer create masterful, domineering heroes without at least some concern about how the story would appear to

---

[6] Romance scholarship influenced a new generation of college-educated romance editors, who hoped to transform the genre. See the scathing critique of these editors' attempted interventions in Krentz (1992b), p. 107.

[7] This applies particularly to series romance, a monthly publication format in which authors have only 200 pages for developing their story and characters. Single-title novels can be longer, sometimes offering Dickensian casts and plots that span generations. However, series romances still account for over a quarter of romance publications (Romance Writers of America 2009) and remain the career entry point for most romance writers. In addition, many writers still prefer using the hero as the major source of conflict.

[8] See the cogent explanation in Krentz (1992b), p. 108.

[9] This redeeming gesture is known in the trade as the "save the cat!" moment. According to Mary Buckham, a romance writer and writing instructor, a romance writer can create "the world's biggest asshole" as her hero as long as he "saves the cat" before readers give up on the book. Personal interview HS with Buckham, 15 July 2008.

editors and other external critics. At the same time, what else could they do? Rather than water down the genre, writers instead turned to a defence of its most troubling aspects.[10] The search for a solid, persuasive, satisfying justification for the romantic power dynamic therefore became a shared goal among the romance writing community.[11] Furthermore, in the face of feminist criticism, it was important for the authors to prove that romance novels, as well as the career of romance writing, were expressions of female empowerment. In other words, the novelists needed to make a case for their domineering hero while simultaneously proving that the overpowered heroine (or perhaps the reader) was really the one in control.

4. The Alpha Hero Arrives

In 1990, romance novelist Jayne Ann Krentz published an article in the RWA trade magazine, *Romance Writers Report*. This article outlined her idea of the alpha male as hero. As a best-selling author who had moved from series romance to single titles – a shift that indicates relative authorial control and status within the romance writing community – Krentz was well positioned as an influential spokesperson for her explanation of the romance hero. She was also motivated and well organised. Within two years, she had assembled a multifaceted rebuttal of feminist criticism, aimed towards an external audience: *Dangerous Men and Adventurous Women: Romance Writers on the Appeal of Romance* (*DMaAW*) (Krentz 1992a) included contributions from twenty-one romance novelists, with Krentz at the helm as editor and contributor of several essays. She presented the book as the result of "a host of conversations that took place over the years among members of the romance writing community" (1992a, p. xi), which suggests that the contributors had been testing and refining their arguments for some time. Published by University of Pennsylvania Press as part of their New Cultural Studies series, the volume fit well with the press's tradition of publishing books with a somewhat scholarly approach for an audience of non-specialist readers. The press hedged its bets by including a positive review from

---

[10] At the same time, the romance novel has changed in ways that appear to address feminist criticism. For example, one important innovation entails offering the hero's point of view. In the early 1980s, most romance novels provided only the heroine's point of view, a stylistic convention that Modleski and other critics had interpreted as helping facilitate the genre's troubling gender dynamics.

[11] The felt need to explain the appeal of romance – to their own satisfaction, as well as to an audience of feminist critics – still lingers. See, for example, Lynn Coddington's "Romance and Feminism," published in *Romance Writers Report* in 2004.

romance scholar Janice Radway on the back jacket. *DMaAW* was a rhetorical tour-de-force still cited among the romance writers community more
than fifteen years later; it has had at least as strong an impact as Modleski's
and Radway's work.

DMaAW provided an effective launching pad for the Alpha Hero. In an
essay entitled "Trying to Tame the Romance: Critics and Correctness," Krentz
described alpha males as "the tough, hard-edged, tormented heroes ... at the
heart of the vast majority of bestselling romance novels. ... These are the
heroes who carry off the heroines in historical romances. These are the heroes
feminist critics despise" (Krentz 1992b, p. 108–9). Note that Krentz defined
the Alpha Hero in two contexts: as he related to romance novels and to feminist critics. Insofar as he would come to be used as the fact that definitively
rebutted feminist criticism, the Alpha Hero was indeed the feminist critics'
enemy. She did not take credit for naming this hero, but suggested merely that
he was "what has come to be known in the trade as the alpha male" (1992b,
p. 107). In another chapter, Laura Kinsale cited Krentz as the source of the
term and quoted an earlier definition of the alpha-male hero: the "retrograde,
old-fashioned, macho, hard-edged man" (1992, p. 39). Kathleen Gilles Seidel,
in the same volume, offered a slightly different origin story: "The term 'alpha
male' came into use, I believe, because some authors were engaged in a struggle with editors about a certain type of hero and needed a vocabulary for the
discussion" (1992, p. 178). Seidel liked the term in part because she saw it as
"the only piece of jargon that has originated from the authors themselves"
(1992, p. 178). None of these stories acknowledge the alpha male as a construct originating in a scientific community.

The influence of evolutionary psychology on these narratives can be a
bit hard for outsiders to see. This results primarily from the way that novelists work with research.[12] As is customary in fiction writing, the *DMaAW*
authors did not mark their research with such scholarly apparatus as footnotes or direct citations. At the same time, while the authors were interested
in the etiology of the term "Alpha Hero" and must surely have recognised
it as deriving from the scientific study of animals, they were more concerned in tracing the intellectual legacy of the term within the romance
writing community. Because the authors in *DMaAW* were conscientiously
citing each other, their essays have a scholarly format.[13] However, insofar

---

[12] The work most fiction authors do to make their research invisible stands in sharp contrast
to Michael Crichton's exceptional use of scholarly research indicators. For a more critical
appreciation of Crichton, see Adams, this volume.

[13] Evolutionary psychologists who later examined *DMaAW* would understandably but inac-
curately assume that anything not cited was not based on research, as we'll see later.

as the academic style of explicit citation was adopted for this volume, the authors utilised it only for the words of other novelists and romance scholars. Despite the camouflage, the novelists' reasoning and their word choices still retained the traces of their external sources: For example, although Williamson's language often sounded persuasively unscientific – consider her discussion of bygone "cave days" – her reference to "species" being "propagated" was probably imported (1992, p. 131).

Krentz, the author who introduced the term "alpha male" to the romance community, most closely echoed the logic of evolutionary psychology. She speculated that romance novels, like other genres, derived their power from ancient myths. These myths endured "because they embody values that are crucially important to the survival of the species" (1992b, p. 113). In Krentz's depiction, romance played on the most powerful of all myths, because "the successful pair bond" was "imperative to ... survival" (p. 112). While the alpha male's precise relationship to these ancient myths was never made explicit, Krentz did argue that romance fiction featuring "sensitive, unaggressive heroes" was not popular with authors or readers (1992b, p. 113). This line of reasoning set the stage for the Alpha Hero as the embodiment of women's inherited sexual preferences. However, Krentz did not ultimately cast her argument as scientific. This is most evident in her discussion of virginal heroines, in which she insisted on virginity as a "metaphor" (1992b, p. 111). An evolutionary psychology approach would not see anything metaphorical about virginity: Instead, concerns about female virginity would be read as the psychological side effect of internal fertilisation (e.g., eggs are fertilised inside a woman's body, a process that her mate cannot witness), thereby fostering male doubts about a child's true paternity (see Salmon, 2005, p. 253).

Though the Alpha Hero was not yet linked explicitly to an evolutionary approach, another author in *DMaAW* was already exploring the implications of evolutionary psychology as a way of explaining the romance novel's appeal. In "By Honor Bound: The Heroine as Hero," Penelope Williamson briefly touched on an argument that, while still contextualised here as fantasy rather than fact, explored the line of reasoning about human reproductive strategies from evolutionary psychology. Williamson explained that a woman's desire for a faithful, monogamous man was

> a fantasy deeply ingrained in the female psyche since the cave days, when the woman relied upon the man to provide food and shelter for her and their children, when his abandonment would mean almost certain death. If she could keep her man tied to her with her sexual allure, she would be assured of a provider and protector in the personal sense, and in the larger sense the species would be propagated. (Williamson 1992, 131)

This explanation looks familiar in two ways: in its logic and in its narrative. Once again we have the drama introduced by Buss and Schmitt (1993, p. 207) a personal and yet epic dilemma with a pregnant woman trying to secure resources to ensure her survival. Although the present-day woman's relationship to this drama was defined as fantasy, the sexual strategies scenario from the "cave days" was unquestioned – those were the days *when* the woman relied upon the man, *when* his abandonment would mean death.

The authors in *DMaAW* were "fictionalising" their new facts, by which I mean that they were adapting the scientific ideas to make them function within the romance community's prevailing discourse about writing. The conceptual toolbox native to fiction writing provided other language for talking about the endurance of narrative – as myth, metaphor, legend and so forth; in other words, the scientific explanation was just one more tool and not given any special precedence. Indeed, although within a few years the Alpha Hero came to be understood as a fact, at this early point he was explicitly cast as fantasy. Seidel (1992) was quite insistent that romances and their characters were fantasies. Kinsale described the romance novel as a type of psychological journey in which an "androgynous" reader identifies with both the heroine and the hero; in this case, then, romance readers can relish the alpha-male hero "because the alpha male hero is themselves" (1992, p. 39). The strong contrast between this reasoning and the gender-deterministic interpretation used by the evolutionary psychologists stemmed from another goal of the romance writers: to demonstrate that their novels, contrary to the opinion of feminist scholars, did not reify gender roles but perhaps even liberated women from those roles.

The literary context, with its fantasies and metaphors, helped these evolutionary ideas move successfully into the new community. The only part still missing was a more convenient package, to bundle the narrative with the Alpha Hero and transform it into a fact, helping it range farther and faster.

5. Fantasy Becomes Fact

By the late 1990s, the increasingly popular Alpha Hero had become explicitly linked to a narrative cluster of facts about human reproductive strategies. The alpha male and human evolutionary history, both offered by *DMaAW* as distinct explanations for romance's appeal, became synthesised in later readings of the book. This is clearly visible in an essay by Amber Botts, "Cavewoman Impulses: The Jungian Shadow Archetype in Popular Romantic Fiction" (1999). By comparing the passage from Williamson cited earlier with Botts's paraphrase of that same passage, we can see how the Alpha Hero, with natural leadership skills and dominance displayed in a group of animals, merged seamlessly with the sexual strategies theory:

> [Williamson] asserts that romance novels reflect a cavewoman impulse deep in the heterosexual female psyche that wants to tie the primary *Alpha Male* to her, in order to ensure the survival of the woman and her children. Williamson argues that romance conventions, in which a heroine attracts the best provider and protector of *the 'herd'* and binds him to her, spring from women's ancient memories... (p. 64, italics added).

Botts's overview, for the most part a scrupulous paraphrase, imported the alpha male into Williamson's narrative, as well as the cross-species comparison popular in evolutionary psychology.[14] Botts did not recognise her own substitution of terms because the Alpha Hero, by the time she wrote this article, was already understood as a precise substitute for the caveman. The Alpha Hero had become the literal embodiment of our evolutionary history.

Meanwhile, the romance community continued to appropriate material from evolutionary psychology to round out the character of the Alpha Hero. Suzette Mako articulated the connection explicitly in a 1998 article on the romance novel, explaining that "alpha types were first identified as they related to animal success in the wild." Their "attractiveness ... is a certification of biological quality, taking into account such factors as bilateral symmetry, ... and weight/muscle distribution as cues to freedom from harmful genetic mutations." Applying this to the romance hero, Mako explained that he was "physically powerful, his goal dominance through acquisition and subsequent protection." He had "healthy levels of the male hormone, testosterone," and he was "a born breeder." In addition to a hyper-masculine body type, he embodied other "healthy 'animal' traits." Mako's description was replete with references to "research," genes, Darwin, fertility and test subjects; whether or not the romance writer was consciously aware of creating an Alpha Hero, Mako posited, the appeal of the alpha model "is understood by our very genes."

In addition to the Alpha Hero himself, romance novelists began drawing on evolutionary psychology for more facts about romance. Best-selling novelist Linda Howard introduced the romance community to Desmond Morris's classic work of sociobiology, *The Naked Ape* (1967), particularly the concept of stages of intimacy. This begins, perhaps not surprisingly,

---

[14] Another case in which receivers gain a bundle of facts transported within a fact might be found in the transfer of technology; see Howlett and Velkar, this volume.

with "eye-to-body contact" and wraps up with "genital-to-genital contact." Howard utilised these stages in one of her own novels, the 1998 *Kill and Tell*, in which an angry heroine analyses the hero's behaviour and recognises the step-by-step deliberation with which he has seduced her.[15] Morris's model, via Howard, was then elaborated in a number of venues in the romance writing community.[16] Howard leaned strongly towards a relationship model drawn from evolutionary psychology. In a 2000 interview, the interviewer presented Howard with a reader's assessment of her heroes:

> A Linda Howard hero is very masculine, experienced, and even jaded. He is a throwback to his hunter-gatherer forbearers whose biological duty was to impregnate as many females as possible to ensure the continuation of his gene pool. He is also the end product of Darwin's natural selection – he is smarter, stronger, faster, better than his fellow males. Yet despite thousands of years of genetic imprinting, his sexual appetite locks onto one woman. (Laurie Likes Books, 2000)

This characterisation clearly employs the facts about the Alpha Hero. While the hero himself is obviously a creation – he is "a Linda Howard hero" – his antecedents are real: He had hunter-gatherer forbearers with a biological duty, and genetic imprinting should make him have a sexual appetite for as many women as possible. Howard embraced this narrative as fact in her response: "To me, the ancestral hunter is part of the best part of men" (Laurie Likes Books, 2000). In other words, the ancestral hunter definitively *was* an inherent part of men.

And he certainly was becoming a staple in many romances, in a way that would have real consequences for the evolutionary psychologists who studied the novels later. Popularly understood as the embodiment of facts about our evolutionary history, the Alpha Hero could also strengthen the apparent foundation of those facts through his literary reproduction: Any successful Alpha Hero who inspired other authors to create Alpha Heroes was in effect siring more evidence of his own existence. In other words, researchers examining all these books, if unaware of the intentionality behind the Alpha Hero's creation, would see thousands and thousands of data points suggesting that women were all drawn to the same kind of man. The facts about sexual strategies that the alpha male carried were thus capable of generating their own support. Our hero was indeed a "born breeder."

---

[15] Thanks to Mary Buckham for pointing me to this helpful reference.

[16] See, for example, Mary Buckham's 2005 article on sexual tension in *Romance Writers' Report*.

6. The Alpha Hero versus The Feminist Critics

Feminist literary criticism was the original goad that prompted romance writers to seek alternative explanations of romance novels' appeal, and, via a somewhat indirect path, led to their discovery of the Alpha Hero. The Alpha Hero carried the facts intended to refute feminist interpretations of romance novels. In *DMaAW*, while the Alpha Hero himself makes an appearance in only a handful of the essays, almost all of the book's authors share his uneasy relationship with feminist criticism. Many identify themselves as feminists, perhaps even more truly feminist than the critics they are addressing; others seem to consider all feminists their enemy, as we see in Seidel's comment: "Feminists talk about sisterhood; I do not know how deeply they feel it" (1992, p. 172). The un-sisterly feminist critics consist of Modleski and two other romance scholars from the 1980s: Radway and Mussell. Radway's and Mussell's work includes many lines that, when extracted, sound hideously offensive. However, Radway's populist "reader response" approach proved ultimately reconcilable with the standards of the romance writing community. Kay Mussell also managed a rapprochement. Both scholars received special thanks from the book's editor for their "distinguished work on the romance" (Krentz 1992a, p. xi). In the essay that engages most thoroughly with romance scholarship and is otherwise highly critical of their work, both Radway and Mussell receive a modicum of praise, Radway for "her idea that romance reading is the one thing that many women do for themselves" (Seidel 1992, p. 178) and Mussell for being "alert and sensitive," a "better reader" than many other critics (Seidel 1992, p. 168). Radway also provides the critical blurb on the back cover, in which she both praises the book and scolds the academic work on romance novels. Modleski gets no such tolerance. Almost every comment in the book addressing "critics" or "feminism" can be read as aimed at Modleski.[17] She might be the one of "the more zealous constituents of the movement" who insisted on seeing men and women in direct competition, perhaps a characteristic of "an earlier, more desperate phase of feminism" (Kinsale 1992, p. 40). She is also a "male-trained female critic," who shares male readers' and reviewers' negative response to the genre (Barlow 1992, p. 52). And, in some ways, she may be the intended audience of the book, or at least part of that audience. Certainly Radway's jacket blurb calls for "feminist literary and media critics" to read the book.

---

[17] The extensive entry in the index on "Critics and criticism" concludes with "*See also* Feminists and feminism" (p. 183).

By the end of the 1990s, many feminist scholars of romance had read the book, and almost all had decided, at least publicly, to embrace romance novels as an authentic expression of feminist liberation.[18] Writing about potentially negative aspects of romance had become more or less verboten, with Modleski as the last unrepentant critic standing. Thus, the romance writing community's triumph over feminism paralleled the triumph of the Alpha Hero, in both the romance writing community and in the general parlance.[19] The romance community had facts on their side.

7. The Transformation of the Alpha

Ironically, once the battle with academic feminism was over, there simply was not as much need for the facts about sexual strategies. Even as the animal behaviour model gained ascendancy in American popular culture, the Alpha Hero's star began to fade within the romance writing community. Writers began to run up against the limitations this fact imposed on their plots and character development. Novelist Karen Harbaugh mocked a hypothetical new writer who, in attempting to use the Warrior archetype, instead creates a formulaic "'alpha male' whose conversation consists of grunts and whose reason for being is nothing more than a wall into which the heroine crashes until – for some Neanderthal reason – he decides to pluck up the feisty little thing and carry her off to his cave. Not good" (2002, p. 42). Arielle Zibrak, an editor from Ballantine Books, asserted, "The way gender roles and relationships are portrayed in romance novels is constantly evolving. … 'Sexy' … no longer means domineering or controlling" (Castell 2005, p. 42). The evolution of gendered behaviour moves rapidly in romance novels.

Now that the facts he embodied no longer had a mission, the Alpha Hero's limitations as a life-mate became more salient. Members of the Romance Writing Book Club, discussing the alpha in 2006, were frustrated with the hazards of trying to plot a book with this character. Morris (2007) remarked that "the line between Alpha and Bully is slender and easily crossed." Several writers complained that their alphas naturally transform into gentler "betas," "the type of man who has best friend qualities and a sense of humour" (Dixielandgrl 2007). More telling, for our purposes, is the Alpha Hero's fall in

---

[18] See the exchange begun in the 1997 "romance" issue of *Paradoxa* (a journal devoted to genre fiction)(Vol. 3 Issue 1–2, 1997), continued with more open hostility in the same journal in 1998 (Vol. 4 issue 9). See also the essay collection *Romantic Conventions* (Kaler and Johnson-Kurek 1999), which adhered closely to the approach laid out in *Paradoxa*'s romance issue.

[19] For further discussion of the alpha male, see Heather Schell, "The Big Bad Wolf."

status from fact to fantasy character, one who either appears only in romance novels or, perhaps, who can seem attractive only in that context. Novelist Leigh Michaels noted that, if there were such an "alpha real life man," "whoo-eee, would we run the other direction or what!" The signs of disillusion in the listserv discussions also appear in romance novels themselves, in which even novelists such as Howard and Christine Feehan, both known for their domineering Alpha Heroes, began to soften their leading men. No longer a staple in mainstream romance, the Alpha Hero survives primarily in the paranormal subgenre, in which, in his dual role of monster and lover, there is no doubt that he is a fantasy character and not a fact.[20]

Nowhere are the disadvantages of the Alpha Hero facts clearer than in Laura Zigman's novel *Animal Husbandry* (1997), perhaps more properly chick lit than romance. Zigman's novel was what she termed a "faction," or a fact-based fiction, drawing on her own traumatic break-up and the subsequent reading project she undertook exploring animal models of masculine behaviour; the book is peppered with the results of that reading, citing Charles Darwin, Richard Dawkins and Desmond Morris, among others. Zigman's proxy is Jane Goodall, a heroine more interested in bovines than primates. Goodall invents an "old cow/new cow" theory to explain philandering men – just as bulls get bored with familiar cows and will no longer mate with them, so, too, do men lose interest in "old cows" and seek out new ones. In this novel, the facts clearly demonstrate that the long-term prospects for any relationship are grim. Pushing the Alpha Hero facts to their most extreme implications creates a mate who will never be faithful if he has the chance to stray. Zigman thus reveals the irreconcilable differences between the sexual strategies facts and the premise of the romance genre. Therefore, when Zigman's novel was transformed into the romantic comedy *Someone Like You* (Goldwyn 2001), the screenwriter had to undermine the fact status of the Alpha Hero. In the film version, Goodall's theory is clearly cast as a comical defence mechanism prompted by heartbreak. While the theory proves to have broad appeal among other female characters in the movie, it so gravely offends Goodall's love interest, Eddie, that it nearly derails her opportunity for real romantic happiness. Not only does Goodall find true love by repudiating her own theory, but Eddie rejects his philandering ways for monogamy. Both gestures reject the factual status of

---

[20] At the same time, the facts that the Alpha Hero embodies are still productive, generative tools in paranormal romances. This gives the Alpha Hero a value in paranormal romances that he has mostly ceased to have outside the subgenre. Another example of facts travelling within and across genres can be found in the case of the transmission of Greek architecture to nineteenth-century America; see Schneider, this volume.

this sexual strategies theory. The book and the film highlight serious limitations of the Alpha Hero fact's usefulness to the genre.

8. The Prodigal Son Returns

In 1999, April Gorry, then a doctoral student in anthropology at UC Santa Barbara, completed her dissertation on the mating behaviour of Western women tourists. She had specialised in evolutionary psychology and human sexuality, and she wanted to investigate the appeal of low-status local men to female tourists. This appeal appeared to counter her field's expectations that women are drawn to high status in mates. As part of her research, she decided to test the validity of her findings by analysing romance novels, with the aim of ascertaining if "certain hard-to-fake, evolutionarily relevant indicators of resource potential have a more direct link to female sexual psychology than do Western signs of social and economic success" (p. 195). Her research would mine romance novels as support for facts about the evolutionary impulses behind female mate selection. Later, other evolutionary psychologists, including her dissertation advisor, would rely on her research on romance novels for facts to prove the difference between adaptive male and female sexual selection strategies.[21] The Alpha Hero facts would thus be discovered anew, in the romance community that had adopted them, and brought home with renewed explanatory power. However, the success of this return journey depended upon the entire outing being unrecognised – for these facts to be valid, each romance author had to be creating the Alpha Hero unconsciously and independently. This section will explore how Gorry discovered the Alpha Hero facts and how some of the founding assumptions of evolutionary psychology made the social history of these facts impossible to see.

Gorry predicted that, just as a man's "indicators of resource potential" would increase his sexual attractiveness, so, too, would his outward show of romantic attachment. Such an appearance of attachment would serve as a "predictor of sustained male investment." This hypothesis would "be confirmed if indicators of romantic attachment … are as reliably attributed to the hero character as traits that signal a man's ability to invest" (pp. 195–6). The results of Gorry's predictions would be unsurprising to a romance reader: Yes, the hero must show signs of romantic attachment to the heroine – signs clearly visible to the reader, even though they may be invisible to the

---

[21] See especially Catherine Salmon and Donald Symons, *Warrior Lovers: Erotic Fiction, Evolution, and Female Sexuality* (2001).

heroine or the hero himself. Because these indicators of romantic attachment could be ascribed to genre conventions rather than to the innate
predisposition of the authors, Gorry needed to verify that there were no
genre conventions, at least none imposed by the industry. She thus asked a
number of romance publishing houses and editors if there were still a "formula" for their industry (p. 205), which was stoutly denied. Such assertions,
taken at face value, suggested to Gorry that the patterns she descried in the
romance novel arose naturally and perhaps persisted despite the efforts of
editors: "Although editors may reject the notion that their books are formulaic, the comments of romance novel writers and the results of my survey
indicate that certain elements are required within the genre" (p. 207).

What this disjunction actually reveals is an inherent pitfall of working
across disciplines. A literary scholar could have explained that all genres
have conventions that develop over time, some of which are perpetuated
through affectionate nostalgia long after their proximate cause is gone;
consider the use of black and white film by some directors who want to
achieve a "film noir" look. Romance conventions had a much more specific origin: Harlequin and other publishers used to send aspiring authors a
"formula" for their books, outlining relatively rigid strictures about format,
character and plot. However, by the early 1980s, the term "formula" had
acquired a bad connotation in the industry, suggesting the type of generic,
cookie-cutter books that prompted Modleski's critique and undermined
the authorial freedom of romance writers. Publishers dropped the term and
discontinued the tip sheets. Harlequin's enormous publishing house can
still maintain its successful blueprint through a combination of industry
dominance (it owns almost every romance publisher in North America,
as well as Mills and Boon), strict editorial control and a regular pool of
manuscript submissions from avid readers who are already very familiar
with the conventions of the genre. "Formula" or not, romance publishers
do indeed realise that their genre has conventions, and a differently worded
question would have yielded another answer. Nonetheless, the Alpha Hero
facts look more robust if the romance formula appears to be a spontaneous,
un-orchestrated pattern rather than one mandated by the industry.

Gorry (1999) asked fifty-one women, primarily members of the RWA,
to recommend "highly popular or well-liked" romance novels, compiling
a list of novels that showed up in at least three sets of recommendations.
Of these, she eliminated ones that predated the codification of the genre in
the early 1970s. She analysed the remaining fifty-six books and found not
only that they seemed to have a formula, as mentioned earlier, but that they
also valorised a certain kind of hero. The study design therefore allowed the

romance writing community to direct Gorry towards the model of romantic hero that they had been actively promoting. To interpret these works, she drew heavily on *Dangerous Men and Adventurous Women*, further ensuring that the romance community's alpha-male facts would be foregrounded.

Gorry's research methods drew heavily on the clear routes of communication within the romance community, which would seem to require recognising romance novels as the product of a shared culture, not isolated texts that tellingly reinvent the same universal patterns. However, as we saw with her handling of the genre's formula, Gorry's interpretative moves were premised on the belief that human behaviour derives from evolution, not culture. She was looking for facts with evolutionary underpinnings and so ignored the possibility that her facts could have any other sort of genesis. This is, I would argue, a serious flaw endemic to some work in evolutionary psychology, and we can see its impact in this case study. Because the discipline has focused its energy on looking for evidence of universality, it promotes a type of study design less able to control for cultural influences.

For example, Gorry needed *DMaAW* for an authoritative account of the goals of romance readers and writers. With the evidence from these essays, she was able to reassemble the Alpha Hero facts – without recognizing the romance community's part in creating and sustaining these facts. How was this possible? First, each essay in *DMaAW* was treated independently, suggesting Gorry did not recognise their publication in the same volume as potentially meaningful; the essays seen separately suggest each author's take on the romance is independent confirmation of the Alpha Hero facts, rather than evidence of a community working to create a consensus.

Second, while she clearly noticed the evolutionary leanings of the authors' arguments, she did not consider that the novelists might have been reading about evolutionary psychology. Instead, the fact that these "writers sometimes sound evolutionary when discussing their subject matter" (Gorry 1999, p. 201) became further proof of the indelibility of the Alpha Hero facts. In this regard, Gorry made the same error that Modleski did: She assumed that her own field could not be an object of study to the very people who were the objects of her own study. The romance community may be somewhat extraordinary in this regard; nonetheless, it reminds us that non-academic communities can also distribute, contest, amend and otherwise contribute to the travel of facts.

Gorry's work was used by other evolutionary psychologists interested in sexuality, most especially by her dissertation advisor, Don Symons, and another of Symons' protégées, Catherine Salmon. Symons and Salmon had co-authored a book in 2001 entitled *Warrior Lovers: Erotic Fiction,*

Evolution, and Female Sexuality. In 2005, Salmon returned to this theme for an essay in *The Literary Animal*, which hoped to apply the insights of evolutionary psychology to literary studies. She offered a concise analysis of porn and romance novels as genres that express the clear, innate difference between male and female sexuality. Gorry's research provided essential evidence, allowing Salmon to present the romance novel as virgin soil, a "vast untapped source of data" (Salmon, p. 244) that transcended culture. Insofar as anyone might be said to shape the romance novel, that influence was limited to the "individuals who make an effort to acquire" the novels, the means by which the free market can discern what "tap[s] into basic … female sexual psychologies" (Salmon, p. 245). This explanation discarded(anything that might indicate conscious agency from members of the romance writing community – the authors, the editors, the marketers and so forth. Salmon's summation of Gorry mentioned nothing of Gorry's method, further distancing the results from the romance writers whose strategic professional goals helped shape everything from the novels Gorry read to the way she interpreted them.

As I suggested in the discussion of Gorry, the truth status of the Alpha Hero facts for evolutionary psychology is based on the facts' freedom from the influence of human culture. If instead it was clearly understood that the romance community had adopted and perpetuated the Alpha Hero facts, then the heroes of romance novels might cease to embody the facts. The novels would no longer look like "a window into our natural preferences" (Salmon, p. 245) – that is, a clear, transparent, unmediated view of our true selves, untainted by culture. Even if the Alpha Hero facts could survive, they would be messier, equivocal facts, tainted with human intent.

Intriguingly, these newly rediscovered Alpha Hero facts were still directed towards disproving the explanation offered two decades earlier by feminist scholars. "Romance novels," Gorry asserted, "are about women winning the battle of the sexes. I say this not in a feminist sense but in an evolutionary one" (1999, p. 200). Note the emphasis on a full-scale war, a connotation echoed later when she mentioned that, "on the feminist front, [romances] are condemned for their not-so-politically-correct portrayals of men and men's attitudes towards the object of their affection" (p. 203). Salmon's work resuscitated Modleski, Radway and Mussell again as targets for the Alpha Hero facts. Salmon asserted that "feminist thinking on pornography and romance novels has often gone astray," most especially in their insistence on inserting "politics" into their analysis (2005, p. 251). Given that "modern society … has restricted female options primarily to control through influencing men, females have two choices. They can exert as much control as possible through men (as often seen in romance novels), or they can work to gain control on their own (one feminist impulse)" (p. 249). Setting aside the oddly dated sound of Salmon's assertion, it placed romance novels in direct opposition to feminism. Though Salmon was careful to define her second option as only "one feminist impulse," her binary formulation leaves no room for additional feminist impulses, only for the impulse depicted in the romance novels she saw as the fact-based answer to the unreasonable political expectations of feminists. For Salmon, feminists' politics-based answers are maladaptive, while the "behaviours… that make us more likely to survive and reproduce" are exemplified in romance novels, through the depiction of techniques "to evaluate the willingness of men to commit [sic] investment" (p. 252). Yet decades have passed since feminist scholars posited romance novels' complicity in damaging gender roles, raising the question of why Salmon still needs to refute them. I would suggest that the feminist scholarship on romance novels has also served as a strangely effective companion to the Alpha Hero facts, providing a rhetorical setting against which the facts show in their best light.

9. A Final Word on What Made These Facts Travel

The cluster of facts about sexual strategies travelled extremely well, both with and without the Alpha Hero as their avatar. He protected their integrity in two quite different communities. They successfully made a round-trip, enacting Ludwik Fleck's idea of the circulation of facts through cycles of propaganda and legitimation.[22] However, as I hoped to demonstrate in this chapter, a fact's ability to travel depends on the needs of the varying communities it serves.[23] Scientific communities may place particular constraints on facts and their origins; in this case, while we might have expected that a scientific fact's travel documents would be thoroughly vetted before it was allowed to cross the borders, we instead found that the fact was claimed as a natural-born citizen, with no birth certificate required. In contrast, a lay community might relish a scientific fact's origins as a sign of legitimacy and highlight the travel. The Alpha Hero fact gained stature in the romance community, albeit indirectly, through his association with evolutionary psychology, which provided the fact with what was seen as

---

[22] For a thorough discussion of Fleck, see Morgan, this volume.
[23] For a parallel and contrasting account of how a cluster of facts gets embodied in one fact, and how that fact then serves different communities in varying ways, see Whatmore and Landström, this volume.

a scientific underpinning. In that regard, it was important that the Alpha Hero not be seen solely as a hometown boy but as a fact that had already been discovered in another arena and could be imported. In contrast, when the Alpha Hero facts returned to evolutionary psychology, romance novels in tow, it was essential that this unauthorised excursion remain a secret. These alpha-male facts in their new home still retain the narrative quality that made them travel so well, they still are attractive facts destined for a good love life. But which is the avatar: the Alpha Hero or the alpha male has been elided, perhaps even reversed, through their travels.

Acknowledgments

Many, many thanks to Ron Iannotti, Ann Pfau, Carol Hayes, Mary Buckham, Dolsy Smith, Jon Adams and all my colleagues in the University Writing Program. I vastly enjoyed the opportunity to work with the members of the Facts group.

Bibliography

Barlow, L. (1992) The Androgynous Writer: Another Point of View. In Krentz, J. A. (Ed.) *Dangerous Men and Adventurous Women: Romance Writers on the Appeal of the Romance.* Philadelphia, U Penn P.

Berger, J. (1973) *Ways of Seeing.* New York, Viking Press.

Botts, A. (1999) Cavewoman Impulses: The Jungian Shadow Archetype in Popular Romantic Fiction. In Kaler, A. K. & Johnson-Kurek, R. E. (Eds.) *Romantic Conventions.* Bowling Green, Bowling Green UP.

Briggs, P. (2008) *Cry Wolf.* New York, Ace.

Buckham, M. (2005) What Is Sexual Tension and How Do You Write It? *Romance Writers Report.*

Buss, D. (1989) Sex Differences in Human Mate Preferences: Evolutionary Hypotheses Tested in 37 Cultures. *Behavioral and Brain Sciences*, 12, 1–49.

Buss, D. & Barnes, M. (1986) Preferences in Human Mate Selection. *Journal of Personality and Social Psychology*, 50, 559–70.

Buss, D. & Schmitt, D. (1993) Sexual Strategies Theory: An Evolutionary Perspective on Human Mating. *Psychological Review*, 100, 204–32.

Castell, D. (2005) Up Close and Personal with Arielle Zibrak of Ballantine. *Romance Writers Report.*

Coddington, L. (2004) Romance and Feminism. *Romance Writers Report.*

Cole, K. (2006) *A Hunger Like No Other.* New York, Pocket Star Books.

De Waal, F. (1982) *Chimpanzee Politics: Power and Sex among Apes.* New York, Harper & Row.

Dixielandgrl (2007) Re: Alpha Male. *The Romance Writing Book Club.* Barnes and Noble.

Douglas, A. (1980) Soft-Porn Culture: Punishing the Liberated Woman. *The New Republic.*

Feehan, C. (2007) *Dark Possession*. New York, Berkley.

Goldwyn, T. (director) (2001) *Someone Like You*. Main performers: Ashley Judd and Greg Kinnear. Released by Fox 2000 Pictures.

Gorry, A. (1999) *Leaving Home for Romance: Tourist Women's Adventures Abroad*. Santa Barbara, UC Santa Barbara.

Harbaugh, K. (2002) The Use and Abuse of Archetypes. *Romance Writers Report*.

Howard, L. (1998) *Kill and Tell*. New York, Pocket.

Kaler, A. K. & Johnson- Kurek, R. E. (Eds.) (1999) *Romantic Conventions*. Bowling Green, Bowling Green UP.

Kinsale, L. (1992) The Androgynous Reader: Point of View in the Romance. In Krentz, J. A. (Ed.) *Dangerous Men and Adventurous Women: Romance Writers on the Appeal of the Romance*. Philadelphia, U Penn P.

Krentz, J. A. (1990) The Alpha Male. *Romance Writers Report*.

Krentz, J. A. (Ed.) (1992a) *Dangerous Men and Adventurous Women: Romance Writers on the Appeal of the Romance*. Philadelphia, U Penn P.

Krentz, J. A. (1992b) Trying to Tame the Romance: Critics and Correctness. In Krentz, J. A. (Ed.) *Dangerous Men and Adventurous Women: Romance Writers on the Appeal of the Romance*. Philadelphia, U Penn P.

Laurie Likes Books (2000) Linda Howard: "Who Knew?" All about Romance website, unidentified author interviewed Howard. The interview date was 11 August 2000, http://www.likesbooks.com/lindahoward.html.

Mako, S. L. (1998) Alpha Amours: The Roles of the Alpha Male Hero and Alpha Female Heroine of Romance. Published online at the website Suite 101. The URL is http://www.suite101.com/article.cfm/romance_genre/5883.

Meyer, S. (2006) *New Moon*. New York, Little Brown.

Michaels, L. (2007) Re: Alpha Male. *The Romance Writing Book Club*. Barnes and Noble.

Modleski, T. ([1982]1990) *Loving with a Vengeance: Mass-Produced Fantasies for Women*. New York, Routledge.

 (1998) My Life as a Romance Writer. *Paradoxa*, 4, 134–44.

Morris, C. (2007) Re: Alpha Male. *The Romance Writing Book Club*, online discussion group, hosted by the bookstore Barnes and Noble, http://bookclubs.barnesandnoble.com/.

Morris, D. (1967) *The Naked Ape*. New York, McGraw-Hill.

Mussell, K. (1984) *Fantasy and Reconciliation: Contemporary Formulas of Women's Romance Fiction*. Westport, CT, Greenwood Press.

Mussell, Kay (Ed.) (1997) Where's Love Gone? Transformations in the Romance Genre. Special issue *Paradoxa*, 3(1–2).

 (1998) Kay Mussell Replies to Tania Modleski. *Paradoxa*, 4, 145–7.

Radway, J. (1984) *Reading the Romance: Women, Patriarchy, and Popular Literature*. Chapel Hill, UNC Press.

Romance Writers of America (2009) Romance Literature Statistics: Overview. http://www.rwanational.org/cs/the_romance_genre/romance_literature_statistics. Accessed 1 June 2009.

Salmon, C. (2005) Crossing the Abyss: Erotica and the Intersection of Evolutionary Psychology and Literary Studies. In Gottschall, J. & Wilson, D. S. (Eds.) *The Literary Animal: Evolution and the Nature of Narrative*. Evanston, IL, Northwestern UP.

Salmon, C. & Symons, D. ([2001]2003) *Warrior Lovers: Erotic Fiction, Evolution, and Female Sexuality*. New Haven, CT, Yale UP.

Schell, H. (2007) The Big Bad Wolf: Masculinity and Genetics in Popular Culture. *Literature and Medicine*, 26, 109–25.

Seidel, K. G. (1992) Judge Me by the Joy I Bring. In Krentz, J. A. (Ed.) *Dangerous Men and Adventurous Women: Romance Writers on the Appeal of the Romance*. Philadelphia, U Penn P.

Snitow, A. B. (1979) Mass Market Romance: Pornography for Women Is Different. *Radical History Review*, 20, 141–61.

Sparks, K. (2005) *How to Marry a Vampire Millionaire*. New York, Avon.

Tiger, L. (1969) *Men in Groups*. New York, Random House.

Ward, J. R. (2007) *Lover Revealed*. New York, Signet.

Williamson, P. (1992) By Honor Bound: The Heroine as Hero. In Krentz, J. A. (Ed.) *Dangerous Men and Adventurous Women: Romance Writers on the Appeal of the Romance*. Philadelphia, U Penn P.

Wilson, E. O. (1975) *Sociobiology: The New Synthesis*. Cambridge, MA, Belknap/Harvard UP.

Wright, R. (1994) Feminists, Meet Mr. Darwin. *New Republic*.

Zaitchik, A. (2003) The Romance Writers of America Convention Is Just Super. *New York Press*. New York.

Zigman, L. ([1997]2001) *Animal Husbandry*. New York, Delta/Dell.

Index